ADVANCES IN THE STUDY OF
PEROMYSCUS (RODENTIA)

ADVANCES IN THE STUDY OF *PEROMYSCUS* (RODENTIA)

Edited by
Gordon L. Kirkland, Jr.
and
James N. Layne

TEXAS TECH UNIVERSITY PRESS
1989

Partial funding for this publication was provided by the Archbold Biological Station and Shippensburg University.

This book was set in 10 on 12 Baskerville and printed on acid-free paper that meets the guidelines for permanence and durability of the Committee on Production Guidelines for Book Longevity of the Council on Library Resources.

Published 20 December 1989

Library of Congress Cataloging-in-Publication Data

Advances in the study of Peromyscus (Rodentia / edited by Gordon L.
    Kirkland, Jr. and James N. Layne.
        p. cm.
    Contents: Systematics and evolution / Michael D. Carleton—
Adaptive physiology / Richard E. MacMillen and Theodore Garland, Jr.
—Reproduction and development / John S. Millar—Population biol-
ogy / Donald W. Kaufman and Glennis A. Kaufman—Social behavior /
Jerry O. Wolff—Peromyscus and Apodemus / W. I. Montgomery.
    ISBN 0-89672-170-1 (alk. paper). — ISBN 0-89672-171-X (pbk.:
alk. paper)
    1. Peromyscus. I. Kirkland, Gordon L.   II. Layne, James
Nathaniel.
QL737.R638A.38   1989
599.32′33—dc19   89-4699
                                                                      CIP

Texas Tech University Press
Lubbock, Texas 79409-1037 USA

CONTENTS

INTRODUCTION

GORDON L. KIRKLAND, JR., AND JAMES N. LAYNE

"The diversity of life is so great that we can never hope to answer all of the questions we might pose for all of the kinds of organisms, so we look to detailed studies of representative taxa for elucidation of the principles governing diversification and maintenance of diversity."

W. Frank Blair (1968)

The genus *Peromyscus* is one of the most representative genera of North American mammals and the most intensively studied. Depending upon the classification employed, the genus includes more than 50 species and 200 subspecies, representing varying degrees of genetic relationship. It has a wide geographic distribution extending from the Atlantic to the Pacific and from the Canadian subarctic to Panama. Species of *Peromyscus* can be found in nearly every terrestrial habitat of North America from the barren, rocky alpine zone of northern mountains to lowland tropical forests. Reflecting the vast geographic area and diversity of environments in which it is found and its apparently rapid evolutionary history, the genus exhibits a considerable range of morphological, behavioral, and physiological variation. Thus, it provides unlimited opportunities for studies of basic questions of population biology, community ecology, adaptive physiology, ecological energetics, and evolutionary biology.

For a number of practical reasons, *Peromyscus* are particularly useful for many kinds of field and laboratory investigations. One or more species are usually available for study in almost any region of North America and are often abundant enough so that adequate samples can be obtained for statistical analyses. *Peromyscus* are comparatively easy to capture, handle, and mark. In addition, most species can be maintained in captivity without difficulty, and some breed freely in the laboratory so that stocks can be established for experimental studies. In short, the genus *Peromyscus* is so valuable a research subject for North American mammalogists that if it did not exist, we would have been obliged to invent it!

Beginning with the taxonomic review of the genus by Wilfred H. Osgood (1909), *Peromyscus* has figured prominently in the literature on North American mammals. Pioneer workers in the field of *Peromyscus* biology were Francis B. Sumner, Lee R. Dice, and

1

W. Frank Blair. Much of the work on the genus in the period from the early 1930s through the 1950s emanated from Dice's laboratory at the University of Michigan.

Many of the early studies of *Peromyscus* were important far beyond their contribution to knowledge of the genus per se. To cite just a few examples—Osgood's (1909) monograph served as the model for future systematic work on North American mammals and contributed to the concept of polytypic species. Sumner's (1932) work on geographic variation in pelage coloration in *Peromyscus* aided understanding of the genetics and selective basis of hair coloration of mammals in general. The interspecific hybridization experiments of Dice and his associates (for example, Dice, 1933, 1937; Dice and Liebe, 1937) provided further insight into the process of speciation in sexually reproducing organisms and extended the botanical concept of cenospecies to animals. The comparative study of three species of *Peromyscus* in California by Thomas T. McCabe and Barbara D. Blanchard (1950) was an important forerunner to studies of the adaptive significance of life history strategies. William H. Burt's field studies on white-footed mice (*Peromyscus leucopus*) and other small mammals in Michigan led to refinement of the concepts of home range and territory (Burt, 1943). Walter E. Howard's (1949) investigation of prairie deer mice (*P. maniculatus bairdii*) on the George Reserve in southern Michigan, which suggested that a genetic component was involved in the dispersal behavior of animals, was the stimulus for expanded interest in the role of genetic differences among individuals in various aspects of animal behavior and ecology. In the 1960s, John A. King conceived the idea of bringing together the previous research on *Peromyscus* to demonstrate the value of the comparative species-level approach to many scientific disciplines and to evaluate, organize, and synthesize the available data as an aid to future research on the genus. The result of this effort was the appearance in 1968 of *Biology of* Peromyscus *(Rodentia)*. The volume comprised 14 chapters written by different authors who covered major topics of *Peromyscus* biology, as follows: "Paleontology" by Claude W. Hibbard, "Classification" by Emmet T. Hooper, "Speciation" by Lee R. Dice, "Habitat and Distribution" by Rollin H. Baker, "Anatomy" by David Klingener, "Ontogeny" by James N. Layne, "Parasites" by John O. Whitaker, Jr., "Endocrinology" by Basil Eleftheriou, "Genetics" by David I. Rasmussen, "Home Ranges and Travels" by Lucille F. Stickel, "Population Dynamics" by C. Richard Terman,

"Behavior Patterns" by John F. Eisenberg, "Psychology" by John A. King, and "Activity" by J. Bruce Falls.

In the 20 years since the appearance of *Biology of* Peromyscus *(Rodentia)*, there has been much additional research on *Peromyscus* and many important advances in our knowledge of the genus. Some of the advances have been made possible by new methods in biochemistry, karyology, or multivariate statistics, but most of the exciting new information is the result of more sophisticated conceptual questions being asked about such things as the role of social behavior in population biology and the evolution of reproductive strategies. Despite significant increases in our knowledge of many aspects of the biology of this genus, much remains to be learned about even the most basic features of the life history and ecology of many species, and data for even the best-studied taxa tend to be restricted to relatively few localities within the range and are usually based on relatively short-term studies. The present volume is an outgrowth of a symposium, "Biology of *Peromyscus* (Rodentia): 1968–1985", presented at the Fourth International Theriological Congress in Edmonton, Canada, in August 1985. The objective of the symposium was to review the major advances in knowledge of *Peromyscus* biology in the decade and a half since the publication of King's book. Although knowledge of *Peromyscus* since 1968 has advanced on all fronts, time constraints prevented coverage of all subject areas represented by the 14 chapters in *Biology of* Peromyscus *(Rodentia)*. Instead, we were forced to select five areas that we felt had been particularly dynamic, and in which there had been significant advances in terms of their contribution both to specific knowledge of the genus, as well as to the understanding of broader ecological and evolutionary processes. The first five chapters in this volume constitute substantially expanded versions of the papers presented in the symposium.

Members of the genus *Peromyscus* have often been ascribed the status of ecological equivalents of Old World wood mice (*Apodemus*). Members of the genus *Apodemus* are commonly referred to as wood mice, in much the same way that members of the genus *Peromyscus* are referred to as deer mice, even though the deer mouse (*P. maniculatus*) is but a single representative. North American mammalogists who have collected either wood mice (*Apodemus sylvaticus*) or yellow-necked mice (*A. flavicollis*) cannot help but be impressed by the obvious morphological, ecological, and behavioral similarities between these Old World species and such representatives of *Pe-*

romyscus as *P. leucopus* and *P. gossypinus*. *Apodemus sylvaticus* and *A. flavicollis*, along with various species of *Sorex, Clethrionomys, Microtus*, and *Pitymys*, provide American mammalogists doing field work in Europe with a genuine sense of home.

Faunal and floral similarities between the Nearctic and Palearctic biogeographic regions have intrigued biogeographers for more than a century and led to the formulation of the Holarctic biogeographic region. The concept of the Holarctic represents an acknowledgment of the common evolutionary history of many of the plants and animals of this circumpolar region. The list of approximately 35 genera of mammals, representing six orders and 14 families, which have Holarctic distributions provides evidence of a long history of faunal interchange between the Palearctic and Nearctic.

Recent taxonomic revisions have led to a burgeoning of the list of mammalian species that have Holarctic distributions. Conspicuously absent from this list is that most taxonomically diverse, geographically widespread, and ecologically ubiquitous of North American mammal genera, *Peromyscus*. Likewise missing is *Apodemus*, which is equally widespread geographically in Eurasia but is not as taxonomically diverse or broadly distributed ecologically as *Peromyscus*.

Thus, in terms of their zoogeography, *Peromyscus* and *Apodemus* have complementary distributions. Both are principally north temperate in distribution but extend into tropical realms, *Peromyscus* into the Neotropical and *Apodemus* into the Oriental. Although both genera are geographically and ecologically widespread, neither has invaded the true arctic zone to any great extent, and the northern limits of both genera correspond to the tree lines in their respective hemispheres. This helps to explain why neither has succeeded in extending its distribution across Beringea so as to become Holarctic in distribution.

Inasmuch as conveners of symposia at the Fourth International Theriological Congress were asked to include some international representation among their invited participants, and given the fact that the oft-reported status of *Peromyscus* and *Apodemus* species as ecological equivalents previously has not been subjected to critical analysis, we felt that the symposium on the biology of *Peromyscus* at ITC/IV would provide an appropriate forum to address this question. In recognition of his many and varied contributions to the literature of *Apodemus*, we asked W. I. Montgomery to undertake the difficult task of preparing a comprehensive and critical review of the similarities and differences between members of these two genera. His resultant contribution, based on a review of over 400 references dealing with the biology of both genera, constitutes the final chapter in this volume.

ACKNOWLEDGMENTS

In addition to thanking the contributors for their efforts and cooperation in preparing papers for ITC/IV and manuscripts for this volume, we wish to acknowledge the support for this project provided by the Archbold Biological Station and Shippensburg University. In addition, Carol J. Kirkland, Patricia M. Krim, and John V. Planz, the latter two graduate students at Shippensburg University, provided invaluable assistance during the final editorial work on this volume.

LITERATURE CITED

BLAIR, W. F. 1968. Introduction. Pp. 1–5, *in* Biology of *Peromyscus* (Rodentia) (J. A. King, ed.). Spec. Publ., Amer. Soc. Mamm., 2:1–593.

BURT, W. H. 1943. Territoriality and home range concepts as applied to mammals. J. Mamm., 24:346–352.

DICE, L. R. 1933. Fertility relationships between some of the species and subspecies of mice in the genus *Peromyscus*. J. Mamm., 14:298–305.

———. 1937. Fertility relations in the *Peromyscus leucopus* group of mice. Contrib. Lab. Vert. Gen., Univ. Michigan, 4:1–3.

DICE, L. R., AND M. LIEBE. 1937. Partial infertility between two members of the *Peromyscus truei* group of mice. Contrib. Lab. Vert. Gen., Univ. Michigan, 5:1–4.

HOWARD, W. E. 1949. Dispersal, amount of interbreeding and longevity in a local population of prairie deer mice on the George Reserve, southern Michigan. Contrib. Lab. Vert. Biol., Univ. Michigan, 43:1–50.

KING, J. A. (ED.). 1968. Biology of *Peromyscus* (Rodentia). Spec. Publ., Amer. Soc. Mamm., 2:1–593.

MCCABE, T. T., AND B. D. BLANCHARD. 1950. Three species of *Peromyscus*. Rood Assoc., Santa Barbara, Calif. 136 pp.

OSGOOD, W. H. 1909. Revision of mice of the genus Peromyscus. N. Amer. Fauna, 28:1–285.

SUMNER, F. B. 1932. Genetic, distributional, and evolutionary studies of the subspecies of deer mice (*Peromyscus*). Bibliog. Genet., 9:1–106.

SYSTEMATICS AND EVOLUTION

MICHAEL D. CARLETON

Abstract.—Research on the systematics and evolution of *Peromyscus* since Hooper's 1968 review has been more profuse and vigorous than that which appeared in the 60 years following Osgood's 1909 revision. This recent research activity evidences three principal trends: 1) proliferation of nontraditional data, especially karyologic and electrophoretic information, that address many facets of *Peromyscus* systematics; 2) revival of interest in Middle American *Peromyscus* that has revealed the complexity of certain species groups and the discovery of new species; and 3) focus on supraspecific systematic questions using more rigorous phylogenetic methods that have kindled debate over the definition and boundaries of the genus as arranged by Hooper. Taxonomic research bearing on species-level and supraspecific decisions is reviewed as a preamble to a revised classification of the genus and related peromyscine rodents, which include: 53 species of *Peromyscus* arranged in 13 species groups; four species of *Habromys*; one of *Podomys*; one of *Neotomodon*; three of *Megadontomys*; three of *Onychomys*; one of *Osgoodomys*; and two of *Isthmomys*. Other subjects discussed include the taxonomic discovery and refinement of species recognition of *Peromyscus*, relatives of *Peromyscus* and the generic scope within Muroidea, and areas for future research.

Growth in systematic knowledge of mice of the genus *Peromyscus* has been punctuated by two influential classifications that have served both as insightful summaries of the state of that knowledge and as catalysts to stimulate and focus subsequent research. Wilfred H. Osgood's (1909) "Revision of the Mice of the American Genus *Peromyscus*" appeared at a time of transition in species concepts and heralded the establishment of the polytypic species viewpoint in systematic mammalogy. Emmet T. Hooper's (1968) chapter on *Peromyscus* classification incorporated much new information from his own research efforts, which shifted questions of relationship involving *Peromyscus* to a higher taxonomic level, bringing fresh explicitness to its generic infrastructure and its hierarchical position within the radiation of New World mice and rats. The attention accorded the classificiation of this single genus of rodents is justified because *Peromyscus* has played a central role in the maturation of systematic mammalogy, especially in North America. Its significance in this regard parallels that of *Drosophila* in the growth of systematic biology as a whole, a comparison that emphasizes both the historical and current importance of studies of the genus *Peromyscus* in shaping our comprehension of the facts and theories of systematic biology and evolution. In recent years, the comparison to *Drosophila* has become even more apt by the formulation of a standard reference system to identify individual chromosomes of deer mice (Committee, 1977). This development, together with the advent of electrophoretic and other biochemical techniques, has thrust *Peromyscus* to the forefront of cytotaxonomic, cytogenetic, and genic studies and made it a favorite for research on population genetics—perhaps rivaled only by *Microtus* and *Mus* in the latter case.

The purpose of this account is to review research on the systematics and evolution of deer mice that has appeared since the publication of the *Biology of* Peromyscus *(Rodentia)* (King, 1968). Research in these fields has been prodigious and continues to be dynamic and vigorous, a situation which leaves me dissatisfied in attempting to provide a retrospective and synthesis seasoned by a longer passage of time. In addition, the chapter's title conceivably could encompass the topics covered in the chapters on paleontology, classification, speciation, habitats and distribution, anatomy, and genetics in the earlier volume, but I have found such coverage to be an intractable goal. Accordingly, the focus adopted herein corresponds most closely to Hooper's (1968) coverage of *Peromyscus* classification and thus emphasizes the taxonomic implications distilled from the literature surveyed, which I have tried to keep current through 1988. For many of the taxa discussed below, published information has been supplemented by my own examinations of museum specimens, particularly the types and series contained in the National Museum of Natural History (USNM) and the American Museum of Natural History (AMNH).

Other prefatory comments on scope and procedure are warranted. The classification set forth by Hooper in the 1968 *Peromyscus* volume forms the framework for my discussion. Wherever the generic scope is critical to interpretation of a statement, I shall make a distinction between *Peromyscus* sensu Hooper (1968) (or *s. l.*) and *Peromyscus* sensu Carleton (1980) (or *s. s.*). Although this review emphasizes contributions since 1968, I have made frequent reference to Osgood (1909) in order to provide the necessary historical perspective to current taxonomic problems and levels of comprehension. Graphs of the taxonomic composition of *Peromyscus* through time (Figs. 3, 5, 6) are based on the following sources: True (1885); Miller and Rehn (1901); Trouessart (1904–5); Osgood (1909); Miller (1924); Ellerman (1941); Miller and Kellogg (1955); Hall and Kelson (1959); Hooper and Musser (1964a); Hooper (1968); Corbet and Hill (1980); Hall (1981); and Honacki *et al.* (1982). For the cumulative plots of new taxa (Fig. 3), the taxonomic level (species or subspecies) is as given by the original describer. In order to provide a common denominator of generic limits over the past century, I have accepted the generic scope established by Osgood (1909) for depicting the chronology of subspecies and species descriptions. I have listed the species included in a given taxon at three points in time: Osgood's original 1909 revision; Hooper's 1968 classification; and Hall's 1981 edition of *The Mammals of North America*. Numbers in parentheses for Osgood (1909) and Hall (1981) refer to subspecies recognized, including the nominate subspecies.

I dedicate this paper to Emmet T. Hooper (Fig. 1), who served in function, although not officially, as major professor. Hooper's 1968 classification of *Peromyscus* is one of his best systematic essays, perhaps second only to his revision of *Reithrodontomys*. His concise discussion of what a systematic classification is and is not would reward a second or third reading. Although not couched in the Hennigian-Popperian language of current fashion, his brief discourse gives an eminently practical view and would seem to meet many of their tenets of phylogenetic inquiry and the nature of scientific hypotheses. In addition, his papers on dental variation in the genus (1957) and on the anatomy of the glans penis (1958) are minor classics of their kind that have substantially reordered our attitudes about characters considered important in the evaluation of kinship and construction of classifications, involving not only *Peromyscus* but also the superfamily Muroidea. Lee and Schmidly's (1977) description of *Peromyscus hooperi* is, therefore, one of the more fitting patronyms associated with the genus.

THEMES OF RECENT SYSTEMATIC RESEARCH

One can document convincingly that the quantity of research on *Peromyscus* in the past 20 years surpasses that which transpired in the three-score years between Osgood's (1909) revision and Hooper's (1968) review. The diversity and provocativeness of this recent research activity issue not only from the introduction of new analytical techniques and the attendant ability to assess untapped informational sources and variation, but also from the infusion of more explicit methodologies for evaluating those data bases, inferring relationships, and deriving classifications. In addition to the proliferation of nontraditional data in taxonomic studies of *Peromyscus*, the topics under investigation evidence a renewal of systematic interest in Middle American taxa and a focus on supraspecific systematic questions.

FIG. 1.—Emmet Thurman Hooper, Curator of Mammals, University of Michigan Museum of Zoology (1938 to 1979).

Proliferation of nontraditional data as taxonomic evidence.—"Non-traditional" relates only to the historical recency of using chromosomal and biochemical data, for by now the preparation of karyotypes and preservation of tissues are as standard field procedures as putting up museum skins. Clearly, the hallmark development over the past two decades has been the tremendous outpouring of karyological and electrophoretic results, addressing nearly every facet of *Peromyscus* systematics and evolution. To underline the vitality of these various research programs, I would note that of the 200 or so references accumulated in the preparation of this review, over 130 of them are post-1968, and almost 70 percent of those use karyotypic and biochemical approaches to questions of *Peromyscus* taxonomy and evolutionary biology. In contrast, Hooper's bibliography contained no references employing such data sources. The dominance of these research tools over this period appears almost incongruous, when we recall that Blair (1950:253), only a few years earlier, had disparaged the utility of cytological and genetic investigations, as then practiced and refined, for evolutionary studies of *Peromyscus*. If ever there was a more serendipitous juxtaposition of a perceptive systematic review and emerging technologies poised to examine those systematic questions, it was Hooper's 1968 paper and the explosion of karyologic and biochemical research that followed.

Although comparatively new as a technique of systematic investigation, karyology has already undergone substantial modification in its approach to questions of *Peromyscus* taxonomy. Earlier studies using standard bone marrow preparations applied a basically phenetic methodology to determination of interspecific relationships and directions of karyotypic change, concentrating on the number and relative lengths of chromosomal arms and on the morphology of the sex chromosomes (Hsu and Arrighi, 1966, 1968; Bradshaw and Hsu, 1972; Lee *et al.*, 1972). During this period, a karyotypic "character" was largely a meristic abstraction of the entire chromosomal complement (for example, diploid number, fundamental number, number of metacentrics), although some phylogenetic reconstructions attempted to identify homologous rearrangements among species for certain large and distinctive autosomes (for example, Hsu and Arrighi, 1968).

More sophisticated application of karyotypic data in mammalian systematics emerged in the middle 1970s, a transition fostered by technological advances in tissue culturing and staining procedures, adoption of a Hennigian approach to character analysis and phylogenetic inference, and the formulation of a standarized nomenclature to identify the 24 individual pairs of *Peromyscus* chromosomes. The ability to discriminate heterochromatic and euchromatic sequences within a chromosome permitted firmer establishment of chromosomal homologies among species and improved recognition of the kinds of chromosomal change that produced observed differences. With the greater resolution of chromosomal structure, a karyotypic "character" acquired a more restricted meaning, usually connoting one chromosome or even a single autosomal arm, with specific rearrangements (for example, pericentric inversions, addition or deletion of heterochromatic short arms, whole-arm translocations) corresponding to character state transformations. Delineation of chromosomal attributes and polarities at this level augmented the number of characters suitable for advancing phylogenetic hypotheses and engendered cladistic analyses of karyotypic data following the methods of Hennig (1966). One perception immediately altered by these new procedures has concerned the magnitude of karyotypic evolution in rodents. The notion of substantial genetic change, as suggested by wide differences in diploid and autosomal numbers (AN), has been supplanted by a more conservative picture of chromosomal evolution, as inferred from the apparent homology of banding patterns among distantly related taxa (Pathak *et al.*, 1973; Mascarello *et al.*, 1974).

For *Peromyscus*, the epochal study is that of Greenbaum and Baker (1978), who implemented a Hennigian analysis to depict the hierarchical relationships among eight species representing three subgenera and to reconstruct an ancestral karyotypic condition for the genus. They identified pairs 1, 22, and 23 as primitively biarmed and all other autosomal pairs as probably acrocentric, with heterochromatin confined to the centromeric region. This is the familiar "primitive condition" frequently referenced in the following Systematic Review, although chromosomal variation discovered recently within *Peromyscus* (*s. l.*) may entail reconsideration of the ancestral complement (Stangl and Baker, 1984*b*; Davis *et al.*, 1987). Greenbaum and Baker's study did much to consolidate a methodological approach to systematic treatment of karyotypic data and inaugurated an important series of cytotaxonomic investigations involving not only *Peromyscus* (*s. l.*) (Yates *et al.*, 1979; Robbins and Baker, 1981; Rogers, 1983; Rogers *et al.*, 1984; Stangl and Baker, 1984*b*) but also *Neotomodon* (Yates *et al.*, 1979), *Baiomys* (Yates *et al.*, 1979), *Onychomys* (Baker *et al.*, 1979; Baker and Barnett, 1981), *Reithrodontomys* (Robbins and Baker, 1980; Hood *et al.*, 1984), *Ochro-*

tomys (Engstrom and Bickham, 1982), *Scotinomys* (Rogers and Heske, 1984), *Nelsonia* (Engstrom and Bickham, 1983), *Neotoma* (Koop *et al.*, 1985), and various genera of South American sigmodontines (Elder, 1980; Baker *et al.*, 1983*a*; Koop *et al.*, 1983, 1984; Elder and Lee, 1985).

In addition to its many taxonomic implications, this research has provided insights to modes of chromosomal evolution within various taxonomic groups. In *Peromyscus* (*s. l.*), for instance, the number of centromeres is unaltered (diploid number constant), euchromatic banding patterns remain conservative, and chromosomal alterations primarily consist of pericentric inversions and heterochromatic additions (Robbins and Baker, 1981). The acquisition of heterochromatic short arms within *Peromyscus* is confined largely to members of three species complexes, the *leucopus* and *maniculatus* groups and *eremicus* and its kin, resulting in high autosomal numbers for these species (Greenbaum *et al.*, 1978*a*, 1978*b*; Robbins and Baker, 1981). This evolutionary polarity now seems well substantiated for these groups, controverting Lawlor's (1974) earlier contention of reductions in chromosomal arms as the principal direction of karyotypic change. In contrast, other New World genera exhibit different modes of karyotypic modification, for example, tandem and centric fusions in *Sigmodon* (Elder, 1980), translocations and centric fusions in *Oryzomys* (Baker *et al.*, 1983*a*), and solely heterochromatic additions in *Onychomys* (Baker and Barnett, 1981). Among some taxa, the extent of euchromatic rearrangements has been so pronounced that relatively few intact chromosomes or arms can be identified with reference to the Committee's (1977) standard; examples include certain species of *Reithrodontomys* (Robbins and Baker, 1980; Hood *et al.*, 1984), *Ochrotomys* (Engstrom and Bickham, 1982), and *Scotinomys* (Rogers and Heske, 1983). Assuming that our current generic constructs mostly correspond to monophyletic groupings, the amount of chromosomal evolution since the divergence of sigmodontine genera has been small relative to the chromosomal alterations that have occurred within many generic lineages (Baker *et al.*, 1983*a*; Koop *et al.*, 1984). Why one kind of chromosomal rearrangement should predominate in one line of descent and not another is poorly understood (see Greenbaum *et al.*, 1986), but the existence of such contrasts has led to frequent invocations of karyotypic orthoselection in the sense of White (1978:49). As cautioned by Patton and Sherwood (1983), this term more accurately relates to a descriptive statement of karyotypic trends rather than a causal process of evolution or speciation, though such a distinction is sometimes ambiguous in published contexts.

Recent publications appear to signal another transition in the use of karyotypic information, at least as its pursuit bears on questions of *Peromyscus* taxonomy. This latest shift seems to reflect the accumulating evidence of homoplastic karyotypic changes that have become apparent as more and more *Peromyscus* species were subjected to chromosomal analyses. Both the number of chromosomes suspected of homoplastic rearrangements and the total number of homoplasies predicted on chromosomal cladograms have increased with the number of species surveyed: three parallel rearrangements involving three different chromosomes for eight species sampled (Greenbaum and Baker, 1978: fig. 9); two parallelisms involving two chromosomes for ten species (Yates *et al.*, 1979: fig. 6); 14 parallelisms involving seven chromosomes for 18 species (Robbins and Baker, 1981: fig. 4); 16 parallelisms involving eight chromosomes for 27 species (Rogers *et al.*, 1984: fig. 2); and 21 parallelisms involving ten chromosomes for 31 species (Stangl and Baker, 1984*b*: fig. 3). The more recent attention to intraspecific polymorphisms, especially of pericentric inversions, also has reordered perceptions about chromosomal variability. Davis *et al.* (1986:648) observed that "As chromosomal studies of *Peromyscus* (particularly those involving banding analyses) have generally involved small numbers of individuals, the available data may not serve as reliable indicators of the extent of intraspecific chromosomal variation." Examples of population-level variation include the hybrid cline involving three pericentric inversions within *P. leucopus* (Stangl, 1986) and the apparent elevational trend involving heterochromatic additions and one inversion polymorphism among populations of *P. maniculatus* (Macey and Dixon, 1987).

As more chromosomal data have become available since the late 1970s, cladistic studies of banded chromosomes have modified their tone and procedure. Whereas Yates *et al.* (1979:40) viewed chromosomal changes to occur as "distinct events," Stangl and Baker (1984*b*) identified three species that exhibit rearrangement polymorphisms, collectively involving nine of the 23 autosomal pairs, at the population level. Whereas Yates *et al.* (1979:40) maintained that "certain types of chromosomal changes are rare (for instance, pericentric inversions) . . . ," independently evolved pericentric changes have now been reported for some seven to 10 chromosomes (Rogers *et al.*, 1984; Stangl and Baker, 1984*b*). Thus, Rogers *et al.* (1984) omitted chromosome 6, a karyotypic "hotspot," from their cladistic analysis because of its predicted tendency to undergo parallel rearrangements. Whereas Yates *et al.* (1979:40) asserted that ". . . there are enough pericentric inversions to establish synapomorphy patterns. . . ," Rogers (1983:621) cautioned that ". . . the resolving power of the G-band technique is such that autapomorphies may be disguised as synapomorphies or symplesiomorphies. . . ." Whereas earlier studies seldom acknowledged the confounding problem of reversals in euchromatic changes, Rogers *et al.* (1984) excluded reversal events but admitted parallelisms in generating their trees, and Stangl and Baker (1984*b*) considered both kinds of homoplastic rearrangements. Whereas Greenbaum and Baker (1978:829) suggested that ". . . in *Peromyscus* chromosomal characters evolve less rapidly . . . and, therefore, are better indicators of evolutionary history. . . ," Rogers *et al.* considered these data most effective at resolving relationships at the species group level and Stangl and Baker (1984*b*: 652) left the issue undecided: ". . . the cladistical analysis of the chromosomal data clearly documents the extensive amount of convergence. However, not until comparable cladistical analyses of classical morphological and genic data are performed, will we be able to determine if convergence is less a problem using these types of data." The papers of Rogers *et al.* (1984) and Stangl and Baker (1984*b*) are candid reviews of the cladistic applications and limits of currently available banding data, and identify problems of character definition and homology, polarity of character state transformations, detection of homoplasy, and decisions about character weighting. Such challenges form the grist of good phylogenetic inquiry using whatever kind of data set.

Although the detection of protein variation by electrophoresis of tissue extracts was a proven technique by the time of Hooper's (1968) review, he cited no studies employing this method. Their omission is understandable because, as noted by Avise (1974), earlier electrophoretic investigations seldom explored the taxonomic import of their results but rather used allelic variation to explore other questions of evolutionary significance, such as hypotheses concerning selective neutrality and genetic load, levels of heterozygosity, and the genetics of speciation. Some of these topics, as they relate to *Peromyscus* and other mammals, are covered in Smith and Joule's (1981) volume *Mammalian Population Genetics*. More general reviews of the application of electrophoretic data to systematic problems include Avise (1974), Straney (1981), and Buth (1984).

The first studies to apply electrophoretic methods to *Peromyscus* focused on single protein analyses, generally serum albumins or transferrins (Brown and Wesler, 1968; Jensen and Rasmussen, 1971). Because of this limited scope and difficulties involving procedural equivalence, such studies presaged limited taxonomic usefulness of the technique. For example, Brown and Wesler (1968: 423) found no apparent relationship between the serum albumin mobilities of 14 species of four subgenera of *Peromyscus* and accepted phylogenetic arrangements of the species. However, improvements of gels, buffer systems, and staining protocol, together with surveys of multiple loci (typically more than 20) using various tissues (usually liver, kidney, heart, testis, and blood), have demonstrated the taxonomic utility of protein electrophoresis. The standardization of the horizontal starch-gel method set forth by Selander *et al.* (1971) served as a paradigm of this research approach and gave impetus to the proliferation of electrophoretic studies of *Peromyscus* that appeared in the 1970s.

As in the evaluation of karyotypic information, methods of analysis and interpretation of electrophoretic data have evolved appreciably since their first application to *Peromyscus* taxonomy. The early single-protein studies typically presented visuals of gels, simple tabulations of alleles and their population frequencies, or both (for example, Brown and Wesler, 1968). The advent of multiloci surveys promoted the use of numerical taxonomic methods for summarizing allelic variation among populations and species and assessing relationships. Most studies of this kind on *Peromyscus* have adopted either Nei's distance coefficient (D) or Rogers' similarity coefficient (S) as a measure of genetic resemblance between samples and used the unweighted pair-group method with arithmetic averages (UPGMA) to amalgamate OTUs. Foremost examples of this investigative approach are the series of biochemical surveys initiated by Selander and colleagues, who have synthesized patterns of allelic variation at various hierarchical levels within *Peromyscus* and demonstrated its taxonomic pertinence (Selander *et al.*, 1971; Smith *et al.*, 1973; Avise *et al.*, 1974a, 1974b, 1979; Robbins *et al.*, 1985). In general, phenetic analyses of electrophoretic data have corroborated the traditional supraspecific associations of *Peromyscus* at the species group level, and to a lesser extent, the subgeneric groupings, but not all seven subgenera are represented among the approximately 27 species examined to date (Avise *et al.*, 1979; Schmidly *et al.*, 1985). Unlike the case of recent karyological studies, cladistic applications of electromorphic traits of *Peromyscus* are markedly fewer, in part because of methodological hurdles involving character definition, polymorphic character states, determination of character polarities, and undetected allelic variation (see Straney, 1981, and Buth, 1984, for discussions). The contributions of Patton *et al.* (1981) and Kilpatrick (1984) constitute notable exceptions to the general lack of cladistic analyses of electrophoretic data up to the present.

The evaluation of genic variation at various hierarchical levels of differentiation within *Peromyscus* has lent insight to the genetics of the speciation process (Zimmerman *et al.* 1978). Ranges of intraspecific genetic similarity (Rogers' S) generally measure 0.85 to nearly unity, with mean values clustering around 0.92 to 0.95 for 11 species of *Peromyscus* representing four subgenera (Table 1). A conspicuous exception is the strong degree of genetic differentiation recorded between population samples of *pectoralis*, which have a mean S near 0.80, a condition attributed to severe range disjunctions and the formation of Pleistocene refugia (Kilpatrick and Zimmerman, 1976). Gene flow now apparently occurs following the establishment of secondary contact of the former isolates. Such surveys of genetic variability within species of *Peromyscus* and other vertebrates have promulgated a 0.85 phenon-level "rule of thumb" for interpreting the specific differentiation of taxa whose status is unknown or problematic, because Rogers' similarity coefficient typically falls above 0.85 for most conspecific population comparisons (Selander and Johnson, 1973; Avise, 1974). The frequent exceptions to this generalization, even among the few well-studied species of *Peromyscus* (Tables 1 through 3), expose the hazards of its injudicious application, as stressed by several authors (Avise *et al.*, 1974*a*; Kilpatrick and Zimmerman, 1975). Cases in point involve *Peromyscus eremicus* and *merriami* or *P. boylii* and *levipes*, pairs of species with similarity values greater than 0.85 where their ranges overlap but whose sympatry underscores the maintenance of their genetic identities (Avise *et al.*, 1974*b*; Rennert and Kilpatrick, 1986, 1988).

Genetic divergence is generally greater at kinship levels more distant than conspecific populations, but there are exceptions and considerable overlap exists (Tables 2, 3). These tabular comparisons are similar to the stratification of genetic differentiation presented by Zimmerman *et al.* (1978). However, I feel the examination of allelic similarities between probable sister species is less arbitrary than the designations of "semispecies" and "sibling" species by Zimmerman *et al.* (1978), as their definitions of those terms

TABLE 1.—*Genetic similarity between conspecific populations of* Peromyscus. *(References: 1, Selander and Johnson, 1973; 2, Smith et al., 1973; 3, Avise et al., 1974a; 4, Avise et al., 1974b; 5, Zimmerman et al., 1975; 6, Kilpatrick and Zimmerman, 1975; 7, Browne, 1977; 8, Avise et al., 1979; 9, Smith, 1979; 10, Price and Kennedy, 1980; 11, Robbins et al., 1985; 12, Rennert and Kilpatrick, 1986; 13, Werbitsky and Kilpatrick, 1987; 14, Nelson et al., 1987; 15, Calhoun et al., 1988).*

Taxa	No. samples[1]	No. loci	Rogers' S mean (R)	Reference
Haplomylomys				
californicus	4	25	.94 (.91–.98)	4
" "	13	31	.90 (.82–.98)	9
eremicus	11	25	.86	4
Peromyscus				
maniculatus Group				
maniculatus	18	22	.93 (.87–.97)	8
" "	13	24	.97 (.95–.99)	15
polionotus	30	32	.95 (.82–.99)	1
melanotis	4	22	.92 (.88–.98)	8
leucopus Group				
leucopus	6	28	.95 (.92–.99)	7
" "	29	14	.91[2] (.83–.99)	10
" "	21	25	.94[2] (.92–.99)	11
" "	21	38	.94	14
gossypinus	11	14	.91[2] (.85–.97)	10
" "	8	25	.95[2] (.92–.98)	11
boylii Group				
boylii	7	21	.95 (.89–.99)	3
" "	5	15	.92 (.86–.99)	5
" "	15	17	.92 (.84–.98)	6
" "	5	30	.95 (.88–.97)	12
attwateri	22	17	.97 (.94–.99)	6
pectoralis	3	21	.79 (.75–.84)	3
" "	4	15	.83 (.78–.89)	5
" "	20	27	.83 (.69–.99)	6
Podomys				
floridanus	4	39	.97 (.96–.98)	2
Megadontomys				
thomasi	4	30	.80 (.72–.90)	13
			Grand mean = .91	

[1] "Samples" usually correspond to single localities but may represent composites of several collecting sites.
[2] Midpoint of range (R).

imply a certain mode of speciation and conclusions regarding reproductive isolation and lack of morphological distinction. For instance, their view (1978:572) that *truei-gentilis* (= *gratus*) and *difficilis-nasutus* are sibling species that ". . . developed reproductive isolation without concomitant morphological differentiation . . ." remains largely unsubstantiated. Viewed from a cladistic perspective, the degree of genetic divergence roughly corresponds to hierarchical levels of relationship (Tables 1 through 3) and does not present the enigmatic contrasts in genetic similarity that Zimmerman *et al.* (1978) discovered for semispecies and sibling species. This pattern of monotonically increasing average genetic distances may obtain regularly for kinship levels successively greater than conspecific populations and sister species, as noted by Schnell and Selander (1981). Such a pattern is observed, for example, in comparisons of genetic similarities of *eremicus* and *eremicus*-like derivatives to that of *eremicus* and *californicus* (Table 3), but statements of relationship at these earlier branching levels are less well supported.

The occurrence of at least limited fertility between certain sister species and even between some non-sister species of the same species-group suggests that attainment of species status, as judged by sympatry and the absence of appreciable introgression, may transpire without the irreversible divergence of gene pools that would make interbreeding impossible and guarantee reproductive isolation. These assertions perhaps assume that the two dozen-odd proteins scored in electrophoretic studies are typical genomic subsets and that the differences quantified somehow index genetic changes relevant to the speciation process, which may not be true in either case. *Peromyscus leucopus* and *P. gossypinus* are good ex-

TABLE 2.—*Genetic similarity between presumed sister species of* Peromyscus. *(References: 1, Avise et al., 1974a; 2, Avise et al., 1974b; 3, Zimmerman et al., 1975; 4, Avise et al., 1979; 5, Price and Kennedy, 1980; 6, Kilpatrick, 1984; 7, Robbins et al., 1985.).*

Sister species	Inter-fertility[1]	Distribution[2]	Rogers' S mean (R)	Reference
Haplomylomys				
eremicus-merriami	?	BS	.92 (.88–.95)	2
guardia-interparietalis	?	A	.92	2
Peromyscus				
maniculatus Group				
maniculatus-polionotus	I	LS	.90	4
leucopus Group				
leucopus-gossypinus	I	BS	.66 (.57–.74)	5
" "	I	BS	.80 (.75–.86)	7
boylii Group				
boylii-attwateri	?	A	.81 (.80–.81)	1
" "	?	A	.78 (.73–.81)	3
" "	?	A	.79 (.71–.84)	6
truei Group				
truei-gratus	?	LS	.86	3
" "	?	LS	.87	4
difficilis-nasutus	?	A	.81	3
" "	?	A	.86	4
		Grand mean = .83		

[1] N = not interfertile in lab; I = at least partially interfertile in lab; ? = laboratory fertility unknown.
[2] A = allopatric; LS = limited extent of sympatry; BS = broadly sympatric.

TABLE 3.—*Genetic similarity between selected non-sister species-pairs of* Peromyscus *from the same supraspecific taxon. (References: 1, Avise* et al., *1974b; 2, Zimmerman* et al., *1975; 3, Kilpatrick and Zimmerman, 1975; 4, Avise* et al., *1979.).*

Species-pairs	Inter-fertility[1]	Distribution[2]	Rogers' S mean (R)	Reference
Haplomylomys				
eremicus-interparietalis	I	A	.87	1
eremicus-dickeyi	?	A	.91	1
eremicus-caniceps	?	A	.77	1
eremicus-californicus	N	BS	.69	1
Peromyscus				
maniculatus Group				
maniculatus-melanotis	I	LS	.72 (.66–.78)	4
maniculatus-sejugis	?	A	.79	4
boylii Group				
boylii-pectoralis	?	BS	.65 (.63–.66)	2
attwateri-pectoralis	?	LS	.63 (.58–.64)	3
polius-pectoralis	?	A	.52	2
polius-boylii	?	BS	.67 (.62–.71)	3
truei Group				
truei-nasutus	I	BS	.86	2
" "	I	BS	.75 (.74–.77)	3
gratus-difficilis	?	BS	.74	2
" "	?	BS	.77	3
		Grand mean =	.74	

[1] N = not interfertile in lab; I = at least partially interfertile in lab; ? = laboratory fertility unknown.
[2] A = allopatric; LS = limited extent of sympatry; BS = broadly sympatric.

amples of species that retain some interfertility, as demonstrated under laboratory conditions (Dice, 1937; Bradshaw, 1968), yet exhibit little morphological or genetic evidence of hybridization in the wild (Engstrom *et al.*, 1982; Price and Kennedy, 1980; Robbins *et al.*, 1985). Such observations on *Peromyscus*, and knowledge of the substantial overlap of genetic similarities at various levels of differentiation, conform to the consensus that has emerged from electrophoretic studies of *Drosophila*, namely, that speciation does not necessarily entail extensive reorganization of gene pools: "Against the yardstick of infraspecific differentiation, speciation does not stand out as a special process" (Throckmorton, 1977:249).

Cytogenetic and molecular techniques other than karyology and electrophoresis have been used infrequently in studies of systematic relationships of *Peromyscus*. Microcomplement fixation analysis of serum albumins (Fuller *et al.*, 1984), DNA-DNA hybridization (Brownell, 1983), and restriction analysis of mitochondrial DNA

(Ashley and Wills, 1987; Nelson *et al.*, 1987) are exceptions that will likely see expanded use in the next decade.

Although cytogenetic and biochemical approaches have characterized much of the recent research on *Peromyscus* systematics, studies relying upon information sometimes stereotyped as "classical morphology" have not been moribund over this same period. It should be emphasized, however, that the procedural and methodological underpinnings that animated the "morphological" studies of Osgood (1909), Hooper and Musser (1964*a*, 1964*b*), and Carleton (1977, 1980) do differ from one another. Similarly, it would be as erroneous to equate both standard karyotypic analyses of the fifties and sixties and euchromatic banding studies of the seventies and eighties as examples of "classical karyology." Some recent contributions drawing upon diverse aspects of the rodent phenotype include Linzey and Layne's (1969) anatomical treatise of the male accessory reproductive glands, Bradley and Schmidly's (1987) survey of phallic diversity in the *boylii* group, Musser's (1969, 1971) taxonomic treatment of various Central American species, and Schmidly's (1972, 1973*b*) and Carleton's (1977, 1979) work with mice of the *boylii* group.

The numerical taxonomic programs developed by the phenetic school of systematic philosophy have encouraged more rigorous assessment of variation and documentation of taxonomic decisions. Multivariate analyses of cranial, dental, and external measurements have been applied in two principal ways regarding questions of *Peromyscus* taxonomy. One has involved discrimination of sympatric species whose morphological distinction is slight, thereby providing less ambiguous means of identification (for example, Choate, 1973; Choate *et al.*, 1979; Engstrom *et al.*, 1982), and the other has examined broader covariation patterns as a yardstick to infer species limits and to evaluate the status of problematic forms (for example, Schmidly, 1973*b*; Carleton, 1977, 1979; Schmidly *et al.*, 1988).

The ability of multivariate analyses to discriminate very subtle size and shape differences of the peromyscine skull has revealed the surprising constancy of such contrasts, particularly at the level of species delimitation (for example, Schmidly, 1973*b*; Carleton *et al.*, 1982; Engstrom *et al.*, 1982; Schmidly *et al.*, 1988). These findings, together with the substantial component of genetic heritability demonstrated for muroid morphometric variables (Bader, 1965; Leamy, 1977; Atchley *et al.*, 1981; Lofsvold, 1986, 1988), encourage the notion that significant dispersion in multivariate space may reflect underlying genetic divergence of the sampled populations. The best support for this critical assumption derives mainly from quantitative genetic studies on inbred laboratory species, for which the breeding regimen and pedigrees are known. Lofsvold (1986, 1988), however, obtained supportive data in studies using the laboratory progeny of wild-caught *P. leucopus*, *P. maniculatus bairdii*, and *P. m. nebrascensis*, specimens that had been preserved from the breeding stocks maintained by Lee R. Dice and his associates. His results suggested a positive but weak correlation between phenotypic and genetic covariance patterns within each taxon and a general correspondence between increasing differences in genetic covariance structure and greater distances in multivariate space for the taxonomic levels examined (populations, subspecies, and species). Whether the genetic and environmental components of cranial metric variation in natural muroid populations conform to these results invites study.

Less attention has been devoted to assessing broad geographic trends among populations of *Peromyscus* using morphometric data. Smith's (1979) study of *Peromyscus californicus* represents one of the few to employ multivariate techniques to collate patterns of geographic variation of craniodental dimensions and to delimit subspecific boundaries.

Smith's (1979) study, which included complementary analyses of electrophoretic characters as well, also serves as an example of the need for multidisciplinary analyses of such a complex genus and the advantage of elevating our taxonomic inferences beyond the provinciality of a single informational base, analytical technique, or systematic methodology. Other examples of this approach are Lawlor's (1971) investigation of the evolutionary origin of *Peromyscus* on islands in the Sea of Cortez and Schmidly's *et al.* (1985) effort to resolve the enigmatic phylogenetic status of *Peromyscus hooperi*. The ascendency of multidisciplinary studies in recent decades partly accounts for the increase of multiauthored papers dealing with *Peromyscus* systematics (Fig. 2). However, this trend may also reflect the nature of funding sources now supporting much systematic work, coupled with the broader diffusion of systematic research into a university setting conducive to the collaboration of professor and student.

Systematic emphasis on Middle American taxa.—A major advance in *Peromyscus* systematics in recent years has been the revisionary at-

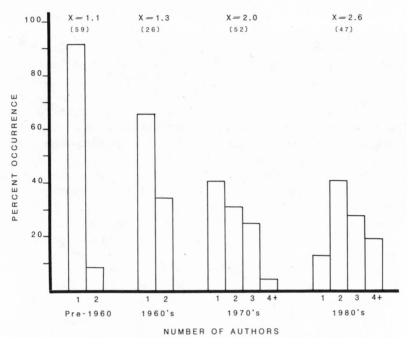

FIG. 2.—Frequency histogram illustrating the increase in number of multi-authored papers dealing with *Peromyscus* systematics over the past three decades.

tention devoted to species-groups and subgenera inhabiting tropical and subtropical realms of Mexico and Central America. Mice of the *boylii* group, in particular, have been examined from morphological, chromosomal, and biochemical perspectives—for example, Avise *et al.* (1974*a*); Schmidly and Schroeter (1974); Carleton *et al.* (1982); Bradley and Schmidly (1987); Rennert and Kilpatrick (1987); Houseal *et al.* (1987); and Schmidly *et al.* (1988). This scrutiny has revealed the complex relationships and unsuspected diversity of species assigned to this assemblage, findings inconsistent with their traditional assignment to a single species group. Huckaby's (1980) synopsis of the *mexicanus* group provided a preliminary look at species limits and patterns of variation in this largest of species groups recognized by Hooper (1968). The immense morphological variation embraced by this group also casts doubt on its monophyletic nature, but the taxa karyotyped thus far enigmatically appear monomorphic in their banding patterns (Stangl and Baker, 1984*b*; Smith *et al.*, 1986).

For a rodent fauna presumed to be among the best documented in the world, the description of four new species of *Peromyscus*, three from Mexico and one from Guatemala, in the latter half of the 1970s is something of a surprise (Carleton and Huckaby, 1975; Robertson and Musser, 1976; Carleton, 1977; Lee and Schmidly, 1977). That these discoveries emerged from collections made in the 1960s and afterwards underscores the continuing need for basic field surveys of poorly known areas. Two of the new species diagnoses forecasted problems with the supraspecific arrangement of *Peromyscus*, one (*P. mayensis*) at the species-group level and one (*P. hooperi*) at the subgeneric level.

Focus on supraspecific systematic questions.—The third major theme of *Peromyscus* research during the last two decades is an outgrowth of the creative tension that has marked the dialogue between cladistic, phenetic, and evolutionary taxonomic philosophies. Thus, confidence in the monophyly of our taxonomic entities, knowledge of their patterns of descent, and the construction of formal classifications to reflect and retrieve that phylogenetic information have become increasingly prominent research goals in systematic mammalogy. And these goals are being addressed by a bewildering array of newly developed systematic tools and information bases, and by new, more rigorous ways of interpreting traditional sources of taxonomic information. As a result, much attention has been directed toward verifying the monophyly of the various subdivisions of *Peromyscus* and resolving their hierarchical relationships, activities which have led to broader questions about the definition and boundaries of the genus itself.

A consensus has emerged that Hooper's generic concept is inadequate, but a basic dichotomy exists on how to nomenclaturally redress that inadequacy. The contradictory recommendations reference different character complexes and perhaps reflect different systematic philosophies, although there are some elements of congruence in the phylogenies hypothesized. One viewpoint, drawing upon cladistic interpretations of banded chromosomes and first advanced by Yates *et al.* (1979), advocates expansion of the generic scope to include *Neotomodon alstoni*. More recently, cladistic argumentation based on broader surveys of inversion data would require that *Peromyscus* should also subsume *Onychomys* (Stangl and Baker, 1984*b*). The other recommendation, put forth by Carleton (1980) using quantitative phyletic anaylses of morphological evidence, favors restriction of the generic definition to include only the subgenus *Haplomylomys* and the nominate subgenus, and elevation of the subgenera *Podomys*, *Habromys*, *Osgoodomys*, *Megadontomys*, and *Isthmomys* to generic status. Either proposal is somewhat disconcerting to the longtime familiar generic concept of *Peromyscus* that we learned from Hall and Kelson (1959) and Hall (1981).

TAXONOMIC DISCOVERY AND REFINEMENT OF SPECIES RECOGNITION
OF *Peromyscus*

The manner in which systematists have formally recognized our accumulated knowledge of the biological diversity of *Peromyscus* mirrors historical changes in our species concept and chronicles fundamental revisions in the treatment of individual, population, and species variation. During the early discovery phase (1840 to 1885) of *Peromyscus*, systematists had not yet grasped the distinction between individual and geographic variation. This was the era of the morphological or essentialist species concept, and almost all new forms were recognized as species (Fig. 3). In fact, the sub-

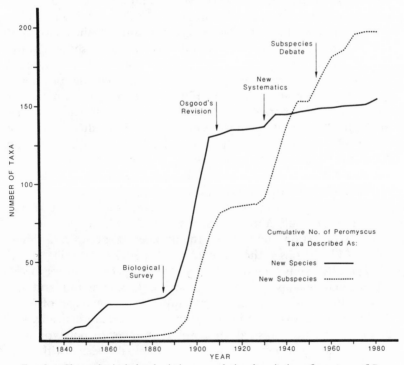

FIG. 3.—Chronological plot depicting cumulative description of new taxa of *Peromyscus* (data from Hall, 1981).

species was not a standard part of our taxonomic filing system during this period—the few subspecies names dating back to this early phase were described as varieties.

The appreciation of geographically variable species populations emerged dramatically with C. Hart Merriam and the inception of the Biological Survey in 1885. These early field studies—the boundary surveys, explorations into Alaska and western North America, Merriam's life zone transects, and Nelson-Goldman's expeditions into Mexico and Central America—uncovered an astonishing complexity of variation among mammals that was not consonant with the fixity of a morphological species concept. Our knowledge of *Peromyscus* grew precipitously during this later discovery period (1885 to 1905), an interval marked by the appearance of Merriam's (1898*b*) descriptions of 20 new species of *Peromyscus* and Osgood's (1904*b*) diagnoses of 30 new forms from Mexico and Guatemala. The trinomial appeared as a regular infraspecific category during this phase, probably reflecting its regular employment by ornithologists and the fact that early workers on mammals were ornithologists first by training and profession. Still, in testimony to the pervasiveness of the morphological species concept, the majority of new taxa diagnosed over this period were named as species, at the rate of about three species to every two new subspecies recognized (Fig. 3). In a very real way, these early scientific activities promoted by the Biological Survey gave identity to systematic mammalogy as a field of study separate from ornithology in North America. The interactions and mutual interests of the many individuals who were a part of the Survey also helped to foster the establishment of the American Society of Mammalogists.

Osgood (Fig. 4) undertook his revision of *Peromyscus* at the end of this fertile descriptive era and had to contend with 130 nominal species and over 60 subspecies ascribed to the genus. Drawing upon the then substantial number (over 27,000) of specimens in museum collections, he reduced the number of species considered valid to 43 and raised the number of subspecies to 100 (Fig. 5). In doing so, he firmly established the polytypic species concept in systematic mammalogy and set a standard of documentation and format for the revisionary studies that appeared in the pages of *North American Fauna* over the next several decades. Paradoxically, Osgood's revision diminished the absolute number of polytypic species because he consolidated several nominal species with geographic representatives into even larger polytypic arrays, as for instance, in his treatment of *leucopus* and *maniculatus*. However, if

Fig. 4.—Wilfred Hudson Osgood, Assistant Biologist, Biological Survey (1897 to 1909), and later Curator of Mammals, Field Museum of Natural History (1909 to 1941).

one looks at the number of polytypic species as a percentage of those species considered valid, then Osgood's conclusions are observed to have dramatically boosted their use in *Peromyscus* taxonomy (Fig. 6).

Predictably, a quiesent period of taxonomic activity followed the appearance of Osgood's treatise in 1909, but the description of new

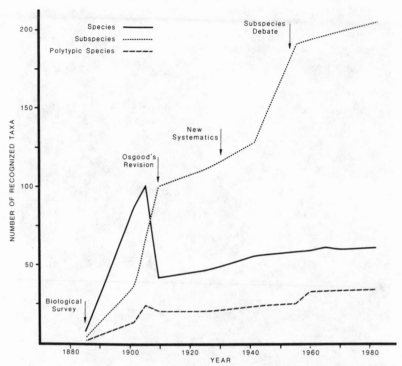

FIG. 5.—Chronological plot of taxa of *Peromyscus* considered valid (see Introduction for literature sources).

Peromyscus did resume two decades later, albeit with a different focus. I have borrowed Huxley's term "New Systematics" to characterize this period of taxonomic investigation of *Peromyscus*, but it's roots are with Fisher's (1930) seminal work *The Genetical Theory of Natural Selection*. In addition to Fisher, this was the time of Haldane, Wright, Dobzhansky, and Mayr and the formulation of the biological species concept. The concern of systematic biologists now focused on the genetics of speciation, the origin of new species, and the role of geographic variation and differentiation in the process of speciation. As a consequence, geographic races or subspecies assumed new significance for understanding presumed underlying patterns of genetic discontinuities and for identifying species *in statu nascendi*. Accordingly, subspecies descriptions became a conspicuous part of revisionary and faunal studies conducted over this period by scientists such as R. H. Baker, W. H. Burt, I. McT. Cowan, G. G. Goodwin, E. R. Hall, D. F. Hoffmeister, and E. T.

Hooper. Despite the emphasis on formal recognition of geographic variants, the percentage of polytypic species of *Peromyscus* increased but slightly throughout the period of the New Systematics (Figs. 5, 6). The unexpected stability in number of polytypic species subsequent to Osgood's revision reflects primarily the fact that most new subspecies were named in species already interpreted as polytypic (for example, *maniculatus, leucopus,* and *truei*), and secondarily that approximately as many new monotypic species were named as were demoted to subspecific status. Still, the number of new subspecific taxa quickly multiplied, surpassing those described or recognized as species (Figs. 3, 5).

Unfortunately, the naming of subspecies became an end in and of itself, precipitating the critical review of this practice by Wilson and Brown (1953). Their paper sparked about a decade of debate on the value of the subspecies category and how to convey, or even whether one should convey, interpopulational variation in formal classifications. A lag period occurred as these new arguments gained credence and were assimilated, with the result that subspecies descriptions reached a plateau by the end of the 1960s, at least as re-

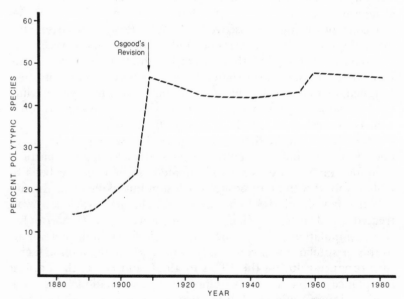

FIG. 6.—Chronological plot of number of polytypic species of *Peromyscus*, expressed as a percentage of the species considered valid (see Introduction for literature sources).

flected in *Peromyscus* taxonomy (Fig. 3). Bowen's (1968) diagnoses of five new geographic races of *P. polionotus* probably represent the last burst of such classical subspecific taxonomy, which will not likely resume.

Kuhn (1970) characterized progress in scientific understanding as a succession of paradigmatic revolutions, in which a prevailing theory initially constrains both the nature of our questions and the interpretations of our data. Gradually the number of factual discrepancies observed and the increasingly frequent recourse to *ad hoc* explanations to preserve the current theory in its original form cause it to be abandoned for another ordinating theory. The ironic feature of these paradigmatic replacements is that scientists often cite the same data and evidence as proof of the explanatory power of both the old and the new theories.

The historical changes in rates of subspecies and species descriptions (Figs. 3, 5, 6) suggest that we have passed through two such periods as regards the assessment of morphological variety and its bearing on the definition and limits of species of *Peromyscus*. Perhaps we stand at the threshold of a third period, wherein we must readjust our definition of species limits to a finer degree than previously appreciated by systematists working on *Peromyscus*. For in the same way (but not degree) that adherence to a morphological species concept during the discovery phase of *Peromyscus* systematics inflated the number of species beyond biological reality, recent revisionary studies suggest that, under the guiding influence of the New Systematics, scientists had occasionally become dogmatic and assigned subtle-but-constant populational contrasts to subspecific-level variation. For instance, in discussing the *nasutus-difficilis* problem, Hoffmeister and De La Torre (1961:2) reasoned that "The changes that occur are those to be expected between populations of one species and not between two species." In selecting this particular quote, my purpose is not to intimate that the issue has been finally resolved with an opposite conclusion but to illustrate the notion that certain kinds of characters and character variation were treated as only of subspecific value; they were ". . . *to be expected* between populations of one species . . ." [italics mine]. Statements to this or similar effect commonly appear in the taxonomic literature produced during the 1930s to 1960s and reveal the implicit acceptance of geographic variation as an explanation for certain kinds of interpopulational differences.

The advent of karyotypic investigations has spurred awareness of finer degrees of discrimination and narrower limits of intraspecific variation. Zimmerman's (1970) karyotypic investigation of *Sigmodon* represents a hallmark example among North American rodents, in which he revealed highly disparate diploid counts between "subspecies" of *hispidus* in the western United States and Mexico. Difference in diploid number and the suspicion of meiotic incompatibility were not the only compelling arguments in raising these karyotypic morphs to species, but also the nearness of their geographic occurrence and lack of intergradation. This is not to say that the morphological differentiation of these forms was unappreciable, as identifying features were later demonstrated by Severinghaus and Hoffmeister (1978), only that the diagnostic cranial features were judged insufficient for species recognition by their first revisor (Bailey, 1902).

Two other examples drawn from recent changes in the taxonomy of the *boylii* species group give evidence of a narrower delineation of morphological limits in identifying species of *Peromyscus*. Together they emphasize the need for caution and recourse to other kinds of information before invoking geographic variation to explain slight differences in craniodental morphologies among populations of *Peromyscus*. The first example involves the western Mexico populations that Osgood (1909) interpreted as an intergrading series of subspecies, principally *rowleyi*, *spicilegus*, and *simulus*. As reviewed below, current research suggests that the populations represented by each of these epithets are genetically discrete from one another: multivariate analyses have detailed the constancy of subtle contrasts in size and shape; instances of sympatry have been recorded in areas where intergradation might be expected; and congruent evidence drawn from phallic anatomy and chromosomal data has supplied other means of species discrimination (Schmidly and Schroeter, 1974; Carleton, 1977; Carleton *et al.*, 1982). The slight cranial differences now shown unambiguously to separate examples of *simulus*, *spicilegus*, and *rowleyi* in multivariate space were qualitatively recognized by Osgood, who cited them in his comparisons and descriptions; yet where he and subsequent systematists considered their value as indicative of subspecies-level variation, other kinds of information now elevate their significance in species delimitation. The difference in treatment partly is due to refinement in character weighting, for Osgood seemingly attributed greater significance to pelage characteristics of deer mice in inferring intergradation, as previously noted by Schmidly (1973*b*) and Carleton *et al.* (1982).

The second example comes from Hooper's (1957) classic paper

on dental variation in *Peromyscus*, in which *P. boylii* is used to exemplify a species with pronounced geographic variability in occurrence of accessory lophs and styles (Fig. 7). This study forcefully disputed the exclusive taxonomic significance historically accorded the presence or absence of these enamel structures, not only in the study of *Peromyscus* but of other cricetids as well, and, in this context, I certainly agree with his general assessment that they had been given undue weight and were no less variable than other cranial parts. However, subsequent studies have disclosed that, instead of a single, dentally plastic species, Hooper's *boylii* samples comprise at least five separate species, namely *boylii* (western U.S. and Durango), *simulus* (Sinaloa), *spicilegus* (Nayarit and Jalisco), *aztecus* (El Salvador), and *attwateri* (Oklahoma) (Schmidly, 1973*b*; Schmidly and Schroeter, 1974; Carleton, 1977, 1979; Carleton *et al.*, 1982). Schmidly (1973*b*), in fact, drew attention to the constancy of accessory cusp development in arguing for the specific distinction of *attwateri* from *boylii*. Although the removal of these localities and species substantially alters the range of dental variation pictured for *boylii* (Fig. 7), the degree of variability may still be exaggerated, for the conspecificity of the remaining population samples remains clouded, those from the western United States and Durango representing *b. boylii* and *b. rowleyi* and those from southern Mexico *b. levipes* (see Rennert and Kilpatrick, 1987; Schmidly *et al*, 1988). Thus, although intraspecific variation in the frequency of accessory enamel structures does occur, as revealed by Hooper (1957) and others (for example, Wolfe and Layne, 1968), the magnitude of such variation is not as great as implied by the composite example of *boylii*.

Our subspecies legacy within *Peromyscus*, as so neatly mapped out in Hall (1981), for example, stems from Osgood's (1909) original revision, supplemented by the pulse of subspecific descriptions from the 1930s through 1950s. Many, if not most, of these geographic races would not survive present standards of systematic analysis and documentation, for their diagnoses were frequently based on qualitative assessments of size and color differences and on limited geographic representation. Given the historical genesis of our subspecies nomenclature and their somewhat anachronistic nature, their employment having been associated with a period of changing species concepts, I find it ingenuous to fault Osgood with misapplying the subspecies concept (Stangl and Baker, 1984*a*: 142), particularly when its approved definition, that of an independently

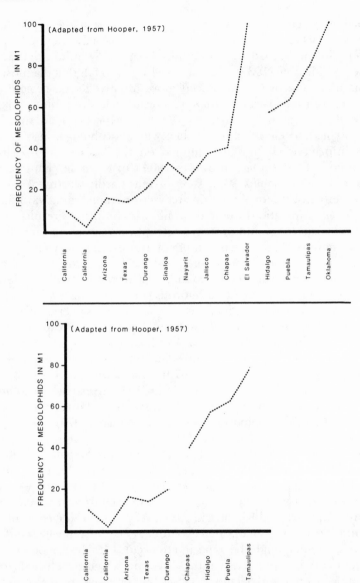

FIG. 7.—Frequency of occurrence of the mesolophid in populations of the *Peromyscus boylii* species group. *Top*—variation as originally depicted by Hooper (1957) for the species *P. boylii*. *Bottom*—variation of *P. boylii* as portrayed after removal of locality samples determined by later investigators to represent different species.

evolving genetic entity, is set forth in a paper published in the 1960s (Lidicker, 1962). The failure of conventionally defined subspecies in this regard was emphasized earlier by Sumner (1932), Dice and Blossom (1937) and Blair (1943*a*), who felt that diagnostic traits that were usually advanced more often reflected ecological responses or trends rather than genetically cohesive units, a point reiterated by Wilson and Brown (1953). To expect concordance of karyotypic and electrophoretic data with morphology, insofar as that informational source is judged on the basis of subspecific boundaries of *P. leucopus* drawn in 1909 (for example, Stangl and Baker, 1984*a*; Robbins *et al.*, 1985), seems hardly worthy of note. Whether careful morphometric analyses of craniodental variation of *leucopus* populations would or would not conform to results suggested by karyology and protein electrophoresis is unknown at this point; however, an interpretation of the evolutionary independence of genic or chromosomal vis-à-vis morphological change based on archaic subspecific boundaries formalized with modern zip-a-tone does not constitute rigorous proof of same.

In cases where thorough multivariate summaries of cranial variation have been performed, there is an impressive correspondance of relationships based on these data with karyotypic or, more often, electrophoretic patterns. Examples include studies of geographic variation of *californicus* (Smith, 1979), *maniculatus* (Caire and Zimmerman, 1975), and *pectoralis* (Schmidly, 1972; Kilpatrick and Zimmerman, 1976), and analyses of the taxonomic status of *attwateri* (Schmidly, 1973*b*; Kilpatrick and Zimmerman, 1975), *comanche* (Schmidly, 1973*a*; Johnson and Packard, 1974; Modi and Lee, 1984), and the *boylii-levipes* complex (Rennert and Kilpatrick, 1987; Houseal *et al.*, 1987; Schmidly *et al.*, 1988). One notable exception involves the lack of congruence between craniometric and electrophoretic data for samples of *maniculatus* on islands off the coast of Maine (Aquadro and Kilpatrick, 1981), yet just such zoogeographic circumstances form an optimal evolutionary context in which independent change might be expected to occur. Clearly, the potential evolutionary independence of these systems is well established (for example, Schnell and Selander, 1981), but, precisely in view of such documentation, the concordance of electrophoretic, karyotypic, and morphological data bases at the population-level within *Peromyscus* is stronger than one might anticipate.

The subspecies taxonomy of *Peromyscus* requires renewed attention if we are to use it to critically evaluate the correlation and significance of various kinds of evolutionary modification and their genesis at the population level. The challenge is forbidding, but the studies of Smith (1979) on *californicus* and Schmidly (1972) and Kilpatrick and Zimmerman (1976) on *pectoralis* offer model examples of careful analysis of patterns of geographic variation. A firmer grasp of such patterns is likewise required to sustain inferences about microevolutionary processes. With some exceptions, our current subspecific classification of *Peromyscus* is insufficient for this purpose.

A correlative dividend of such revisionary attention will be to further identify valid biological species. At this stage of our knowledge, I find it difficult to accept that species of *Sigmodon* are cryptic species, as implied by Patton and Sherwood (1983), or that *Peromyscus truei* and *gentilis* (= *gratus*) are sibling species "indistinguishable by typical means," as maintained by Zimmerman *et al.* (1978)—unless one construes "cryptic" to mean by the criteria set forth in late nineteenth or early twentieth century revisions and "typical means" to signify gestalt judgements of museum skins and skulls. The fineness of our taxonomic focus in many muroid genera is still too coarse to allow conclusions that certain species are indeed morphologically indistinguishable. Certain forms of the *boylii* species group, namely *levipes* and *baetae*, may qualify as the best examples of cryptic species within *Peromyscus*, but their status requires further substantiation (see Schmidly *et al.*, 1988). In order to unequivocally identify such cases within *Peromyscus*, in the same manner as that documented for mole rats of the genus *Spalax* (Nevo, 1979), then we must further refine our species-level systematics beyond that available now.

The probable number of species in the genus *Peromyscus* warrants comment in light of the foregoing review and the specific predictions tendered by both Osgood (1909:24) and Hooper (1968: 39). Writing 60 years apart, it is remarkable that each man converged on 40 as the probable number of valid species. Osgood listed 43 species, but he considered many of these to be subspecies "masquerading" as full species; thus, he speculated that "The number of bona fide species scarcely exceeds forty, and of these some half dozen eventually may be reduced in rank." By the time of Hooper's review, the generic scope of *Peromyscus* had been restricted by the elevation of *Baiomys* and *Ochrotomys* to genera. Still his classification numbered 57 nominal species. Hooper expressly adhered to the biological species concept in the sense of Mayr's

(1963) definition and predicted that future revisionary studies would reduce that count to 40, not to ". . . exceed 50 unless the scope of the genus is enlarged or another concept of the species is employed."

The estimate of 40 by both students of *Peromyscus* clearly reflects a polytypic species viewpoint and the expectation that many monotypic forms they listed as species would one day be reclassified as geographic variants. Yet the systematic research of the past 20 years has not exhibited such a reductionary trend, even though our usage of the species category in *Peromyscus* still essentially conforms to a biological species definition, to the extent that our mostly inferential and secondhand knowledge can appraise such. In this review, I recognize 53 species (Table 4) in *Peromyscus (sensu stricto)*, with another 11 species distributed among the five taxa Hooper (1968) identified as subgenera. It is difficult to imagine where within the genus, of either broad or narrow scope, significant synonymies of species might occur, with the possible exception of revised opinions regarding insular taxa in the Gulf of California. Instead, the need for careful alpha systematic investigations of many species (for example, *eremicus, maniculatus, crinitus, melanophrys, boylii*, and *mexicanus*) portends a slight increase over the current estimate of 53. The results of such revisions, together with the possibility of yet other new species discoveries, would bring the number of species into the range of 55 to 60, or near 70 using the generic boundaries defined by Hooper (1968).

SYSTEMATIC REVIEW

As noted by Hooper (1968), the species composing the nominate subgenus, although highly variable as a whole, form a morphological continuum. Those set apart from this continuum, or central cluster, have been segregated as different subgenera. Thus Osgood (1909) arranged 29 of the 40 species he recognized in the subgenus *Peromyscus*, and Hall (1981) placed 41 of the 59 species he listed in it. In both classifications, the remaining species are distributed among five or six other subgenera, each of which embraces appreciably less morphological, ecological, and geographical variety compared to the core subgenus. Within the subgenus *Peromyscus*, discontinuous patterns of variation of a finer degree have in turn been expressed through recognition of species groups, a practice introduced by Osgood. Although the identity and composition of these species groups and satellite subgenera have changed, in some cases markedly, between the eras of Osgood and Hall, systematists have

TABLE 4.—*Classification of peromyscine rodents (Muridae: Sigmodontinae). * = allocation of species to species group is tentative. (?) = conspecificity of included populations is suspect.*

Genus *Peromyscus* Gloger
  californicus group
    californicus (Gambel)
  eremicus group
    eremicus (Baird) (?)
    guardia Townsend
    interparietalis Burt
    dickeyi Burt
    pseudocrinitus Burt
    eva Thomas
    caniceps Burt
    merriami Mearns
    pembertoni Burt
  hooperi group
    hooperi Lee and Schmidly
  crinitus group
    crinitus Merriam (?)
  maniculatus group
    maniculatus (Wagner) (?)
    polionotus (Wagner)
    sejugis Burt
    oreas Bangs
    sitkensis Merriam
    melanotis Allen and Chapman
    slevini Mailliard*
  leucopus group
    leucopus (Rafinesque)
    gossypinus (Le Conte)
  aztecus group
    aztecus (Saussure) (?)
    spicilegus Allen
    winkelmanni Carleton
  boylii group
    boylii (Baird)
    levipes Merriam (?)
    stephani Townsend
    attwateri Allen
    simulus Osgood
    madrensis Merriam
    pectoralis Osgood * (?)
    polius Osgood *
  truei group
    truei (Shufeldt)
    gratus Merriam
    bullatus Osgood
    difficilis (Allen)
    nasutus (Allen)

Genus *Peromyscus* Gloger (*continued*)
  melanophrys group
    melanophrys (Coues) (?)
    perfulvus Osgood
    mekisturus Merriam *
  furvus group
    furvus Allen and Chapman
    ochraventer Baker *
    mayensis Carleton and Huckaby *
  megalops group
    megalops Merriam
    melanurus Osgood
    melanocarpus Osgood
  mexicanus group
    mexicanus (Saussure) (?)
    gymnotis Thomas
    guatemalensis Merriam
    zarhynchus Merriam
    grandis Goodwin
    yucatanicus Allen and Chapman *
    stirtoni Dickey *
Genus *Habromys* Hooper and Musser
  lepturus (Merriam)
  lophurus (Osgood) (?)
  simulatus (Osgood)
  chinanteco (Robertson and Musser)
Genus *Podomys* Osgood
  floridanus (Chapman)
Genus *Neotomodon* Merriam
  alstoni Merriam
Genus *Megadontomys* Merriam
  thomasi Merriam
  nelsoni Merriam
  cryophilus (Musser)
Genus *Onychomys* Baird
  leucogaster (Wied-Neuwied)
  torridus (Coues)
  arenicola Mearns
Genus *Osgoodomys* Hooper and Musser
  banderanus (Allen)
Genus *Isthmomys* Hooper and Musser
  flavidus (Bangs)
  pirrensis (Goldman)

continued to represent variation and hypothesized relationship within the genus using similar taxonomic conventions. Perhaps this traditional view of *Peromyscus* taxonomy conforms to our preconception of a common evolutionary pattern, one of adaptive radiation with specialized offshoots arising from a successful and dynamic stem group.

Historically then, the construction of supraspecific taxa of *Peromyscus* has been essentially a phenetic method, relying upon body proportions, cranial size and shape, dental morphology, and pelage color. Our attitudes regarding the relative taxonomic importance of some of these character suites have undergone revision; for instance, Osgood gave greater weight to chromatic variations as evidence of intergradation than have modern workers, who have emphasized differences in cranial conformation in reversing some of his decisions (for example, Schmidly, 1973*b*; Carleton *et al.*, 1982). Although phenetic in operation, systematists have proceeded under the guiding assumption that these kinds of morphological similarities reflect community of descent (Hooper, 1968). Concepts such as phylogenetic relatedness and monophyly continually have been important goals in developing classifications of *Peromyscus*.

Nevertheless, the recent awareness of the need to bring added rigor to the formulation of our taxonomic hierarchies and the phylogenetic framework on which they rest, coupled with the proliferation of different sources of systematic data by which to assess relationship, have ushered in a period of frank reappraisal of our traditional taxonomic constructs involving *Peromyscus* and its relatives. One conspicuous sign of this on-going reassessment is the debate over the subgenera encompassed by the genus (for example, Yates *et al.*, 1979; Carleton, 1980; Rogers, 1983; Stangl and Baker, 1984*b*). Yet the possibility of needed classificatory change is more fundamental than altering the taxonomic rank of forms peripheral to the main cluster of species. The integration of this new information also questions the phyletic cohesiveness of the core subgenus itself and the adequacy of certain specific groupings. Results of both electrophoretic and karyologic investigations, for example, intimate that the subgenus *Peromyscus* is a polyphyletic entity (Avise *et al.*, 1979; Rogers, 1983; Stangl and Baker, 1984*b*; Schmidly *et al.*, 1985). The phylogenetic ramifications and eventual taxonomic adjustments issuing from these and other studies have yet to be fully absorbed. The following taxonomic review principally focuses on recent research results that address the validity of currently recognized species and the question of a taxon's monophyly.

Subgenus Haplomylomys

Osgood (1909)	Hooper (1968)	Hall (1981)
crinitus (3)	eremicus	eremicus (14)
eremicus (10)	merriami	eva (2)
goldmani	caniceps	merriami (2)
californicus (2)	guardia	guardia (3)
	interparietalis	interparietalis (3)
	collatus	collatus
	dickeyi	dickeyi
	pembertoni	pembertoni
	stephani	californicus (5)
	californicus	

The number of forms assigned to *Haplomylomys* has more than doubled since Osgood (1904a) diagnosed it, an increase principally reflecting the taxonomic discoveries of *Peromyscus* from islands in the Gulf of California (Burt, 1932). Although Hooper (1968) observed that any of eight mainland species (*eremicus, merriami, californicus, crinitus, maniculatus, pectoralis, boylii, truei*) may have served as the source stock for evolution of the island taxa, subsequent investigations have implicated only four of those (*eremicus, merriami, maniculatus*, and *boylii*), with *eremicus* or an *eremicus*-like form playing the most important role. Insular species believed to bear a descendent relationship to *eremicus* include *guardia, interparietalis, dickeyi*, and *pseudocrinitus* (Fig. 8).

Although Burt (1932) named *interparietalis* as a subspecies of *guardia*, Banks (1967) elevated it to a species based on size and cranial distinctions and viewed both as probable derivatives from *eremicus*-like parentage. He speculated that they originated by separate invasions from the mainland rather than isolation following submersion of the ridge formerly uniting the northern (Angel de la Guardia) and southern (Isla Salsipuedes, Islas San Lorenzo Norte and Sur) island groups, which are inhabited respectively by *guardia* and *interparietalis*. Brand and Ryckman (1969) and Lawlor (1971b) supported the specific distinction of *guardia* and *interparietalis* and the interpretation of their origin from an *eremicus* progenitor. Because *guardia* possesses certain invariant characters, such as a unique hemoglobin band and a relatively complex phallus, Lawlor's study revealed it as more divergent from *eremicus* than *interparietalis*. However, in a broader electrophoretic survey, Avise *et al.* (1974b) disclosed that the genetic differentiation of *guardia* and *interparietalis* is unremarkable in comparison to distances among

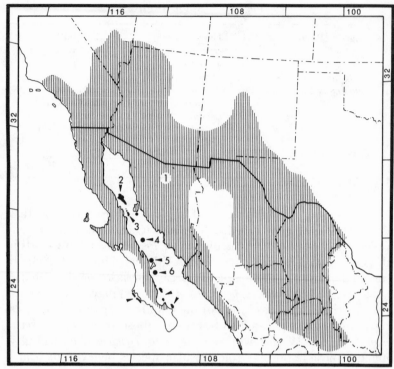

FIG. 8.—Distribution of some species of the subgenus *Haplomylomys*: 1, *P. eremicus*; 2, *P. guardia*; 3, *P. interparietalis*; 4, *P. dickeyi*; 5, *P. pseudocrinitus*; 6, *P. caniceps*. Arrows without numbers indicate insular populations of *eremicus*.

mainland populations of *eremicus*, though the authors retained the two as species. In contrast to previous zoogeographic reconstructions, they considered the weak allelic divergence as supportive of their common origin from a single stock fragmented by rising sea level. *Peromyscus dickeyi*, on Tortuga Island, is also similar morphologically and genetically to *eremicus* and presumably derivable from it (Avise *et al.*, 1974*b*; Lawlor, 1983). The little-known species *pseudocrinitus* on Coronados Island was provisionally allocated to the *crinitus* species group of the subgenus *Peromyscus* by Hooper (1968), but Lawlor (1971*a*) reported that it possesses a phallus like that of *eremicus* and later (1983) listed it among the island species having affinity to *eremicus*.

Two other nominal forms from islands in the Gulf of California are believed to have been derived from close mainland relatives of *eremicus*: *caniceps* from *eva* and *pembertoni* from *merriami* (Fig. 9). *Pe-*

romyscus eva had been arranged as a subspecies of *eremicus*, following Osgood's (1909) revision, until Lawlor (1971*a*) drew attention to its different cranial, external, and phallic traits and documented instances of sympatry with *eremicus*. The geographic range of the resurrected form approximately corresponds to Nelson's (1921) Cape Faunal District, which encompasses the southern half of the Baja California peninsula, and it was judged to have differentiated from an *eremicus*-like ancestor. Still, as noted by Lawlor (1971*a*), certain problematic specimens, such as those from the Sierra de la Giganta, combine features of both *eremicus* and *eva* and require further clarification of their biological status in zones of contact. Specimens of *caniceps*, known only from Monserrate Island off the southern Baja peninsula, possess a long slender baculum resem-

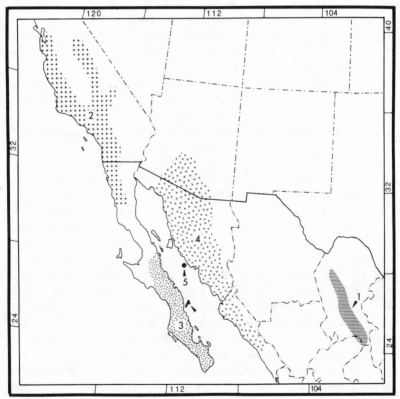

Fig. 9.—Distribution of 1, *P. hooperi* and other species of the subgenus *Haplomylomys*: 2, *P. californicus*; 3, *P. eva*; 4, *P. merriami*; 5, *P. pembertoni*. Arrow without number indicates insular subspecies of *eva*.

bling that of *eva* (Lawlor, 1971a), which Lawlor (1983) regarded as its probable mainland progenitor. Avise *et al.* (1974b) established that *caniceps* is distinct biochemically from peninsular *eremicus*, but examples of *eva* were unavailable for comparison. Hall (1981) listed *caniceps* in the *crinitus* group without explanation.

Recent studies have affirmed that *merriami*, which Osgood (1909) relegated to full synonomy with *eremicus*, deserves specific distinction. Despite the subtlety of their morphological differences, *merriami* and *eremicus* have been shown to occur in broad sympatry over southern Arizona and northwestern Mexico (Hoffmeister and Lee, 1963; Lawlor, 1971a; Hoffmeister and Diersing, 1973). Their close resemblance extends even to their biochemical profiles, for Avise *et al.* (1974b) in an assay of 25 allozymic loci reported a Rogers' similarity coefficient of 0.88, a figure typically accepted as within the range of geographic races of a single species. In view of the sympatry of *eremicus* and *merriami* and their obvious behavior as "good" biological species, the authors cautioned against simplistically applying genetic resemblance coefficients as a yardstick to infer species limits. Both Lawlor (1971a) and Avise *et al.* (1974b) regarded the morphological and genic resemblance of *merriami* and *eremicus* as evidence of close relationship, not convergence. The major question involving infraspecific variation of *merriami* is whether to retain *goldmani* as a formal subspecies, a taxonomic decision judged unnecessary by Lawlor (1971a) but favored by Hoffmeister and Diersing (1973). The insular species *pembertoni* has not been collected since Burt (1932) described it from San Pedro Nolasco; the mouse may be extinct, as extensive efforts to collect it have been unsuccessful (Lawlor, 1983). Based on bacular characters and cranial proportions, Lawlor (1971a; 1983) considered *pembertoni* a large-bodied descendant of *merriami*.

Adjustments in the composition of *Haplomylomys* following Osgood's (1909) revision also have involved deletions of species from the subgenus. In addition to the earlier reallocation of *crinitus* to the subgenus *Peromyscus* (Hooper and Musser, 1964b), Lawlor (1971b) identified *stephani* from San Esteban Island as a member of the *boylii* species group of *Peromyscus*. Furthermore, he placed *collatus* (Turner Island) as a subspecies of *eremicus*.

Unlike the changes in taxonomic treatment of their insular relatives, *eremicus* and *californicus*, the core species of the subgenus (Figs. 8, 9), have remained relatively stable in the light of recent systematic evaluations. In their electrophoretic survey of mainland and insular populations of *eremicus*, Avise *et al.* (1974b) discovered little correspondence between allelic variation and recognized sub-species but noted a contrast between populations west of the Colorado River and those east of it. The similarity coefficient (Rogers' S = 0.86) derived for these two moieties approximates the lower limit of values generated for conspecific populations and recommends further investigation of populations in the area of the Colorado River. The insular species *interparietalis, guardia,* and *dickeyi* aligned genetically with the western segment of *eremicus*. Smith (1979) reduced the number of subspecies in *californicus* from five to two, having identified a sharp step-cline in genic and craniometric characters between northern (*c. californicus*) and southern populations (*c. insignis*).

The trenchant feature in Osgood's (1904a) diagnosis of *Haplomylomys* was the comparative simplicity of the molar teeth, resulting from the usual absence of accessory lophs and styles. In spite of the devaluation of dental characters in *Peromyscus* taxonomy, particularly following Hooper's (1957) critical review of dental variability, the unity of the subgenus has largely persisted in the light of new morphological data. Nevertheless, some taxonomic realignments, such as the removal of *crinitus* and *stephani* to the subgenus *Peromyscus*, were necessitated by new information (Hooper and Musser, 1964b; Linzey and Layne, 1969; Lawlor, 1971a, 1971b). Linzey and Layne (1969), Lawlor (1971b), and Carleton (1980) regarded many of the morphological traits exhibited by members of *Haplomylomys* to be primitive for the genus whether defined broadly or narrowly.

Confidence in the monophyletic origin of forms assigned to the subgenus *Haplomylomys* receives additional support from electrophoretic investigations but not karyologic studies. In phenetic analyses of allelic data, all sampled members of *Haplomylomys*, including *eremicus, merriami, caniceps, dickeyi, guardia, interparietalis,* and *californicus*, composed a single large cluster distinct from other species groups and subgenera (Avise *et al.*, 1974a, 1974b, 1979; Schmidly *et al.*, 1985). In these studies, *californicus* consistently emerged at the first bifurcation within the *Haplomylomys* complex. Knowledge of chromosomal bands currently exists for only three species, *californicus, eremicus,* and *merriami* (Greenbaum and Baker, 1978; Yates *et al.*, 1979; Robbins and Baker, 1981). *Peromyscus eremicus* and *merriami* share 20 heterochromatic short-arm additions but no rearrangements of euchromatic material, whereas *californicus* displays an inversion at chromosome 2 and only centromeric heterochromatin. As a result, cladistic analyses of banded chromosomes portray *californicus* as originating separately from *merriami* and *eremicus* on the tree, conveying a polyphyletic interpretation of *Haplomylomys* (Robbins and Baker, 1981; Rogers, 1983; Rogers *et al.*, 1984; Stangl and Baker, 1984*b*). However, if the subgenus is truly monophyletic despite the karyotypic evidence, Stangl and Baker (1984*b*) reasoned that chromosome 2 must either have experienced a reversal to the acrocentric state in the *eremicus-merriami* lineage or have undergone an independent rearrangement to the biarmed condition found in *californicus*.

Subgenus Peromyscus

Indicative of the difficulties with the current taxonomy of *Peromyscus* are the recent descriptions of two new species, *mayensis* and *hooperi*, in which the authors refrained from attempting supraspecific assignment of their new form. Because both species were eventually assigned to the subgenus *Peromyscus*, they are covered here prior to the review of the species groups.

Carleton and Huckaby (1975) named *Peromyscus mayensis* from high-elevation cloud forest in the Cordillera de los Cuchumatanes of Guatemala. Certain characteristics of the new species suggested its affiliation with the *mexicanus* species group, in particular *furvus*, but they preferred to withhold allocation until supraspecific groupings within the subgenus *Peromyscus* had been reevaluated.

The affinity of *Peromyscus hooperi* proved to be more problematic. Lee and Schmidly (1977) discovered *hooperi* in a restricted zone of grassland transitional between Chihuahuan desert shrub and montane chaparral, along the eastern flanks of low mountains in Coahuila and northeastern Zacatecas (Fig. 9). The diagnostic traits of *hooperi* so evenly recall features of both *Haplomylomys* and *Peromyscus* that Lee and Schmidly (1977) declined subgeneric assignment. Later, Schmidly *et al.* (1985), drawing upon numerous data sources, allocated *hooperi*, with reservation, to the subgenus *Peromyscus* as the sole member of a new species group.

Subgenus Peromyscus: maniculatus *Group*

Osgood (1909)	Hooper (1968)	Hall (1981)
polionotus (5)	*polionotus*	*polionotus* (16)
maniculatus (35)	*maniculatus*	*maniculatus* (67)
sitkensis (2)	*sejugis*	*sejugis*
melanotis	*slevini*	*slevini*
	sitkensis	*sitkensis* (2)
	melanotis	*melanotis*

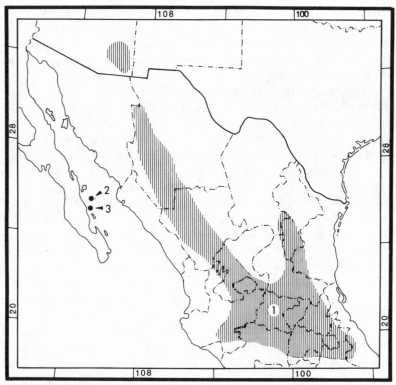

FIG. 10.—Distribution of some species of the *maniculatus* species group: 1, *P. melanotis*; 2, *P. slevini*; 3, *P. sejugis*.

Next to the *leucopus* group, the *maniculatus* species group has remained the most stable of the species aggregations identified by Osgood (1909). The addition of two species named from Gulf of California islands and a doubling of the number of recognized subspecies, many of them also insular, mark the principal changes in taxonomic composition of the group since Osgood's revision.

At the time of Hooper's (1968) review, *Peromyscus sejugis* and *slevini* were poorly known forms described from oceanic islands in the Gulf of California off the southern part of the Baja Peninsula (Fig. 10). Our knowledge of *sejugis* has markedly improved in the interim, but the systematic position of *slevini* remains obscure. In his original description, Burt (1932) associated his new form *sejugis*, known only from Santa Cruz and San Diego Islands, with the *maniculatus* species group, noting its larger size compared to the subspecies *coolidgei* from the peninsula. This relationship has been further substantiated by phallic morphology (Hooper and Musser, 1964*b*) and biochemical variation (Avise *et al.*, 1974*b*, 1979), and Lawlor (1983) considered *sejugis* to have arisen from a *maniculatus*-like ancestor. *Peromyscus sejugis*, however, is not simply an insular variant of mainland *maniculatus*. Of the four *maniculatus*-group species sampled by Avise *et al.* (1979), the degree of genetic differentiation of *sejugis* from *maniculatus* matched that of *melanotis* and surpassed that of *polionotus*.

Although Santa Catalina Island, the type locality of *slevini*, lies just north of the islands of Santa Cruz and San Diego, the affinity of its endemic species does not seem especially close to either *sejugis* or *maniculatus*. The large size and comparatively simple molars of *slevini* persuaded the original describer, Maillaird (1924), to link it with *californicus* in the subgenus *Haplomylomys*. Burt (1934) disputed this assignment and transferred *slevini* to the subgenus *Peromyscus*. In doing so, he did not expressly make a species-group determination for *slevini* but only mentioned the proportional similarities of its skull to that of *maniculatus*, in which species group it has remained although Hooper (1968) considered the affiliation tentative. The resemblance in cranial shape to *maniculatus* is not definitive. In particular, a supraorbital shelf is well expressed in *slevini* and imparts an elongate appearance to the braincase, reminiscent of certain Mexican species, rather than the circular form typical of the *maniculatus* group. Like *sejugis*, Lawlor (1983) also considered *slevini* a derivative of a *maniculatus* ancestral stock, but this putative kinship needs critical reexamination using a variety of systematic approaches. Perhaps *slevini* will prove an exception to the prevalent pattern of relationships described for mammals in the Gulf of California, where insular taxa generally are thought to have originated from the nearest mainland source (Lawlor, 1971*b*, 1983).

The systematic status of *polionotus*, as summarized by Hooper (1968), was not viewed as problematic. The species was believed to bear closest relationship to *maniculatus* and presumably evolved from it as a peripheral isolate (Fig. 11), in particular from short-tailed grassland forms like the subspecies *pallescens* and *bairdii*. Although this perception has not been strongly disputed by recent evidence, neither has it remained quite so clearcut. Electrophoretic data have revealed *polionotus* as the nearest kin of *maniculatus* (Avise *et al.*, 1979; Schmidly *et al.*, 1985). In the study of Avise *et al.* (1979), however, the sample of *polionotus* was not uniquely separated from OTUs of *maniculatus*, which included representatives of numerous

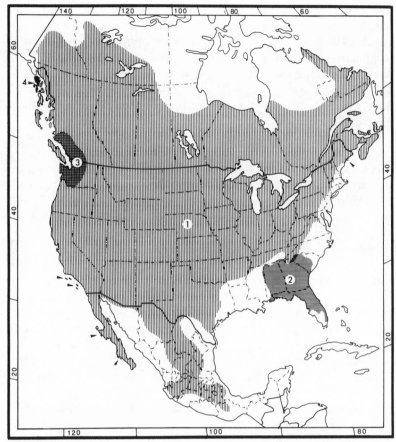

FIG. 11.—Distribution of some species of the *maniculatus* species group: 1, *P. maniculatus*; 2, *P. polionotus*; 3, *P. oreas*; 4, *P. sitkensis*. Arrows without numbers indicate insular populations of *maniculatus*.

subspecies, and instead clustered with *m. nubiterrae*, one of the long-tailed forest races. Earlier investigations of banded karyotypes depicted *polionotus* as the sister species to *maniculatus* (Greenbaum *et al.*, 1978*b*; Yates *et al.*, 1979), but with the addition of species to karyotypic cladograms, *polionotus* appears as a more distant relative of *maniculatus* than either *oreas* or *sitkensis* (Robbins and Baker, 1981; Rogers *et al.*, 1984; Stangl and Baker, 1984*b*).

Bowen (1968) evaluated morphological variation among Gulf Coast populations of *polionotus*, principally examining coat-color variation and determining inheritance of pelage traits from laboratory breeding experiments. As a result, he recognized five additional subspecies, most of them restricted to coastal dunes and sandbars and exhibiting varying degrees of coat-color dilution in comparison to inland populations. In Bowen's interpretation, mice distributed along the coast experienced a complex history of isolation, differentiation, and secondary contact with inland populations, fostered by post-Pleistocene changes in sea level and coastal geomorphology.

Furthermore, he postulated a convoluted fissure zone of incipient speciation dividing certain of the coastal populations and those in central Florida as one nascent species from the remaining coastal and mainland forms as another. The correspondence of his Pleistocene biogeographical account to the origination of the incipient species is obscurely drawn. The unique nature of the selective backdrop afforded by the beach-dune environment was underscored by Selander *et al.* (1971), who reported that genetic heterozygosity levels among populations inhabiting the Gulf Coast barrier beaches and islands were about one-third to one-half those recorded for mice on the adjacent mainland. They attributed the uncommonly low levels of genetic variation to the homogeneity of the beach habitat and to the isolation of these marginal populations and consequent loss of variety through genetic drift. Although their results may suggest that Bowen's involved evolutionary scenario has some merit in broad terms, if not in detail, the systematics of *polionotus* warrants critical review.

The range of *melanotis* had been understood to encompass higher elevations in the pine-fir zone and intermixed grasslands of Mexico, extending throughout the Cordillera Transvolcanica and northward along the Sierra Madre Oriental and Occidental (Fig. 10). Bowers *et al.* (1973) and Bowers (1974), however, documented its range to include upper elevations in the Chiricahua, Pinaleno, and Santa Catalina Mountains of southern Arizona. Although the presence of *melanotis* on these southern Arizona mountaintops is wholly plausible from a zoogeographic standpoint, these mice had been confused with *maniculatus rufinus* since Osgood's revision. Rasmussen (1970), in an electrophoretic study, earlier had identified certain of these populations as anomalous, but he erroneously attributed their genetic singularity to drift in small isolated populations of *maniculatus*. Using standard karyotypic preparations, electrophoresis of blood samples, multivariate analyses of cranial measurements, and interbreeding tests, Bowers and colleagues (1973, 1974) convincingly demonstrated that the southern Arizona mice represent northernmost populations of *melanotis*. Despite numerous breeding trials, no mice from the Arizona populations mated successfully with contiguous representatives of *maniculatus* but were interfertile with Mexican examples of *melanotis*, although Clark (1966) reported one hybrid litter from a cross between *melanotis* and *maniculatus bairdii*. Bowers' (1974) assessment of intraspecific variation provided no grounds for subspecific divisions within *melanotis*.

Despite the vast research efforts devoted to various aspects of the biology of *maniculatus*, our taxonomic comprehension of the species has, ironically, improved little since Osgood's (1909) revision. The status of *maniculatus*-like forms was exceedingly complex at the time of Osgood's study. Of the taxa that he ultimately brought together as the species *maniculatus*, 24 stood as nominal species; several, such as *michiganensis* and *canadensis*, already contained geographic representatives; and many described forms had been incorrectly associated with *leucopus*. That he brought a credible and stable classification to this nomenclatural morass alone would have marked his 1909 work as a seminal contribution to North American mammalogy. Osgood consolidated the numerous described forms exhibiting great individual and geographic variability into one species with 35 subspecies, which constituted the nucleus of his *maniculatus* species group. This taxonomic framework has persisted to the present, adjusting to the creation of over 30 additional subspecies. With its fold-out distribution map serving as Osgood's multicolor frontispiece, *maniculatus* epitomized both the polytypic model of a sound revisionary study and prima facie evidence of the evolutionary process, with numerous interdigitating geographic races, various degrees of population differentiation, and purported zones of intergradation. The species continues to deserve our attention because it remains foremost among *Peromyscus* in current research on intraspecific variation and speciation.

Discussions of the intraspecific relationships of *maniculatus* have centered historically upon the well-documented existence of two distinctive morphs, the short-tailed, grassland-inhabiting subspecies and the long-tailed, forest-dwelling forms. These external morphologies mirror fundamental contrasts in cranial size and shape as well (see, for example, Koh and Peterson, 1983). Hooper (1968) broadly sketched the range of the long-tailed forms as an inverted "U" upon the North American continent, distributed across boreal biotopes in Canada and extending southwards along the Appalachian Mountains in the east and the Rockies in the west. Populations of short-tailed *maniculatus* generally are distributed to the south, occupying the central prairies and Mexican Plateau and the western deserts and intermontane grasslands. Distributional anomalies involving both the western and eastern prongs of long-tailed subspecies highlight the complexity and resulting uncertainty of the status of the diverse forms Osgood consolidated as *P. maniculatus*.

Osgood (1909) recognized one problematic area in western Washington and British Columbia, where *oreas* and *austerus* occurred sympatrically and seemed to retain their morphological identities. Subsequent research (summarized in Hooper, 1968) revealed that, elsewhere in its range, *oreas* also occurs with two other shorter-tailed forms, *artemisiae* and *gambeli*, prompting Sheppe (1961) to regard it as a full species. Investigators using chromosomal data also have viewed *oreas* as a species distinct from *maniculatus* based on the former's consistently higher autosomal numbers (AN = 85–88 versus AN = 74–76) (Robbins and Baker, 1981; Rogers *et al.*, 1984; Gunn and Greenbaum, 1986). Exclusively using specimens of known karyotype, Allard *et al.* (1987) demonstrated the consistent multivariate discrimination of *oreas* and *m. austerus* in northwestern Washington.

Hooper (1968:45) opposed species recognition of *oreas* because it "would solve no biological problem." Although the status of *oreas* may not be resolved in some final sense by treating it as a species, such action acknowledges the many documented instances of its sympatry with short-tailed races of *maniculatus* and its morphological and chromosomal differentiation from them. Furthermore, such recognition encourages added scrutiny of material from critical geographic regions that might be otherwise installed under the nearest applicable subspecific epithet. What is now required, as noted by Hooper, is further elucidation of the range of *oreas* (Fig. 11) and its relationship to other long-tailed populations, in particular to the subspecies *macrorhinus* found farther north along the British Columbia coast. The numerous insular races also warrant attention as possible relatives of *oreas*.

Unraveling the affinities of populations distributed among the islands off Alaska, British Columbia, and Washington is a crucial but daunting prerequisite for understanding the level of speciation not only of *oreas* but also of *sitkensis* (Fig. 11). Other than *sitkensis*, which he retained as a species, Osgood (1909) recognized few taxa of the *maniculatus* group on these islands. The taxonomic and zoogeographic complexity of this region changed dramatically in the following decades, which witnessed a two-fold increase in the number of subspecies of *maniculatus*, almost two-thirds of them named from islands off western North America.

The complexity evident in the proliferation of epithets for insular representatives seems to be borne out in the karyological information published to date. Thomas (1973) found both low (74–78) and high (84–86) autosomal numbers among karyotypes of mice from two small island groups near Vancouver Island. The high-AN animals he interpreted as examples of *sitkensis* and consequently reassigned certain subspecies (*doylei, insolatus, saxamans,* and *triangularis*) to that species, a decision that greatly extends the southern limits of *sitkensis* and gives it an even more disjointed and enigmatic distribution. Thomas' definition of *oreas* was somewhat ambiguous, for he also identified low- and high-AN karyotypes from the mainland as belonging to *oreas* but formally associated only the low-AN numbers with it. In view of the consistent autosomal differences they encountered, Gunn and Greenbaum (1986) surmised that Thomas had confused examples of *oreas* and *maniculatus austerus*. Pengilly *et al.* (1983) obtained karyotypic samples from the type locality of *sitkensis*, where they recorded even higher autosomal numbers (86–91); they refrained from drawing conclusions on its taxonomic status. The contradictory karyotypic findings raise several questions: do two kinds of *oreas* exist on the adjoining mainland; do the island mice Thomas aligned with *sitkensis* in fact bear closer kinship to the high-AN *oreas*; and does *sitkensis* fit with the long-tailed southern island mice or is it another species altogether. A detailed taxonomic review of the island *Peromyscus* of this region and determination of their origin from continental populations offer exciting research opportunities that should be pursued.

The systematic and distributional picture involving the eastern prong of long-tailed *maniculatus* is far less complicated but still not fully resolved. Indeed, were it not for the prevalent belief that intergradation of short-tailed and long-tailed populations occurs someplace farther west, the two ecophenotypic forms would likely be regarded as separate species in view of their morphological distinctiveness and habitat segregation. Much of their present-day range contact has transpired with the habitat changes that accompanied settlement and deforestation of eastern North America (Hooper, 1942; Baker, 1968). The original range of *bairdii* probably encompassed the prairie region of the Upper Mississippi Valley; Osgood (1909) recorded it only as far east as Central Ohio. Apparently within the past century and a half, its range has extended northward into central and western Michigan, eastward to West Virginia, southcentral Pennsylvania, and southcentral New York, and across the Appalachians into northern Virginia and western Maryland (Paradiso, 1969). As a result, the distributions of *bairdii* and the long-tailed subspecies *gracilis* and *nubiterrae* have come to intermingle in the northeastern United States, yet their morphological and ecological characteristics seem to persist in these areas of secondary overlap.

In view of the morphological distinctiveness, ecological separation, and records of sympatry of the various subspecies of *maniculatus*, the low allelic differentiation thus far substantiated within the species is unexpected. Using starch-gel electrophoresis, Avise *et al.* (1979) examined allelic variation from 71 localities throughout the continental range of *maniculatus*, their samples representing 16 subspecies. Of 21 loci scored, only six were measurably polymorphic and most populations shared alleles at both monomorphic and polymorphic loci, resulting in a uniformly high coefficient of genetic similarity (mean = 0.934) among the samples. No discernible patterns of heterozygosity were revealed, nor did examples of long-tailed and short-tailed races segregate in any meaningful way in cluster analysis of the data. In a survey of four subspecies distributed over the western United States and Baja California, Calhoun *et al.* (1988) similarly documented high levels of genetic similarities among their population samples (mean = 0.973) and the absence of biochemical differentiation along traditional subspecific divisions. Working with electrophoretic samples over a still finer geographic scale, Aquadro and Kilpatrick (1981) obtained comparable values of heterozygosity and genetic divergence for populations of *maniculatus abietorum* on small islands and the adjacent mainland of Maine; however, lower variability was generally recorded for the insular samples.

Avise *et al.* (1979:187) attributed the conservative genic divergence within *maniculatus* to the ". . . relatively recent separation of populations, coupled with a genetic inertia resulting from a selected cohesion of the genome." Calhoun *et al.* (1988:43) elaborated a similar theme, remarking that ". . . morphological attributes may respond to external adaptive pressures, but karyotypic and biochemical polymorphisms are maintained by an internal or genetic environment." Assuming that one can facilely compartmentalize an organism or the selective regime into external and internal components, one might expect substantial agreement of allelic data with patterns derived from variation in nucleotide sequences of mitochondrial DNA. However, the extensive mitochondrial-DNA variation uncovered by Lansman *et al.* (1983), using samples drawn across the North American range of *maniculatus*, exhibited discordancies with formal subspecific boundaries as well as the allozyme data.

Chromosomal data, mostly based on standard preparations, suggests greater geographic differentiation within *maniculatus*, with races possessing consistent modal differences in number of autosomal arms (Bradshaw and Hsu, 1972; Bowers *et al.*, 1973). The karyotypic data has not been evaluated extensively from the perspective of banding homologies, but variation in autosomal number results from both pericentric inversions and short-arm heterochromatic additions, and polymorphisms of both kinds have been reported at the population level (Robbins and Baker, 1981; Robbins *et al.*, 1983; Stangl and Baker, 1984*b*; Gunn and Greenbaum, 1986; Macey and Dixon, 1987). Macey and Dixon (1987) noted a correspondence between the incidence of certain pericentric and heterchromatic polymorphisms and elevation among *maniculatus* populations in eastern Colorado.

More than any other species of *Peromyscus, maniculatus* has served as a natural model for ideas on the evolutionary origin of new taxa, owing to such traits as its wide distribution, morphological and karyological variability, and broad ecological tolerances. Blair (1950) emphasized *maniculatus* in developing his hypothesis that "offspring" species are distributed at the periphery of the "parent" species, interrelating the degree of geographic and ecological variation in a species to the size of its range and to the likelihood that marginal populations will become isolated. The *maniculatus* species group reflects this pattern remarkably well, with the more restricted distributions of *polionotus, melanotis, sejugis,* and *sitkensis* arrayed peripherally to the larger, basically continental range of *maniculatus* (Figs. 10, 11). Based on karyotypic variation, Bowers *et al.* (1973), and later Greenbaum *et al.* (1978*b*), supported Blair's fundamental thesis and extended it to incorporate Brown's (1957) hypothesis of centrifugal speciation, in which the central portion of the range of a species is thought to be more evolutionarily dynamic, with descendent species arising as peripheral isolates and evolving at slower rates. Of the peripheral relatives of *maniculatus*, only *melanotis* and *polionotus*, which have more plesiomorphic karyotypes, meet the predictions of this argument, but *sitkensis* and *oreas*, which have a derived karyotypic condition (Pengilly *et al*, 1983; Gunn and Greenbaum, 1986), do not. In his discussion of the interplay of geography and ecology in taxon cycles, Glazier (1980) highlighted *maniculatus* as an example of the generalization that widespread species tend to be habitat generalists, have higher reproductive potentials, and are more successful at colonizing marginal environments. The broad distribution and variability of *maniculatus* was viewed from yet another perspective by Robbins *et al.* (1983), who noted a positive correspondence between the geographic area occupied by a species and the magnitude of its chromosomal evolution since divergence from a common ancestor.

In reviewing the literature on the *maniculatus* species group, and on *maniculatus* in particular, one is struck by the necessity to resolve the phylogenetic component of the group before theories on speciation, biogeography, genetic variability, and anagenetic processes of evolution can assume a more credible basis. Our confidence in such theories demands, if not presupposes, a sound framework of relationship and pattern of descent. This interdependency of evolutionary theory and classificatory soundness is exemplified by the misidentification of the montane isolates in southern Arizona: Rasmussen (1970) attributed their genetic uniqueness to drift among populations of *maniculatus*, a hypothesis later replaced by one of evolutionary stasis through disclosure of their identity as another species, *melanotis* (Bowers *et al.*, 1973).

Systematic comprehension of *maniculatus* warrants further refinement, especially in two areas. First, the purported intergradation of long-tailed and short-tailed subspecies requires unambiguous documentation, drawing upon multiple data bases and emphasizing regional studies and detailed transects across suspected zones of contact and hybridization. Persuasive investigations of subspecific transitions are limited to short-tailed forms, for instance, the intergradation of *bairdii* into *osgoodi* (= *nebrascensis*) across North Dakota (Dice, 1940*a*) and the circular overlap of *ozarkiarum* and *pallescens* in Texas and Oklahoma (Caire and Zimmerman, 1975). The status of long- and short-tailed deer mice populations, where they meet, is characterized more by the maintenance of their distinctness than by concrete evidence of introgression. Second, our tendency to pose the taxonomic dilemma of *maniculatus* as a species consisting of two contradistinctive sets of populations, "the" long-tailed subspecies versus "the" short-tailed subspecies, may mask the reticulate genealogical complexity of these organisms and hinder appreciation of their interrelationships and level of differentiation. Several lines of evidence suggest that this traditional view is an oversimplification. The instances of reported sympatry indicate a more complicated picture of distribution and affinity than a simple long-tailed versus short-tailed dichotomy. For example, where *artemisiae* occurs with *oreas* in Washington and British Columbia, it is the shorter-tailed of the two (Sheppe, 1961); whereas, in western Montana, *artemisiae* was typified as a long-tailed forest form and so contrasted to the sympatric *osgoodi* (Murie, 1933). Furthermore, the patterns of variation revealed by the biochemical and karyological data are inconsistent with the notion that the two morphological stereotypes are equivalent to two evolutionary clades (Lansman *et al.*, 1983; Calhoun *et al.*, 1988). Lastly, in view of the profound biotic modifications in western North America as a result of past orogenic and glacial episodes, and the effect these events had on the diversification of other mammals in this region, it seems remarkable that *maniculatus*, with its low vagility, small deme sizes, and broad altitudinal range, has remained as genetically cohesive as the electrophoretic and interbreeding studies suggest. Still, experience with other species of *Peromyscus* cautions that these kinds of data are not final arbiters of species status and must be considered in the context of intensive field studies and thorough revisions using a broad array of information.

The cohesiveness (and inferred monophyly) of the *maniculatus* group generally has been borne out in molecular and cytogenetic studies of the past decade. In a dendrogram (UPGMA) of genetic similarities based on 19 allozymic loci, the four species of the *maniculatus* group sampled (*maniculatus, melanotis, polionotus, sejugis*) formed a single cluster among the 20 species examined, which included representatives of the subgenera *Haplomylomys* and *Podomys* as well (Avise *et al.*, 1979). The four species united at a genetic similarity value of 0.78 and formed a pair group with the *leucopus* species group. Other phenetic treatments of electrophoretic data have produced similar results, supporting the association of species Osgood and subsequent authors have identified as the *maniculatus* group (Avise *et al.*, 1974a, 1974b; Schmidly *et al.*, 1985).

Equally corroborative are the phylogenetic interpretations of banded chromosomes, which have been described for *maniculatus, melanotis, oreas, polionotus,* and *sitkensis* (Greenbaum and Baker, 1978; Greenbaum *et al.*, 1978a; Pengilly *et al.*, 1983; Yates *et al.*, 1979; Robbins and Baker, 1981; Gunn and Greenbaum, 1986). Members of the *maniculatus* group are defined by a pericentric inversion of chromosome pair 20 (also shared with one race of *leucopus*), in addition to inversions at 2, 3, 6, and 9, and therefore are viewed as highly derived karyotypically (Rogers *et al.*, 1984; Stangl and Baker, 1984*b*). Such studies have arranged *melanotis* as the first species to separate within the group because it lacks certain additional pericentric alterations and the proliferation of short-arm heterochromatin exhibited by the other species. Based on a pericentric inversion of chromosome 19, Gunn and Greenbaum (1986) hypothesized a sister-group relationship for *oreas* and *sitkensis*, as distinct from *maniculatus*, which seems plausible from the standpoint of their morphology and geography.

Subgenus Peromyscus: leucopus *Group*

Osgood (1909)	Hooper (1968)	Hall (1981)
leucopus (13)	*leucopus*	*leucopus* (17)
gossypinus (4)	*gossypinus*	*gossypinus* (7)

Except for a modest increase in the number of recognized subspecies, the composition of the *leucopus* group has remained unchanged since Osgood's (1909) revision. The broad geographic range of *leucopus* largely encompasses the smaller one of *gossypinus*, although the latter extends to the coast in the southeastern United States and throughout peninsular Florida (Fig. 12). Where the two species co-occur, their physical similarity can prove confusing, the distinction primarily resting on the slightly larger size of *gossypinus* (Osgood, 1909).

In view of their close morphological resemblance, their broad distributional sympatry, and knowledge of the fecundity of interspecific laboratory crosses (Dice, 1937; Bradshaw, 1968), hybridization under natural conditions has been long suspected. Such expectations have not been realized, however, whether examined from the standpoint of traditional external and cranial features (Dice, 1940*b*; Engstrom *et al.*, 1982; McDaniel *et al.*, 1983) or from the modern perspective of starch-gel electrophoresis (Price and Kennedy, 1980; Robbins *et al.*, 1985). Engstrom *et al.* (1982) applied the discriminatory power of multivariate statistical programs to cranial dimensions of *gossypinus* and *leucopus* from eastern Texas and found no evidence of hybridization; in fact, their analyses definitively associated two specimens that McCarley (1954) had identified as wild-caught hybrids with one or the other species. Price and Kennedy (1980) scored 14 structural loci from 31 samples of *leucopus* and 11 of *gossypinus* from eastern Arkansas, western Tennessee, and northern Mississippi, including localities where the two oc-

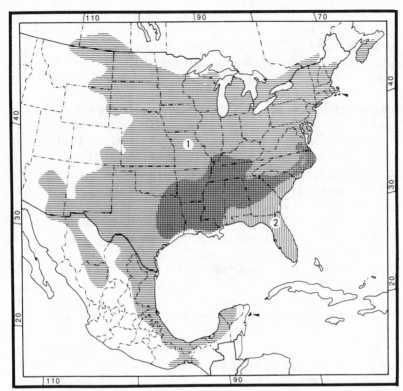

Fig. 12.—Distribution of species of the *leucopus* species group: 1, *P. leucopus*; 2, *P. gossypinus*. Arrows without numbers indicate insular populations of *leucopus*.

curred sympatrically as well as singly. Samples of each species displayed clear genetic separation, segregating at a similarity level around 0.82; furthermore, they uncovered no differences in levels of individual heterozygosity or amount of population polymorphism in areas of sympatry compared with those of allopatry, suggesting the absence of introgression. Zimmerman *et al.* (1978) reported a genic identity of 0.84 for the two kinds and considered this degree of divergence as exemplifying semispecies, forms which have not acquired complete reproductive isolation after the establishment of secondary overlap. Browne (1977) compared genetic variation of island populations of *leucopus* to that of mainland localities and documented a reduction in allelic variability among the island samples, which he attributed to a founder effect. An examination of electrophoretic variation of isolated populations distributed over a small geographic area revealed no appreciable differentiation (Tolliver *et al.*, 1987), which the authors attributed to ocassional dispersal between the preferred habitats.

More biologically interesting, perhaps, than the confirmation of the reproductive integrity of *leucopus* and *gossypinus* is the discovery of a rift between populations of *leucopus*, as suggested by electromorph differences in salivary amylase (Aquadro and Patton, 1980) and by chromosomal variants (Baker *et al.*, 1983*b*). Or rather a rediscovery, for Osgood (1909:115) had perceived a subdivision among the populations he assigned to *leucopus*: "The species . . . *leucopus* . . . is naturally divisible into three subspecies in the northeast [*leucopus, noveboracensis, aridulus*] and several others (*texanus,* etc.) in the southwest." That he thus would interpret the affinity of the geographic samples then available to him is apparent from an inspection of his distribution map (Osgood 1909: fig. 2), which shows a large gap in the region of Kansas, Oklahoma, and eastern Texas, separating the populations in the northeast from those in the southwest. In fact, until Osgood's revision, the southwestern segment had been recognized as a species (*texanus*) distinct from that (*leucopus*) in the northeast (see, for example, Miller and Rehn, 1901).

The provocative aspect of Baker's *et al.* (1983*b*) finding is the magnitude of the karyotypic differences between the two moieties, which they term chromosomal races. These include three euchromatic alterations involving chromosomes 5, 11, and 20. Such contrasts typically, but not always, mark full species of *Peromyscus*, but this is not the case for the two segments of *leucopus*, which interbreed freely with no apparent depression of hybrid fertility (Stangl and Baker, 1984*a*, 1984*b*; Stangl, 1986). An earlier electrophoretic survey covering much of the same geographic area failed to discern the same dichotomy (Robbins *et al.*, 1985). Robbins *et al.* (1985) uncovered more genic differentiation among populations having the northeastern karyotype than between those in the northeast and those in the southwest, although the latter did form an almost homogeneous cluster. However, a more detailed geographic transect across the contact zone of the two chromosomal races revealed a strong correspondence between allozyme and karyotypic variation (Nelson *et al.*, 1987).

The intergradation between the northeastern and southwestern moieties of *leucopus* now constitutes the best-studied example of a hybrid cline within *Peromyscus*. Stangl (1986) convincingly argued that the zone originated as a result of introgression following secondary contact, in view of the geographic concordance of three clines involving pericentric inversions of separate chromosomes. Clinal profiles in frequency of marker alleles and mitochondrial-DNA haplotypes are likewise congruent with the chromosomal pattern (Nelson et al., 1987). Stangl (1986) postulated that the establishment of secondary contact probably transpired within the past 9000 years, following the subdivision of *leucopus* into two sets of populations due to Wisconsin-age climatic changes. The present-day distributions of these two races broadly correspond to the occurrence of major plant biomes, the Great Plains Grasslands in the west and the Eastern Deciduous Forest in the east, and the hybrid zone is approximately situated at the oak-savannah ecotone between them (Stangl, 1986). The status of these *leucopus* populations, and the considerable information already available about them, offers a dynamic evolutionary system that invites still further investigation. Laboratory studies of the meiotic behavior of hybrid individuals, the extent of their interfertility, and the inheritance of racial traits would benefit understanding of hybrid tension zones, the interaction of genetic and environmental factors leading to maintenance of step-clines, and introgression under secondary contact.

Electrophoretic and karyologic surveys involving both *leucopus* and *gossypinus* provide further confirmation of the closeness of their relationship and the notion that they have diverged only recently from a common ancestor. Cluster analyses of allelic data have repeatedly associated the two in the same linkage group, generally at a level of differentiation comparable to that noted for other species groups in the subgenus *Peromyscus* (Avise et al., 1974b, 1979; Zimmerman et al., 1978; Schmidly et al., 1985). In like manner, heterochromatin segments in the short arms of chromsomes 21, 22, and 23 affiliate the two species in cladistic interpretations of banding information (Robbins and Baker, 1981; Rogers et al., 1984; Stangl and Baker, 1984b).

Subgenus Peromyscus: crinitus *Group*

Osgood (1909)	Hooper (1968)	Hall (1981)
Not recognized,	*crinitus*	*crinitus* (8)
crinitus (3) part of	*pseudocrinitus*	*pseudocrinitus*
(*Haplomylomys*)		*caniceps*

Although Osgood (1909) regarded the subgeneric assignment of *crinitus* as equivocal, he favored its allocation to *Haplomylomys* largely because it possesses only four mammae instead of the six that he regarded as normal for *Peromyscus*. Subsequent research has revealed that absence of the pectoral pair of mammae is not unknown in the subgenus *Peromyscus* (for example, some species of the *mexicanus* group, Huckaby, 1980), rendering this character less pivotal. Studies of the glans penis of *crinitus* confirmed its alliance with members of *Peromyscus* (Hooper, 1958), but consideration of its intermediate and unique features persuaded Hooper and Musser (1964*b*) to erect the *crinitus* group to contain it and two nominal species, *caniceps* and *pseudocrinitus*, of questionable status and affinity.

The basic phyletic affinity of *crinitus* to species included in the subgenus *Peromyscus* has been generally supported in later systematic investigations. Linzey and Layne (1969) documented that male specimens of *crinitus*, like those of the subgenus *Peromyscus*, lack preputial glands, in contrast to the conspicuous preputial glands found in *eremicus* and *californicus*. Electrophoretic assays also align *crinitus* with species groups in the subgenus *Peromyscus*, in particular with members of the *maniculatus* group (Zimmerman *et al.*, 1978; Schmidly *et al.*, 1985). This association is consistent with the phallic resemblances between *crinitus* and the *maniculatus* group remarked upon by Hooper (1958) and Hooper and Musser (1964*b*), but those authors additionally noted similarities to certain species of the *boylii* group. As reported by Greenbaum and Baker (1978), chromosomal banding suggests that *crinitus* retains a karyotypic morphology identical to that advanced as the ancestral condition in the genus (*s.l.*). As a result, karyological information offers little resolution concerning the phylogenetic position of *crinitus*. In such studies, investigators have arranged *crinitus* as a basal evolutionary branch in an unresolved polychotomy that also includes species of *Haplomylomys*, *Osgoodomys banderanus*, and certain forms of *Peromyscus boylii* (Robbins and Baker, 1981; Rogers *et al.*, 1984; Stangl and Baker, 1984*b*).

Hooper (1968) slightly modified the composition of the *crinitus* group and tentatively referred *caniceps* to *Haplomylomys*. Lawlor's research (1971*a*, 1971*b*; 1983) on rodents from islands in the Gulf of California supports this reassignment. As discussed above, Lawlor treated both *caniceps* and *pseudocrinitus* as only marginally differentiated insular species that originated from mainland parental stocks of *eremicus* (*pseudocrinitus* on Isla Coronados) and *eva* (*caniceps* on Monserrate), both forms of the subgenus *Haplomylomys*.

That the *crinitus* group will remain monotypic seems doubtful.

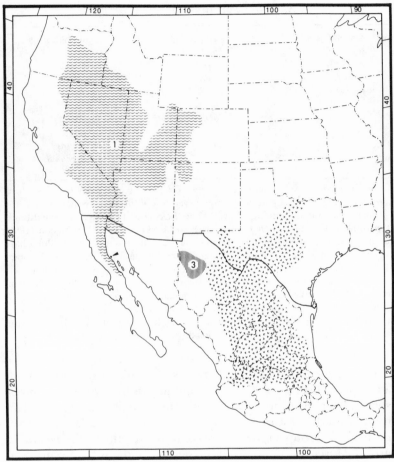

FIG. 13.—Distribution of: 1, *P. crinitus*; 2, *P. pectoralis*; 3, *P. polius*. Arrow without number indicates insular subspecies of *crinitus*.

The species as now recognized embraces appreciable morphological diversity and covers a tremendous elevational range, essentially from sea level to over 10,000 feet (Fig. 13) (Hall and Hoffmeister, 1942). Moreover, its patchy local distribution and preference for rocky habitats suggest a history of range fragmentation and differentiation among isolates. The degree of divergence of subspecies populations inhabiting various regions of the Great Basin and the purported intergradation of long-tailed (for example, *disparilis* and *delgadilli*) and short-tailed subspecies (for example, *auripectus*, *crinitus*, and *stephani*) merit reassessment.

Subgenus Peromyscus: boylii *Group*

Osgood (1909)	Hooper (1968)	Hall (1981)
pectoralis (3)	*pectoralis*	*pectoralis* (3)
boylii (9)	*boylii*	*boylii* (12)
oaxacensis	*polius*	*attwateri*
hylocetes	*evides*	*polius*
	aztecus	*stephani*
	hondurensis	*evides*
	oaxacensis	*aztecus*
	hylocetes	*oaxacensis*
		winkelmanni
		hylocetes

Nomenclatural note.—The species *boylii* derives its name from Charles Elisha Boyle, a young Ohio physician who journeyed to California in 1848 as part of the goldrush migrations (Jennings, 1987). There he collected the specimens that Baird (1855) later described as *Hesperomys boylii*. In his generic revision, Osgood (1909) used, without comment, the species name *boylei*, and these alternative spellings have since appeared in later taxonomic treatments of the species (for example, Hooper, 1968, versus Hall, 1981). In his role as first revisor, Osgood's choice of spelling merits special consideration. However, the International Code of Zoological Nomenclature (1985) makes clear that Baird's original latinization of Boyle as *boylii* is acceptable (Article 31a) and that Osgood's use of *boylei*, although the currently preferred formulation for a patronym (Appendix D: Recommendation III), constitutes an incorrect subsequent spelling (Article 33c,d).

During the past 15 years, there has been arguably more systematic research conducted on the *boylii* group than upon any other species assemblage of *Peromyscus*. The taxonomic adjustments stemming from this research activity have raised the number of species in the *boylii* group so that it is second in size only to the *mexicanus* group. Moreover, these recent taxonomic findings have reordered our perception of the status and affinity of the member taxa in a manner unanticipated by either Osgood (1909) or Hooper (1968).

The guiding paradigm on *boylii* group research had been one of a complex Rassenkreis, with hybridization and intergradation occurring among some of the component populations and infertility and geographic overlap among others. Osgood (1909:159) characterized the situation as a circuitously intergrading mosaic of subspecies: "Perhaps here is another example of two subspecies [i.e., *oaxacensis* and *levipes*] of the same group occurring together, for *levipes* appears to intergrade with *spicilegus*, *spicilegus* with *evides* and *aztecus*, and quite probably *aztecus* with *oaxacensis*." Osgood thus arranged most of the aforementioned forms as subspecies of a highly variable *boylii*. Although Hooper had himself weakened Osgood's interpretation of species boundaries, he (1968:53–54) echoed the notion of varying degrees of interfertility, suggesting that ". . . (a) the isolation of eastern and western arms of the range of *boylii*-like forms in Mexico has been removed and (b) the large gap between them has been filled by morphologically different populations from the south and now are incompletely interfertile with those western and eastern series of populations." Our gradual abandonment of these thematically similar hypotheses mostly has been influenced by more detailed studies of museum specimens and documentation of distributions, activities which have disclosed the consistency of subtle morphological distinctions, uncovered fresh bases of comparison, and provided additional instances of sympatry. As a result, the view of a complexly intergrading chain of morphologically dissimilar yet conspecific populations has been supplanted by one of several discrete species whose distributional limits are characterized by zones of sympatry or contiguous allopatry, not introgression of subspecific traits. The nomenclatural consequences of our revised perception are reflected in the elevation of seven of Osgood's (1909) nine subspecies of *boylii* to species rank: *attwateri*, *aztecus*, *evides*, *levipes*, *madrensis*, *simulus*, and *spicilegus*.

Western Mexico has proven to be a critical region for unraveling some of the complexities of the *boylii* group. In this area, according to Osgood's revision, populations of *boylii* were distributed from the Mexican Plateau (*rowleyi*), along the flanks of the Sierra Madre Occidental (*spicilegus*), onto the coastal plain of Sinaloa and Nayarit (*simulus*), and occupied the near-shore Tres Marias Islands (*madrensis*).

Hooper (1955, 1968) drew attention to the occurrence of two "morphological types" of *boylii* in coastal Sinaloa and Nayarit. Although all fell within the range mapped for *boylii simulus*, his specimens from foothills just east of San Blas, Nayarit, the type locality of *simulus*, more closely resembled *spicilegus*. Subsequently, Carleton (1977) documented the sympatry of *simulus* and *spicilegus* in this region, and other instances of their co-occurrence have been recorded elsewhere in Nayarit (Carleton *et al.*, 1982). As suspected by Baker and Greer (1962), samples of *simulus* and *spicilegus* give no indication of intergradation at contact areas in Sinaloa or Nayarit. Instead cranial, phallic, and chromosomal characteristics indicate *simulus* is markedly differentiated from *spicilegus* and perhaps not even closely related to it (Carleton, 1977; Carleton *et al.*, 1982; Bradley and Schmidly, 1987). *Peromyscus simulus* appears confined to lowland tropical forest and thorn scrub on the coastal plain of

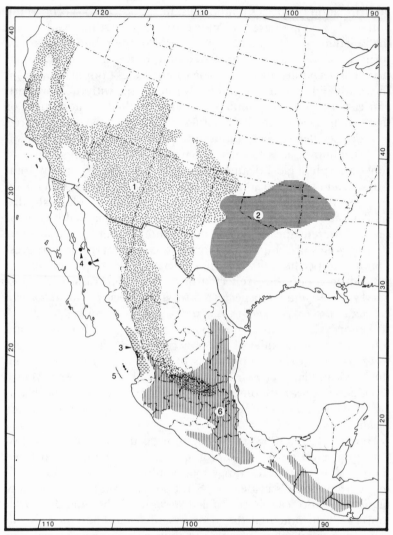

FIG. 14. Distribution of some species of the *boylii* species group: 1, *P. boylii*; 2, *P. attwateri*; 3, *P. simulus;* 4, *P. stephani*; 5, *P. madrensis*; 6, *P. levipes*. Arrow without number indicates insular subspecies of *boylii*.

Nayarit and Sinaloa, generally below elevations of 200 meters (Fig. 14). The southern limit of its distribution, just south of San Blas, is probably known, for at this latitude foothills of the Sierra Madre Occidental closely border the seacoast; however, its northern range limit, presently the vicinity of Mazatlan, Sinaloa, may be subject to future adjustment. As currently delimited, *simulus* is entirely allopatric to the nearest forms to which its morphological and karyological traits suggest phylogenetic affinity: *madrensis* on the Tres Marias Islands (Fig. 14) and *boylii rowleyi* in the higher elevations of the Sierra Madre Occidental and adjacent Mexican Plateau.

Both Merriam (1898a) and Osgood (1909) considered *spicilegus* as the nearest ally of the large-bodied *madrensis*, the former treating it as a species and the latter as a subspecies of *boylii*. Nevertheless, recent studies have not sustained this relationship. Carleton (1977) resurrected *madrensis* to species status because samples of it did not fall within patterns of variation observed among mainland populations of either *spicilegus* or *boylii*. In morphometric analyses using larger sample sizes of *madrensis*, Carleton *et al.* (1982) further substantiated its singular morphological differentiation from all other Nayarit *Peromyscus* and its purported close relative *spicilegus*. Based upon agreement of cranial and phallic proportions and upon similarity in autosomal number, they proposed the derivation of *madrensis* from a *simulus*-like ancestor, although the two represent opposite extremes of the size range of *boylii*-group forms in Nayarit.

Our awareness of the specific distinction of *spicilegus* has been hampered by Osgood's (1909) original definition of its range. Whereas Allen's (1897) description of *spicilegus* was limited to specimens from the type locality in western Jalisco, Osgood's revision extended its distribution to include not only the western lowlands of Jalisco and Nayarit, but also the flanks of the Sierra Madre Occidental and western fringes of the Mexican Plateau. His delineation of *boylii spicilegus* included western segments of true *boylii* as well as *spicilegus*, a low to intermediate elevation species. His error was particularly unfortunate because it misled subsequent investigators to affirm intergradation of *boylii rowleyi* and *spicilegus* when, in fact, they had not sampled *spicilegus* proper (for example, Kilpatrick and Zimmerman, 1975; Schmidly and Schroeter, 1974).

Hooper (1955) forecast the difficulties associated with Osgood's concept of *spicilegus* in western Mexico when he reported it and *boylii* in sympatry and contiguous allopatry at localities in Jalisco. Further distributional records of both kinds are now known for areas in Durango, Sinaloa, and Nayarit (Baker and Greer, 1962; Carleton, 1977; Carleton *et al.*, 1982), and have reduced both the altitudinal and geographic range of *spicilegus* as initially depicted by Osgood. In addition to the distributional evidence, the genetic isolation of *spicilegus* from its contiguous neighbors, *simulus* and *boylii*,

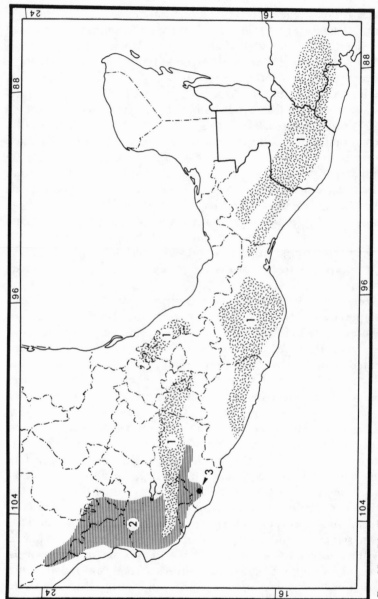

FIG. 15.— Distribution of some species of the *boylii* species group: 1, *P. aztecus*; 2, *P. spicilegus*; 3, *P. winkelmanni*.

is supported by cranial, phallic, and chromosomal attributes, which instead affiliate it with *aztecus* and related forms (Hooper, 1968; Carleton, 1977; Carleton *et al.*, 1982; Bradley and Schmidly, 1987). In raising *spicilegus* to a species, Carleton (1977) extended its southern distributional boundary to encompass populations in Michoacan that had been previously identified as *evides* (Osgood, 1909; Hooper, 1961). In summary, these modifications reveal that *spicilegus* inhabits the humid tropical zone at intermediate elevations (mainly from 500 to 1600 meters) from southern Sinaloa and western Durango in the north, southward through Nayarit, Jalisco, southwestern Zacatecas, and Colima, to western and central Michoacan (Fig. 15). The extent of its range still needs refinement as does its relationship to the *aztecus* complex.

The *aztecus* complex consists of an association of morphologically similar forms that occupy highlands in southern Mexico and Central America as far as southeastern Honduras. As their distributions are currently known, all are allopatric and occur on separate mountain systems: *aztecus* in the southern portion of the Sierra Madre Oriental; *evides* in the Sierra Madre del Sur of Guerrero and Oaxaca; *hylocetes* in the Transverse Volcanic Range; and *oaxacensis*, the most widely distributed, in the central highlands of Oaxaca, Chiapas, Guatemala, Honduras, and El Salvador. Osgood's (1909) comparative accounts indicate that he considered the four, and *spicilegus*, closely related; however, he retained only two as full species (*hylocetes* and *oaxacensis*) and placed the others as subspecies of *boylii*, though he did so with some skepticism, especially in the case of *aztecus*. The seeming unevenness of his treatment underscores his reliance upon distributional data in formulating his taxonomic decisions, for he detected the sympatry of *boylii levipes* with both *hylocetes* and *oaxacensis* but lacked such evidence for *aztecus* and *evides*. Such documentation has now been mustered for *b. levipes* and *aztecus* in Veracruz, Puebla, and Hidalgo (Alvarez, 1961; Hall and Dalquest, 1963; Musser, 1964; Carleton, 1977) and for *b. levipes* and *evides* in Guerrero and Oaxaca (Musser, 1964; Carleton, 1977). Accordingly, both *aztecus* and *evides* have been ranked as species along with *oaxacensis* and *hylocetes* (Hooper, 1968).

The allopatric status of these taxa and their general morphological resemblance, however, prompted doubts about the soundness of this arrangement. Hooper (1968) questioned whether they represent disjunct populations of one species and singled out *oaxacensis* and *hylocetes* as forms that may prove to be conspecific. Carleton

(1977) concurred with his assessment on the basis of their morphological overlap as revealed by craniometric analyses and penile characteristics and later (1979) formally placed *oaxacensis* and *hylocetes*, together with *evides*, as subspecies of *aztecus*. Two other taxa, *hondurensis* Goodwin (1941) and *cordillerae* Dickey (1928), were included as part of this synonymy. Musser (1969) had previously corroborated Hooper's (1968) impression that *hondurensis* was simply a southern example of *oaxacensis*, and Carleton (1979) listed it as a full synonym of *P. aztecus oaxacensis*. Carleton (1979) further pointed out that *cordillerae*, one of two subspecies of *boylii* that Dickey (1928) named from El Salvador, also agrees morphologically with *aztecus* and retained it as a subspecies thereof. These reallocations portray *aztecus* as being a formerly widespread montane species (Fig. 15) whose now fragmented parts have diverged weakly to moderately as a result of isolation and adaptation to slightly different biotic zones.

Carleton's (1979) revisions were based on the fundamental cranial, phallic, and ecological similarities exhibited by these nominal species, particularly the overlapping patterns of variation within and between them in relation to elevation. Populations inhabiting more subtropical conditions at lower elevations (that is, *aztecus*, *cordillerae*, and some *evides*) are smaller, with less inflated bullae, sparsely haired tails, and more brightly colored pelage; whereas, those from higher elevations (that is, *oaxacensis* and *hylocetes*), generally in the pine-oak belt, are larger, with moderately inflated bullae, more somber pelage, and more densely furred and sharply bicolored tails. Patterns of variation among these montane taxa require further confirmation using other kinds of data and additional samples. Available karyological data, although limited to standard preparations for *oaxacensis* and *hylocetes* are consistent with an interpretation of conspecificity (Schmidly and Schroeter, 1974; Lee and Elder, 1977). Of these five taxa, *aztecus* appears to be the most divergent phenetically (Carleton, 1979; Bradley and Schmidly, 1987), and perhaps ecologically, and its status relative to the other forms especially deserves reconsideration. More thorough sampling in the rugged mountains of Oaxaca is needed to better understand the level of differentiation between *oaxacensis* and *evides* in the south and between *oaxacensis* and *aztecus* in the north.

In evaluating the status of taxa that Carleton (1979) judged to be members of the species *aztecus*, two other forms merit attention as probable kin. *Peromyscus spicilegus* is obviously affiliated with this complex (Carleton, 1977; Bradley and Schmidly, 1987), yet in the mountains of Jalisco and Michoacan, *spicilegus* occupies the lower slopes of the Cordillera Transvolcanica, altitudinally separated from the larger *a. hylocetes*. Until controverted by other evidence, this distributional relationship, together with the cranial and phallic differences, supports the continued recognition of *spicilegus* as a species distinct from *aztecus*. Another species whose affinity to the *aztecus* complex invites clarification is *winkelmanni*, described by Carleton (1977) from near Dos Aguas, Michoacan (Fig. 15). In proportions of its glans penis, *winkelmanni* resembles *spicilegus*, but in size and color, it recalls a robust version of *a. hylocetes*. Presently recorded only from the vicinity of the type locality, *winkelmanni* may inhabit other sections of the coastal Sierra de Coalcoman in Michoacan.

Lawlor (1971*b*) demonstrated the *boylii*-like traits of *stephani*, known only from San Esteban Island in the Gulf of California (Fig. 14) and previously classified in the subgenus *Haplomylomys*. His decision to retain *stephani* as a species has been borne out by electrophoretic surveys, in which the genetic distance of *stephani* from *boylii* approximates that of *attwateri* from *boylii* (Avise *et al.*, 1974*a*, 1974*b*, 1979).

Not all recent nomenclatural changes in the *boylii* group have solely involved Mexican populations. Lee *et al.* (1972) reported that examples of *boylii attwateri* from Texas and Oklahoma consistently differed from neighboring populations of *b. rowleyi* in having a higher autosomal number, 56 as compared to 52. Their proposal that *attwateri* should be recognized as a species was endorsed by Schmidly (1973*b*), who detailed cranial and dental traits unique to *attwateri* that agreed with the karyological data. The status of *attwateri* as a full species has been similarly confirmed from phenetic and cladistic analyses of biochemical data (Avise *et al.*, 1974*a*; Kilpatrick and Zimmerman, 1975; Kilpatrick, 1984) and from cladistic assessments of differentially-stained chromosomes (Robbins and Baker, 1981; Rogers *et al.*, 1984; Stangl and Baker, 1984*b*). *Peromyscus attwateri*, as its distribution is presently known (Fig. 14), is allopatric to populations of *boylii*, but the two forms have been recorded from nearby localities in westcentral Texas (Schmidly, 1973*b*). Sampled populations of *attwateri* are more genetically homogeneous compared to most *boylii*, presumably an effect of its isolation on the Edwards Plateau during the late Pleistocene (Kilpatrick and Zimmerman, 1975; Kilpatrick, 1984). Kilpatrick (1984) estimated its time of divergence as 170,000 years before present, having separated as a peripheral isolate from a *boylii*-like ancestor.

Even after the deletion of most of Osgood's (1909) "definable forms," the residual populations comprising *Peromyscus boylii* exhibit substantial variation, notably as observed in standard karyotypic preparations. Much of this karyological variety is confined to populations in Mexico, particularly those known as *boylii levipes* or ones previously misidentified as *spicilegus* (Carleton *et al.*, 1982; Davis *et al.*, 1986; Houseal *et al.*, 1987). On the other hand, populations in the United States and northern Mexico (subspecies *boylii*, *rowleyi*, and *utahensis*) are chromosomally uniform, all displaying an autosomal number of 52 (Lee *et al.*, 1972; Schmidly and Schroeter, 1974; Houseal *et al.*, 1987). These data raise questions about the nature of population contact zones at the southern end of the Mexican Plateau and on its eastern and western boundaries where flatter terrain gives way to the intricate topography of the two Sierra Madres.

Hooper (1968) suspected that *rowleyi* and *levipes* merge near the southern end of the Mesa Central but lacked concrete evidence of this. This opinion has been reiterated in morphological (Schmidly, 1973*b*; Carleton, 1977), karyological (Schmidly and Schroeter, 1974), and electrophoretic (Kilpatrick and Zimmerman, 1975) studies, but the evidence in each case is inferential, based upon similarity criteria or, in the case of the chromosomal data, polymorphisms presumably indicative of hybridization. The karyotypic picture is complicated by variation in the number of biarmed pairs observed among samples referrable to *levipes*. Specimens from some areas in Chiapas, Oaxaca, and Hidalgo are karyotypically inseparable from examples of *rowleyi* (Schmidly and Schroeter, 1974); whereas, others, namely those from Michoacan, Queretaro, and San Luis Potosi, possess additional biarmed pairs yielding autosomal numbers of 54 to 60 (Lee *et al.*, 1972; Schmidly and Schroeter, 1974). Some of these disparate counts are recorded from localities in close geographic proximity, for example, in Queretaro and Hidalgo. Yet electrophoretic assays throughout the range of *boylii*, including samples of *utahensis*, *rowleyi*, and *levipes*, have typically yielded high coefficients of genetic similarity suggestive of an integrated genome (Avise *et al.*, 1974*a*; Kilpatrick and Zimmerman, 1975; Kilpatrick, 1984).

Recent investigations have partially reconciled the lack of concordance among these different data sets, at least in regard to the status of *rowleyi* and *levipes*. Certain populations in Queretaro and western Hidalgo have been found to bear closer affinity to *rowleyi* and other subspecies of true *boylii* (*utahensis* and *glasselli*) as judged from their electromorphic pattern (Rennert and Kilpatrick, 1986, 1987), karyotype (Houseal *et al.*, 1987), and craniodental morphology (Schmidly *et al.*, 1988). This discovery extends the range of *boylii rowleyi* about 250 kilometers to the southeast of its previously known occurrence in Aguascalientes. At Jonacapa, Hidalgo, individuals of *b. rowleyi* were trapped together with others that possess intermediate autosomal numbers (56–60). With respect to each of the above data sources, the latter agree with recently collected topotypic material of *levipes* from Cerro la Malinche, Tlaxcala. On this basis, *levipes* has been proposed as a species distinct from *boylii* (Rennert and Kilpatrick, 1987; Houseal *et al.*, 1987), and Schmidly *et al.* (1988) later formalized this arrangement.

Even if one accepts that the monomorphic AN = 52 populations on the Mesa Central (*rowleyi*) represent the southernmost distribution of true *boylii*, the homogeneity and taxonomy of the remaining populations still remains clouded. In western Mexico, the affinity of certain *boylii*-like populations mistakenly associated with *spicilegus* is problematical. When Carleton (1977) raised *spicilegus* to a species and redefined its distribution, he noted that many of the *boylii* samples Osgood (1909) identified as *b. spicilegus* are not clearly assignable, on geographic or morphological grounds, either to *rowleyi* or *levipes* and refrained from making subspecific designations.

Some of these populations contrast in number of autosomal arms with contiguous representatives of *boylii rowleyi*. Lee *et al.* (1972) reported that specimens from western Durango, in the vicinity of Coyotes and La Ciudad, have two additional pairs of large biarmed elements (AN = 56) compared to those of typical *rowleyi* (AN = 52). Similar chromosomal morphologies have been recorded from other localities southwest of Durango City (Schmidly and Schroeter, 1974; Kilpatrick and Zimmerman, 1975; Boles, 1984). Although this biarmed complement resembles the gross idiogram of *P. attwateri*, banded chromosomal comparisons reveal that certain of the biarmed pairs are not homologous rearrangements (Boles, 1984). This finding and the allelic data presented by Kilpatrick (1984) suggest that the enigmatic form is more closely related to other *boylii* than to *attwateri*. The evidence on the degree of its reproductive isolation from adjacent *rowleyi* populations is contradictory. The presence of some chromosomal polymorphism and heteromorphic pairing within the western Durango samples has been attributed to hybridization of the contradistinctive cytotypes (Schmidly and Schroeter, 1974), as has the occurrence of certain alleles (Kilpatrick and Zimmerman, 1975); however, the determination of chromosomal homologies inferred from banding data prompted Boles (1984) to dispute this interpretation. The distribution and status of these AN = 56 populations thus remain unresolved. The road between Durango City and Mazatlan, Sinaloa, appears to offer the best region for concentrated field efforts aimed at explicating the nature of the contact zone.

Another anomalous karyotypic variant documented for scattered *boylii*-like populations in western Mexico contains 10 or 11 pairs of biarmed chromosomes, yielding a higher autosomal number (AN = 66–68) than previously known for the species. Carleton *et al.* (1982) recorded this complement from Ocota in eastcentral Nayarit, a locality on a small mountain range isolated from the main ridges of the Sierra Madre Occidental. This formula matches that reported by Schmidly and Schroeter (1974) for a specimen from Los Reyes, Michoacan, which they identified as *evides* but which Carleton *et al.* (1982) referred to *boylii levipes*. Houseal *et al.* (1987) reported additional instances of this high-AN formula for populations in Jalisco and Michoacan and considered it to represent an undescribed taxon, presumably a new species. The significance of this chromosomal variant, the extent of its occurrence in western Mexico, and the propinquity of the mice in Nayarit, Jalisco, and Michoacan are all topics for future investigation.

Along the eastern rim of the Mexican Plateau, the genealogical picture among *boylii*-like populations is hardly improved over that in western Mexico. Osgood (1909) placed his samples from this area in *boylii levipes,* but Alvarez (1961) designated the smaller and more brightly colored mice in eastern Coahuila, Nuevo Leon, and the Sierra San Carlos, Tamaulipas, as a new subspecies, *ambiguus.* Schmidly (1973*b*) slightly modified the range of *b. ambiguus* to include northern San Luis Potosi and the Sierra de Tamaulipas and drew attention to the moderately pronounced morphological separation of these populations compared to those of *levipes* to the south. The autosomal numbers characteristic of *ambiguus* (AN = 58–60) exceed those of *b. rowleyi* but appear the same as that documented for the topotypic sample of *levipes* and others in Queretaro, Hidalgo, Puebla, and Veracruz that otherwise appear to be *levipes* (Schmidly and Schroeter, 1974; Kilpatrick and Zimmerman, 1975;

Houseal *et al.*, 1987). The close relationship of *ambiguus* and *levipes sensu stricto* has also drawn support from electrophoretic (Rennert and Kilpatrick, 1987), phallic (Bradley and Schmidly, 1987), and morphometric (Schmidly *et al.*, 1988) investigations, and Schmidly *et al.* (1988) recognized *ambiguus* as a subspecies of *P. levipes*. Nonetheless, samples identified as *ambiguus* seem to be appreciably removed biochemically and craniometrically from others in eastern Mexico (Kilpatrick, 1984; Rennert and Kilpatrick, 1987; Schmidly *et al.*, 1988). Whether two cryptic species intermingle in portions of the northern Sierra Madre Oriental and whether the epithet *ambiguus* properly applies to one of those forms are matters that deserve further scrutiny.

Another problematic form in eastern Mexico whose status requires clarification is *baetae*. In fact, determination of its level of relationship to *levipes* is pivotal to understanding the species limits and distribution of *boylii*-like populations located farther south in Mexico and Central America. Thomas (1903) described *baetae* as a species from Mount Orizaba, Veracruz, but Osgood (1909) relegated it to full synonomy under *boylii levipes*. In 1961, Alvarez resurrected *baetae* as a subspecies of *boylii* to identify the larger, darker mice found in a localized area of the eastern Sierra Madre Oriental in Veracruz. The divergence of these mice from typical *levipes* is borne out by their lower autosomal count (AN = 52–54) and weak to moderate genetic and morphometric distances (Rennert and Kilpatrick, 1987; Houseal *et al.*, 1987; Schmidly *et al.*, 1988). Individuals of both *levipes* and *baetae* have been reported from one locality in northwestern Puebla, prompting their recognition as cryptic species. Based on biochemical and karyotypic criteria, the range of *baetae* is viewed as including populations in Guerrero and Oaxaca (Rennert and Kilpatrick, 1987; Houseal *et al.*, 1987), and presumably those farther south in Central America (Bradley and Ensink, 1987). In contrast, Schmidly *et al.* (1988), noting the morphometic segregation of these samples from *baetae*, delimited its range to the eastern slopes of the Sierra Madre Oriental in Veracruz, Puebla, and Hidalgo. As mapped by Schmidly *et al.* (1988), the occurrence of *baetae* adheres to humid forest at intermediate to high elevations, a biotope which has apparently been important to the differention of other peromyscines such as *P. furvus*, *P. a. aztecus*, *Megadontomys nelsoni*, and *Habromys simulatus*.

The unsuspected diversity among Mexican populations of *boylii* unveiled by research efforts over the last three years clearly requires further synthesis. Illumination is particularly urgent regarding 1) the status and geographic extent of *baetae* in comparison with *levipes* and 2) resolution of the degree of differention and relationship of the many karyotypic morphs so far identified. In light of these uncertainties, I have adopted a pragmatic course and arranged *boylii* (including *glasselli*, *rowleyi*, and *utahensis*) as a northern species distinct from the more southern *levipes* (including *ambiguus*, *baetae*, and *sacarensis*), while acknowledging the possibly composite nature of the latter (Fig. 14).

Ironically, Hooper's (1968:52–53) insight that "The status of the populations known as *P. b. levipes* is at the core of the problem of the interrelationships of the *boylii*-like populations in the highlands of Mexico" holds equally valid today, even though the morphological boundaries and taxonomic content of *boylii* have been significantly restricted since his review. One's impression is that Pleistocene climatic oscillations generated a complex tableau of fragmentation, differentiation, and secondary contact among populations of *boylii* distributed throughout the mountains flanking the Mexican Plateau, but that the vicariant impact of these climatically induced biotic shifts was less disruptive for populations inhabiting the central portion of the Plateau. The likelihood of such range disjunctions is readily apparent today in the distribution of *boylii* forms on mountain spurs isolated from the major cordilleras. The localization of karyotypic variation within *boylii*-like populations lends some credence to this view, as does the pattern of reduction of genetic variation among them (Kilpatrick, 1984). In several respects, *Peromyscus boylii* and its allies, excluding the *aztecus* complex, may offer a better evolutionary microcosm for exploring speciation theories than the *maniculatus* group, which heretofore has been the focus of such studies in *Peromyscus*.

Where *pectoralis* and *polius* fit in relationship to the forms discussed above remains obscure in spite of appreciable growth in our systematic knowledge of them. A tentative answer emerging from recent studies is that the two do not belong with the *boylii* group as currently recognized.

Peromyscus pectoralis occurs over the central Mexican Plateau and adjoining Sierra Madre Oriental, northward to central Texas, southeastern New Mexico, and southern Oklahoma (Kilpatrick and Caire, 1973; Schmidly, 1972) (Fig. 13). Within this area, they seem to occur as disjunct populations, preferring rocky terrain where other *Peromyscus* are scarce (Kilpatrick, 1971). Although the close phenotypic resemblance of *pectoralis* to *boylii* has raised suspicion that the two interbreed, neither morphological nor biochemical studies have uncovered such evidence (Kilpatrick and Zimmerman, 1976; Schmidly, 1972). Instead the genetic dissimilarity of *pectoralis* and *boylii* questions their placement in the same species group. Clustering analyses of genic data have variously linked *pectoralis* with species of *Haplomylomys* or the *truei* group (Avise *et al.*, 1974*a*, 1979; Zimmerman *et al.*, 1975), and in studies in which *pectoralis* associated with another *boylii*-group member, it did so at a comparatively low level of similarity (Kilpatrick and Zimmerman, 1975; Schmidly *et al.*, 1985). The karyotypic information, although restricted to few species of the *boylii* group, also leaves the phylogenetic alliance of *pectoralis* as equivocal. *Peromyscus pectoralis* displays inversions of chromosomes 2, 3, and 9 (Robbins and Baker, 1981); consequently, parsimonious interpretations of G-banded data place it among species of the *mexicanus* group (Rogers *et al.*, 1984; Stangl and Baker, 1984*b*).

Perhaps due to their patchy ecological occurrence, population samples of *pectoralis* exhibit unusually wide variation in levels of heterozygosity and genetic similarity (Avise *et al.*, 1974*a*; Kilpatrick and Zimmerman, 1975; Zimmerman *et al.*, 1975; Kilpatrick, 1984); in fact, the low genetic similarity coefficients among certain segments of *pectoralis* led Avise *et al.* (1974*a*) to question whether it was a composite of two or more species. However, Kilpatrick and Zimmerman (1976), on the basis of a broader electrophoretic survey, attributed the low genetic similarity values to limited gene flow across narrow zones of intergradation. In addition, they interpreted the pronounced genetic discontinuities within *pectoralis* as evidence of a refugium effect, whereby the Pleistocene distribution of *pectoralis* was contracted to three regions before population expansion and secondary contact. The hypothesized locations of these Pleistocene refugia are concordant with character shifts and subspecies ranges as mapped by Schmidly (1972): the Edwards Plateau in southcentral Texas (*p. laceianus*), the central portion of the Mexican Plateau (*p. pectoralis*), and tropical habitats in Tamaulipas (*p. collinus*).

Peromyscus polius is recorded from scattered localities in the mountains of northwestern Chihuahua (Fig. 13), generally in rocky areas above 6000 feet (Anderson, 1972). This ecological setting conforms to that of *difficilis*, which it also resembles in size, but there is little overlap in distributions of the two species, leading Anderson (1972) to suggest some kind of competitive displacement. Osgood (1909) placed *polius* in the *truei* group, where it remained until Hoffmeister

(1951) provisionally reassigned it to the *boylii* group, although mentioning its possible affiliation with the *melanophrys* group. Anderson (1972) questioned whether the phenetic and ecological similarity of *polius* and *difficilis* might not also signify their close phylogenetic affinity and argued that resolution of its species-group status must await firmer comprehension of relationships among species of both the *truei* and *boylii* groups. Phenetic evaluations of allelic variation consistently depict *polius* as a uniquely differentiated species, whose affinity either is ambiguous or weakly associated with the *boylii* group (Kilpatrick and Zimmerman, 1975; Zimmerman *et al.*, 1975, 1978; Schmidly *et al.*, 1985). Cladistic analysis of biochemical characters also portrays *polius* as peripheral to most forms, although it was considered to share more recent ancestry with *boylii-attwateri* than *pectoralis* (Kilpatrick, 1984). The scope of most of the above-cited studies is inadequate to assess alternative hypotheses of relationship of *polius*. Determination of its species-group membership will require broader systematic comparisons, drawing upon other information such as chromosomal banding and reproductive tract morphology.

Whereas the phyletic integrity of most species-group taxa of *Peromyscus* has received moderate to strong corroboration from research conducted since Hooper's review, the evidence for monophyly of the *boylii* group has been severely eroded. On the basis of cranial, phallic, and distributional characteristics, Carleton (1977) sorted *boylii*-group species into two assemblages, each of which appears to constitute a separate clade and whose morphological differentiation from one another approaches that of other recognized species groups. In a more thorough analysis of phallic traits, Bradley and Schmidly (1987) discerned a similar division of species. The variation in autosomal number is also suggestive of a dichotomy, a higher-versus lower-AN group, but too few of the critical species have been banded to ascertain chromosomal homologies. Using current specific nomenclature, the dichotomy consists of a *boylii* complex (*boylii, attwateri, stephani, simulus,* and *madrensis*) and an *aztecus* complex (*aztecus, spicilegus,* and *winkelmanni*). Morphological criteria, particularly phallic traits, associate *pectoralis* and *polius* with the *boylii* complex, but the biochemical and karyological information cast doubt on their affinity to *boylii* proper. Nor does the unity of the group receive support from chromosomal inversion data, the few species banded (*boylii glasselli, b. rowleyi, attwateri,* and *pectoralis*) being widely separated in the cladograms produced (Rogers *et al.*, 1984; Stangl and Baker, 1984*b*).

Osgood's (1909) remarks on the *boylii* group reveal that he considered it exceptionally difficult to distill the reticular patterns of morphologic variation and geographic distribution into a sound classification. Aided by 80 years of hindsight and attendant research upon the group, one better appreciates the challenge he then faced. Although the research contributions of the past decade have resolved many of the problems that puzzled Osgood, this same body of research cautions that the genealogical complexities of the group are still unfolding.

Subgenus Peromyscus: truei *Group*

Osgood (1909)	Hooper (1968)	Hall (1981)
truei (6)	*truei*	*truei* (15)
difficilis (3)	*difficilis*	*difficilis* (8)
nasutus	*bullatus*	*bullatus*
bullatus		
polius		

Evidence accumulated from karyotypic and electrophoretic investigations suggests that the presumed conspecificity of populations identified as *truei* and *difficilis* bears reconsideration. In each case, these data sets point to the existence of northern and southern moieties that are reproductively isolated from one another.

The occurrence of two karyotypic morphs differing in number of autosomal arms (62 versus 54) among populations of *truei* had been known for some time (Hsu and Arrighi, 1968), leading Lee *et al.* (1972) to suspect its species integrity. Modi and Lee (1984) discovered localities of sympatry of the two cytotypes in New Mexico. G-banding demonstrated that the AN = 62 form has experienced pericentric inversions in chromosomes 1, 2, 3, 6, 9, and 15; whereas, the AN = 54 form exhibits an inversion only in chromosome 2. In light of the kinds of interspecific chromosomal variation documented for species of *Peromyscus* (for example, Rogers *et al.*, 1984; Stangl and Baker, 1984b), the euchromatic differences revealed by Modi and Lee (1984) are substantial and largely invariant over broad geographic areas. These authors therefore recognized the southern AN = 54 complex as *P. gratus* Merriam (including *erasmus*, *gentilis*, and *zapotecae*), a species distinct from *truei* (remainder of the northern subspecies, including *t. comanche*). The distribution and extent of sympatry of the two species remain to be accurately mapped (Fig. 16). Zimmerman *et al.* (1978) likewise con-

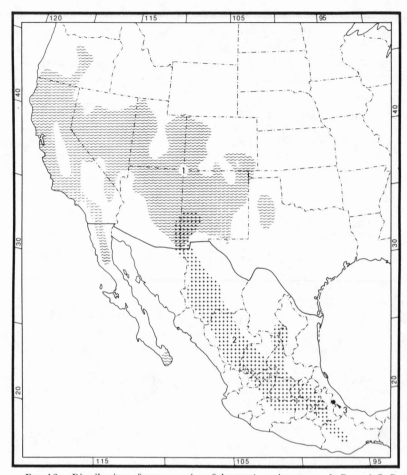

Fig. 16.—Distribution of some species of the *truei* species group: 1, *P. truei*; 2, *P. gratus*; 3, *P. bullatus*.

sidered *truei* to be a composite of two gene pools, although they listed the southern form as *gentilis*. Unlike the chromosomal dissimilarities, the divergence in protein alleles of these forms, which Zimmerman *et al.* called sibling species, is not so remarkable (Nei's identity index = 0.948), being about equal to that recorded for other subspecies within *Peromyscus*.

In addition, Zimmerman *et al.* (1978) identified northern and southern segments of *difficilis* as a sibling species pair (Nei's identity index = 0.943), removing those northern populations previously known as *nasutus* (Fig. 17). In contrast to the taxonomic situation in *truei, nasutus* had been recognized as a distinct species since Osgood's (1909) revision, until Hoffmeister and de la Torre (1961) advocated its synonymy with *difficilis*. Existence of a chromosomal difference (Hsu and Arrighi, 1968; Zimmerman *et al.*, 1975), the northern form *nasutus* possessing an added biarmed pair (AN =

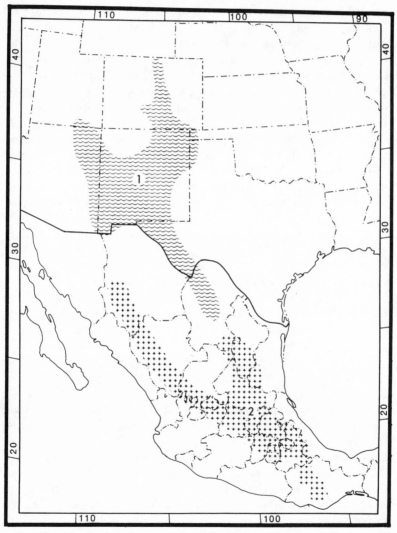

FIG. 17.—Distribution of some species of the *truei* species group: 1, *P. nasutus*; 2, *P. difficilis*.

58) compared to Mexican *difficilis* (AN = 56), suggested that the two might not intergrade as argued by Hoffmeister and de la Torre (but Robbins and Baker, 1981, reported their sample from Durango as having AN = 58). On the basis of electrophoretic studies, Zimmerman *et al.* (1975, 1978) and Avise *et al.* (1979) further questioned the intergradation of *nasutus* and *difficilis* and recommended elevating *nasutus* to its former specific status. To date, the evidence against intergradation of *difficils* and *nasutus* is less compelling than that for *truei* and *gratus*, as it lacks the breadth of geographic sampling and the demonstration of sympatry.

Diersing (1976) retrieved *boylii penicillatus* Mearns from synonymy under *boylii rowleyi*, where Osgood (1909) had erroneously placed it, and reassociated it as *difficilis penicillatus*. Reidentification of this form, known from the Franklin Mountains, Texas, and northwestern Coahuila, narrows the gap between the ranges of *nasutus*-like and *difficilis*-like populations, compared to the wide geographic hiatus between the samples available to Osgood (1909) and Hoffmeister and de la Torre (1961). Moreover, the specimens assigned to *penicillatus* karyotypically and morphometrically resemble examples of *nasutus*, although Diersing (1976) chose to recognize them all as subspecies of *difficilis*. Whether the level of differentiation between *difficilis* and *nasutus* is ultimately judged equivalent to that between *truei* and *gratus*, the congruence of their geographic ranges and coalignment of the presumed contact zones intimate an evolutionary response to the same vicariant factors.

The status of Blair's (1943*b*) *Peromyscus comanche* seems finally to have been resolved. Although one of the few mammalian species descriptions presenting breeding information suggesting the new form's reproductive isolation, Blair's data were incomplete, lacking critical reciprocal crosses to other species in the group, notably *truei*. In his revision of *truei*, Hoffmeister (1951) relegated *comanche* to subspecific level under *nasutus*, and later (Hoffmeister and de la Torre, 1961) to a subspecies of *difficilis* when that form and *nasutus* were synonymized. Lee *et al.* (1972), however, reported that the standard karyotype of *comanche* was inseparable from that of *truei*, and Schmidly's (1973*a*) morphometric evaluation also supported the transfer of *comanche* to *P. truei*. Modi and Lee (1984) have confirmed that individuals of *comanche* and *truei* from New Mexico are indeed reproductively compatible in the lab, although the two populations are separated from each other by over 100 miles of unsuitable habitat and have been so for perhaps 10,000 years (Blair,

1950; Schmidly, 1973*a*). Only the electrophoretic study of Johnson and Packard (1974) questioned the taxonomic status (although not its closer affinity to *truei*) of *comanche*, which they resurrected as a species. Nevertheless, as noted by Modi and Lee (1984), their conclusion is at variance with their own genic data and the genetic distances observed among their samples.

Hooper (1968) believed that *bullatus* (Fig. 16) would prove to be a subspecies of *truei*, but the status of this form has not been addressed. The proportions of the otic capsules of the holotype are truly outsized for either *truei* or *difficilis*.

Mice assigned to the *truei* group form a unified assemblage morphologically, geographically, and ecologically (Hoffmeister, 1951; Hooper, 1968). The cohesiveness, and inferred monophyly, of the group generally draws support from the electrophoretic but not chromosomal banding data assembled to date.

Investigations of allelic variation involving the *truei* group have generally calculated genetic distances among OTUs either from Nei's identity index or from Rogers' similarity coefficient and have depicted levels of similarity in phenograms (UPGMA). In one study, examples of *difficilis* (and *nasutus*) clustered nearer *pectoralis* than *truei* (Zimmerman *et al.*, 1975), but the taxonomic coverage (6 species) and number of polymorphic loci (11) employed were limited. Electrophoretic assays using more species (14 to 26) and additional loci (18 to 23) have uniformly represented *truei-gratus* and *difficilis-nasutus* as members of a pair-group well differentiated from others of comparable taxonomic ranking in the subgenus *Peromyscus* (Zimmerman *et al.*, 1978; Avise *et al.*, 1979; Schmidly *et al.*, 1985).

In marked contrast to electrophoretic data, euchromatic banding patterns picture such a substantial phylogenetic hiatus between *difficilis* and *truei* as to question the validity of their species-group alliance. The two forms differ in five rearrangements, namely pericentric inversions in chromosome 7 in *difficilis* and in 1, 3, 9, and 15 in *truei* (Robbins and Baker, 1981; Stangl and Baker, 1984*b*). Consequently, cladistic interpretations of chromosomal evolution place *truei* as an independent line within a large clade mostly containing other species of the subgenus *Peromyscus* (Robbins and Baker, 1981; Rogers *et al.*, 1984), or as the sister species of *Habromys* (= *Peromyscus* s.l.) *lepturus* within the same clade (Stangl and Baker, 1984*b*). These same studies derive *difficilis* more basally from their trees, either in association with *attwateri* as part of an unresolved trichotomy (Robbins and Baker, 1981; Stangl and Baker, 1984*b*) or with *att-*

wateri and *californicus* as members of the same clade (Rogers *et al.*, 1984). Disagreement over the phylogenetic placement of *difficilis* reflects a character-weighting decision involving the numerous pericentric inversions recorded for chromosome 6, which were included by Robbins and Baker (1981) and Stangl and Baker (1984*b*) and excluded by Rogers *et al.* (1984). In light of Modi and Lee's (1984) data for *gratus*, which departs from the hypothetical ancestral configuration only by an inversion of chromosome 2, the karyologic divergence within the *truei* group becomes even more difficult to reconcile with other kinds of systematic data indicating the group's homogeneity.

Subgenus Peromyscus: melanophrys *Group*

Osgood (1909)	Hooper (1968)	Hall (1981)
melanophrys (3)	*melanophrys*	*melanophrys* (6)
xenurus	*mekisturus*	*mekisturus*
mekisturus	*perfulvus*	*perfulvus* (2)

Although Osgood (1945) expressly denied any close phyletic affinty of his new form *perfulvus* to *melanophrys* (and consequently allied it with the *boylii* group), recent karyological and biochemical evidence convincingly sustains Hooper's (1955, 1968; Hooper and Musser, 1964*b*) arrangement of the two in the same species group. With six pairs of biarmed autosomes (1, 2, 3, 9, 22, and 23), *perfulvus* and *melanophrys* appear inseparable karyotypically (Lee and Elder, 1977), even to the pattern of their euchromatic bands (Stangl and Baker, 1984*b*). The most singular aspect of their chromosomal resemblance, however, is the male's possession of a minute biarmed Y-chromosome. Zimmerman (1974) postulated a complex sex-determining mechanism for *melanophrys*, but Lee and Elder (1977) identified a conventional X-Y mechanism in *perfulvus*, implying that the X-chromosome is a large acrocentric. In studies of genic variation, *perfulvus* and *melanophrys* exhibit strong similarity and consistently cluster together at a similarity level comparable to that observed for the *maniculatus, leucopus,* and *truei* species groups (Zimmerman *et al.*, 1978; Schmidly *et al.*, 1985).

An evaluation of immunological distances derived from microcomplement fixation analysis of serum albumins constitutes the only evidence against a cognate relationship for the two species (Fuller *et al.*, 1984). In one-way comparisons of immunological distances, *perfulvus* scored slightly closer to *P.* (*Osgoodomys*) *banderanus* than to *melanophrys*; however, the authors noted that the confidence limits for the one-way comparisons were insufficient to select the nearest neighbor of *perfulvus* between those two. In reciprocal immunological comparisons, *perfulvus* was uniformly the most highly divergent of the five species of *Peromyscus* (*s.l.*) examined. Interestingly, it formed a pair-group with *banderanus* (*melanophrys* not included in the reciprocal trials), whose range broadly overlaps that of *perfulvus* in the arid lower tropical zone in midwestern Mexico (Fig. 18). In light of the adaptive context attributed to alpha-chain hemoglobin variation in *maniculatus* (Chappell *et al.*, 1988; Snyder *et al.*, 1988), exploration of the physiological role of the serum albumins associating *perfulvus* and *banderanus* would be worthwhile.

Whereas the species-group association of *melanophrys* and *perfulvus* draws support from recent investigations, the cladistic posi-

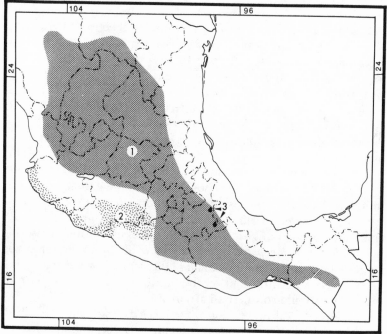

Fig. 18.—Distribution of species of the *melanophrys* species group: 1, *P. melanophrys*; 2, *P. perfulvus*; 3, *P. mekisturus*.

tion of the *melanophrys* group relative to other *Peromyscus* is less clear, although it is undoubtedly affiliated with forms of the nominate subgenus. Stangl and Baker (1984*b*) placed the *melanophrys* group in an unresolved polychotomy numbering eight stems, which also included *pectoralis* and six species of the *mexicanus* group. The character defining this clade is a pericentric inversion in chomosome 6, a rearrangement which has seemingly occurred multiple times in peromyscine evolution (Robbins and Baker, 1981; Rogers *et al.*, 1984; Stangl and Baker 1984*b*). A phenogram generated from allelic variation among 14 species of the subgenus *Peromyscus* linked the *melanophrys* group with the *leucopus* group, which in turn formed a pair-group exclusive of other species representing the *crinitus*, *maniculatus*, *boylii*, and *truei* species groups (Zimmerman *et al.*, 1978). In a broader electrophoretic survey, including 27 species of *Peromyscus* (*s. l.*), the *melanophrys* group again demonstrated affinity to the same five species groups, although it clustered with them at a low similarity value and was set somewhat more apart from members of the *mexicanus* group (Schmidly *et al.*, 1985). The last study lends some credence to Hooper's (1968:59) impression that the *melanophrys* group is ". . . in a sense transitional between temperate *boylii*-like forms and tropical *mexicanus*-like species."

Systematic knowledge of *P. mekisturus* remains meager, as the form is known only from two specimens, Merriam's (1898*b*) holotype from Chalchicomula, Puebla, and Hooper's (1947) record from Tehuacan, Puebla (Fig. 18). Specimens of *melanophrys* also are documented from both localities (Osgood, 1909; Baker, 1952). Of the three species composing the *melanophrys* group, *mekisturus* is most divergent, characterized by more orthodont incisors, lack of a supraorbital shelf, shorter and wider mesopterygoid fossa, and greater development of an incisor capsule on the dentary. Still, the number of features possessed in common (such as the long tail, truncate rostrum, spacious sphenopalatine vacuities, sloping occiput, and molar configuration) recommends its continued association with the *melanophrys* group. In view of the exceptional tail length (about 160 percent head and body length) of *mekisturus* and its relatively long fifth digit on the hindfoot, efforts to rediscover the species should consider an arboreal niche.

The level of differentiation of *melanophrys micropus*, described by Baker (1952) from near Guadalahara, Jalisco, should be reexamined. In its small skull and weakly developed supraorbital shelf, *micropus* stands apart from other geographic examples of *melanophrys*.

Subgenus Peromyscus: mexicanus *Group*

Osgood (1909)	Hooper (1968)	Hall (1981)
banderanus (3)	*ochraventer*	*ochraventer*
yucatanicus (2)	*stirtoni*	*stirtoni*
allophylus	*yucatanicus*	*yucatanicus* (2)
mexicanus (3)	*allophylus*	*mexicanus* (7)
+	*mexicanus*	*gymnotis*
lepturus Group (part)	*furvus*	*furvus*
furvus	*latirostris*	*mayensis*
altilaneus	*melanocarpus*	*melanocarpus*
guatemalensis	*zarhynchus*	*zarhynchus*
nudipes	*grandis*	*grandis*
+	*altilaneus*	*altilaneus*
megalops Group	*guatemalensis*	*guatemalensis* (2)
melanocarpus	*nudipes*	*nudipes* (3)
zarhynchus	*megalops*	*megalops* (4)
megalops (3)		

The *mexicanus* group, the largest of the seven species groups recognized by Hooper (1968), rivals the *boylii* group in its systematic complexity and the need for basic revisionary work. This speciose assemblage, whose morphological limits were expanded by Hooper and Musser (1964*b*), represents an amalgamation of three groups created by Osgood (1909), namely the *mexicanus, megalops,* and a portion of the *lepturus* species groups. The unity of the group derives principally from our meager knowledge of it and the common distribution of member species in subtropical and tropical habitats of Central America, rather than from a clear demonstration of phylogenetic relationship. As is true for the genus as a whole (Glazier, 1980), the greatest number of *mexicanus*-group species is concentrated in the highly dissected mountains of Oaxaca. Additional sampling, particularly transect surveys, from this region would enhance our grasp of the diversity and complexity of the *mexicanus* group. The study by Huckaby (1980) constituted a first attempt to delineate species boundaries and summarize distributions in the *mexicanus* group, but his contribution must be viewed as providing an improved taxonomic framework for the problems yet to be resolved.

The core of the *mexicanus* group consists of five species centered in Guatemala and southern Mexico whose morphological recognition involves almost imperceptible size and color gradations. In qualitative features of the cranium and dentition, mammae number, stomach morphology, glans penis, and male accessory gland

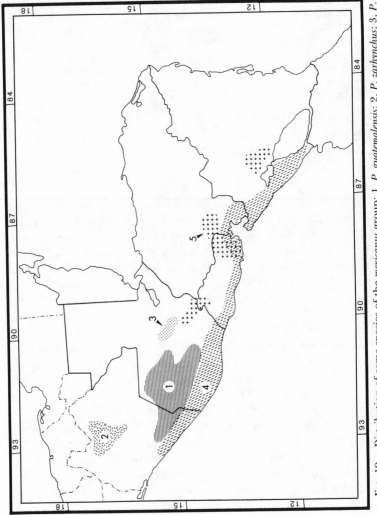

Fig. 19.—Distribution of some species of the *mexicanus* group: 1, *P. guatemalensis*; 2, *P. zarhynchus*; 3, *P. grandis*; 4, *P. gymnotis*; 5, *P. stirtoni*.

complement, the five are inseparable and presumably closely related (Hooper and Musser, 1964*b*; Linzey and Layne, 1969; Carleton, 1973; Huckaby, 1980). Consequently, the case for their specific status rests upon a combination of distributional evidence and degree-of-difference criteria. Arranged from smallest to largest, they are *gymnotis, mexicanus, guatemalensis, zarhynchus,* and *grandis.*

Osgood (1909) considered *Peromyscus gymnotis* as a subspecies of *mexicanus,* but Hooper (1968) questioned its conspecificity with those populations. With Hooper's reservation in mind and having found no evidence of intergradation, Musser (1971) reinstated *gymnotis* as a species and placed Osgood's (1904*b*) Chiapan form *allophylus* in synonymy with it. Huckaby (1980) approved both taxonomic decisions and further documented the range of *gymnotis* in low to intermediate elevations of southern Chiapas and southwestern Guatemala, where it abuts the ranges of *guatemalensis* and *mexicanus* (Fig. 19). Although Huckaby thought that the presence of *mexicanus* constrained the northern and southern geographic limits of *gymnotis,* Jones and Yates (1983) recorded the latter from the Pacific lowlands of western Nicaragua almost to the border of Costa Rica. The emerging distributional picture of *gymnotis* as an endemic of the Pacific coastal plain and adjacent foothills of Middle America cautions that it may occur among samples from El Salvador and the Isthmus of Tehuantepec believed to represent *mexicanus.*

Geographically and altitudinally, *Peromyscus mexicanus* has the widest range of the 14 species Hooper (1968) assigned to the *mexicanus* group, with populations extending from northern Veracruz and southeastern Guerrero to central Nicaragua (Fig. 20). Considerable diversity in size and pelage color, the latter seemingly correlated with humidity and temperature, occurs within the species and is reflected in the number of described taxa associated with it. Faced with the absence of broad geographic trends and the complexity of localized variations, Huckaby (1980) dispensed with formal subspecific divisions of *mexicanus,* an action underscoring the difficulties inherent in conveying reticulate patterns of variation through conventional subspecies usage. However, until the relationships and conspecificity of the included forms are better understood, retention of subspecies may be defended, an unhappy solution but a pragmatic one given our present knowledge.

Such a recommendation acknowledges the numerous systematic reallocations that have involved *mexicanus* over the past 15 years. Taxa transferred into *mexicanus* include *banderanus angelensis* Osgood, *b. coatlanensis* Goodwin, *b. sloeops* Goodwin, and *guatema-*

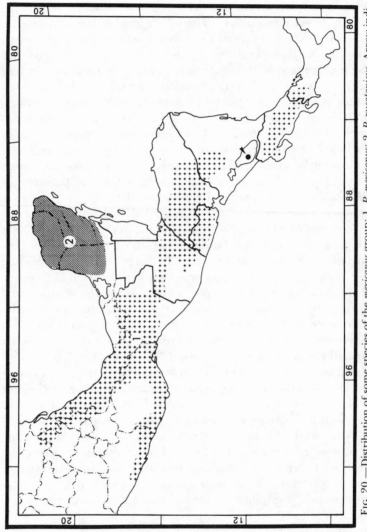

FIG. 20.—Distribution of some species of the *mexicanus* group: 1, *P. mexicanus*; 2, *P. yucatanicus*. Arrow indicates insular population of *mexicanus* in Lake Nicaragua.

lensis tropicalis Goodwin by Musser (1969); and *nudipes* Thomas, *n. orientalis* Goodwin, *n. hesperus* Harris, and *megalops azulensis* Goodwin by Huckaby (1980). Although these forms were improperly associated by their original describers, it is doubtful that they comprise a single, interbreeding array of populations as their synonymy under *mexicanus* implies. In particular, populations of *angelensis* along the lower, dry Pacific-facing slopes of Oaxaca are highly distinctive and contrast in external and cranial morphology with the contiguous higher-elevation races of typical *mexicanus* (that is, *mexicanus* and *totontepecus*). *Peromyscus coatlanensis* fits with *angelensis* in most respects; *sloeops* and *tropicalis* resemble *mexicanus* proper. The holotype of Goodwin's (1956) *azulensis*, described from mountains in the Isthmus of Tehuantepec, is unusually large compared to most *mexicanus*, which it otherwise resembles as indicated by Huckaby (1980).

Hooper (1968) concluded that specimens of *nudipes* are anatomically linked with both *guatemalensis* and *mexicanus*, but Huckaby's (1980) morphometric analyses disclosed a stronger similarity of *nudipes* to the latter. This finding, and the fact that samples of *nudipes* were interspersed randomly among those of *mexicanus* in multivariate space, persuaded him to view populations of *nudipes* as the southernmost extension of the species *mexicanus*. Like so many other taxonomic decisions regarding allopatric populations of *Peromyscus* that bear different names, this conclusion requires corroboration using other data bases. The conspecificity of *mexicanus* and *nudipes* seems zoogeographically plausible, inasmuch as a record of *mexicanus saxatilis* from volcanic peaks in Lake Nicaragua (Jones and Yates, 1983) has greatly diminished the geographic hiatus between it and the nearest populations of *nudipes* in Costa Rica and suggests their former continuity. I am unable to find any differences between the specimens used to substantiate the sympatric occurrence of *mexicanus saxatilis* and *nudipes* in the Cordillera Central of Costa Rica, as shown by distribution maps in Hall and Kelson (1959) and Hall (1981). Until patterns of variation and relationships of Costa Rican and Nicaraguan populations are better understood, the locality report of Jones and Yates (1983) should be considered the southern terminus of *mexicanus saxatilis*.

Peromyscus guatemalensis, zarhynchus, and *grandis* comprise another association of allopatric forms whose slight but demonstrable morphological differences pose largely subjective taxonomic judgements. Huckaby (1973) initially treated the three as fragmented populations of a single species, *zarhynchus*, but later (Huckaby, 1980) altered his conclusion and maintained them as separate species. All three occupy montane habitats, but as presently delimited, their ranges are disjunct (Fig. 19), with *guatemalensis* in intermediate- to high-elevation wet forests of southern Chiapas and westcentral Guatemala, *zarhynchus* in high-elevation cloud forests of central Chiapas, and *grandis* in intermediate-elevation humid forests of eastcentral Guatemala. This distributional pattern intimates an earlier continuity, but the two peripheral forms, *zarhynchus* and *grandis*, are more similar morphologically and chromatically than either is to the central, more broadly distributed *guatemalensis*. Yet the more similar pair inhabit different vegetational and elevational zones. Unable to choose critically between a hypothesis of their independent origin from one or more lowland species versus one of secondary isolation and differentiation of a once continuous montane species, Huckaby ultimately elected to maintain each as a species. In light of our current uncertainty about their status, this continues to be the preferred course.

Peromyscus guatemalensis also requires more refined distributional, ecological, and taxonomic analyses in order to improve our understanding of its relationship to *mexicanus*. In Guatemala, the range of *guatemalensis* borders that of *gymnotis* along its southwestern boundary and that of *mexicanus* at its southeastern and northwestern limits. In both cases, *guatemalensis* occupies the higher, moister, and cooler environments; it has been trapped sympatrically with *gymnotis* but not *mexicanus* (Huckaby, 1980). The separation of *guatemalensis* from the much smaller *gymnotis* is not problematic, but discrimination between *guatemalensis* and *mexicanus* samples can prove difficult, identification relying upon the relatively larger size and grayer pelage of *guatemalensis*. The confusion is heightened by the apparent tendency of *mexicanus* populations to acquire *guatemalensis*-like traits in highlands where *guatemalensis* is absent, for example in the mountains to the south and east of Guatemala City (Huckaby, 1980). Populations from this region deserve further study because I have discovered specimens of both kinds in USNM that were collected at the same localities just west of Guatemala City.

The sympatric occurrence of *mexicanus* and *guatemalensis* resurrects the question of the systematic status of the enigmatic form *altilaneus*, known only by the type specimen from Todos Santos in the Cordillera de los Cuchumatanes, Guatemala (Osgood, 1904*b*). In his revision of 1909, Osgood mentioned the resemblance of *altilaneus* to both *mexicanus* and *guatemalensis* and especially stressed its appearance as a diminutive *guatemalensis*. The form had been retained as a species of unresolved status in subsequent classifications until Carleton and Huckaby (1975) surmised that the holotype might be a composite of a skin of *guatemalensis* and skull of *gymnotis*. This speculation is wrong: the cranium of *altilaneus* is too robust for an example of *gymnotis*. It is, however, well within the range of variation described for *mexicanus*. Huckaby (1980) subsequently argued that Osgood's *altilaneus* is probably based on an inordinately small individual of *guatemalensis* and placed it as a subjective synonym of that species. In view of the sympatry of *guatemalensis* and *mexicanus* elsewhere in Guatemala and the fact that *guatemalensis* was also described from Todos Santos (Merriam, 1898*b*), consideration should be given to the possibility that *altilaneus* is actually a junior synonym of *mexicanus* and the situation represents another instance of sympatry between it and *guatemalensis*.

Osgood (1909) viewed *megalops* and *melanocarpus* as sufficiently distinct to place them together in a separate species group. He included *zarhynchus* in the same group, though his comparisons of it mentioned neither of those species but rather emphasized its similarity to *guatemalensis*. In any case, a close relationship of *melanocarpus* and *megalops* to *zarhynchus* has not been sustained by subsequent evidence (Huckaby, 1980).

Although limited in distribution to mountains in Guerrero and Oaxaca, *megalops* had been divided into four geographic races, two of which, *azulensis* and *melanurus*, have proven to be improperly classified. As noted above, Huckaby (1980) drew attention to the *mexicanus*-like characters of Goodwin's (1956) *megalops azulensis*; his referral of *azulensis* to *mexicanus* has restricted the range of *megalops* to highlands west of the Isthmus of Tehuantepec, mostly at elevations above 2000 meters in the Sierra Madre del Sur (Fig. 21). Although Osgood (1909) diagnosed *melanurus* as a subspecies of *megalops*, he considered it markedly differentiated from typical *megalops* and had reservations about its placement there. His reservations were well founded, for *melanurus* has been reported to differ from *megalops* in gastric anatomy (Carleton, 1973) and to co-occur with it at Juquila, Oaxaca (Huckaby, 1980). Huckaby (1980) raised *melanurus* to a species, its documented range confined to Oaxaca and fringing that of *megalops* on lower, Pacific-facing slopes (Fig. 21).

In the same paper, Huckaby removed *mexicanus putlaensis* to *melanurus*. Goodwin (1964) named *putlaensis* from San Vicente, Oaxaca, which remained the only record of its occurrence until

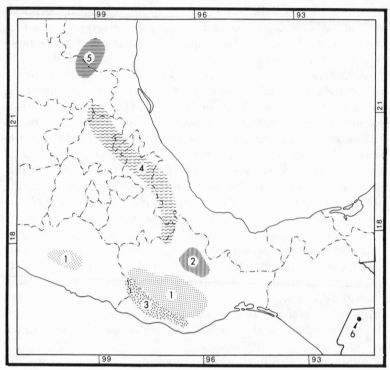

FIG. 21.—Distribution of some species of the *mexicanus* group: 1, *P. megalops*; 2, *P. melanocarpus*; 3, *P. melanurus*; 4, *P. furvus*; 5, *P. ochraventer*; 6, *P. mayensis*.

Ramirez-Pulido *et al.* (1977) reported it far to the west in the Costa Grande of Guerrero. Nevertheless, whether *putlaensis* is conspecific with *melanurus*, or with *mexicanus* as originally described, warrants firmer demonstration. Based on specimens in AMNH, *megalops*, *melanurus*, and *putlaensis* apparently occur in close proximity in the mountains of the Putla District, Oaxaca, although actual sympatry has not been demonstrated.

Despite Hooper's (1968) suggestion that *melanocarpus* may be a disjunct member of some other species, its singular morphological attributes preclude any change in its taxonomic status at the present time (Huckaby, 1980). *Peromyscus melanocarpus* inhabits cloud forest in Oaxacan mountains to the north of the range of *megalops*, its distribution apparently confined to isolated populations in the Sierra de Juarez and the Zempoaltepec range (Fig. 21). This occurrence is congruent with that of *Habromys lepturus*. Osgood's (1909) conclusion that *melanocarpus* and *megalops* are closely related has not been contradicted by subsequent studies. However, the elevation of *melanurus* to species rank necessitates its consideration as a possible cognate relative; *melanocarpus* and *melanurus* exhibit the same intermediate grade of gastric specialization (Carleton, 1973). The montane distributions of the two are separated by intervening populations of *megalops*.

Hooper (1968) suggested that *furvus* and *latirostris*, both found along the eastern slopes of the Sierra Madre Oriental, have the appearance of allopatric races of the same species. Hall (1971) concurred and so transferred *latirostris* to *furvus*, but he judged the amount of geographic variation insubstantial for the retention of subspecies, as did Huckaby (1980). As noted by Huckaby, however, individuals of *latirostris* are the most distinctive of the populations now assigned to *furvus*, which include those previously segregated as *angustirostris*, the latter referred to *furvus* by Musser (1964). The resultant range of *furvus* traces a flange of moist subtropical forest that extends at intermediate elevations from northern Oaxaca to southern San Luis Potosi and eastern Queretaro (Fig. 21). It is generally found upslope from environments inhabited by *mexicanus*, which reaches its northern extreme in the same vicinity. Huckaby (1980) considered the deep gorge of the Rio Santo Domingo-Quiotepec effectively to limit the southern distribution of *furvus*. On the basis of its discoglandular stomach morphology, mammae count, complex dentition, and smooth interorbital region, *furvus* stands apart somewhat from the *megalops* and *mexicanus* complexes (Hooper, 1968; Carleton, 1973; Huckaby, 1980). In addition, Carleton and Huckaby (1975) suggested a closer affiliation of *mayensis* (Fig. 21) to *furvus* among species in the *mexicanus* group.

The remaining three species (*ochraventer, yucatanicus, stirtoni*) that Hooper placed in his *mexicanus* group are not manifestly associated with any of the species discussed so far. *Peromyscus ochraventer* is known from a few localities in the foothills of southern Tamaulipas and northeastern San Luis Potosi (Fig. 21), generally in more temperate but still quite moist forest. Hooper (1968:59) considered it to cluster nearer the *mexicanus* complex than *furvus*, as Baker (1951) had originally proposed in his diagnosis of *ochraventer*. Yet in cranial and dental traits, gastric anatomy, and number of mammary glands, specimens of *ochraventer* agree with those of *furvus* (Carleton, 1973; Huckaby, 1980). However, as the development of some of these characters may be plesiomorphic for *Peromyscus*, an interpretation of close relationship is weakened. The glans penis in examples of *ochraventer* is small compared to that of most species

of the *mexicanus* group (Hooper and Musser, 1964*b*), which led
Huckaby (1973) to reassign it to the *boylii* group. In his review
of the *mexicanus* group, Huckaby (1980) retained *ochraventer* as a
member species without comment.

Peromyscus yucatanicus inhabits semideciduous forest in the northern half of the Yucatan Peninsula (Fig. 20), where it is isolated
from all other species of the *mexicanus* group. Additional locality
records have not reduced the hiatus between it and the nearest
populations of *mexicanus* in eastern Veracruz (Huckaby, 1980). In
spite of the external similarity of the two species, earlier assessments (Osgood, 1909; Lawlor, 1965; and Hooper, 1968) that *yucatanicus* has closest kinship with *mexicanus* are open to dispute. Trenchant differences in anatomy of the phallus, stomach morphology,
and supraorbital construction indicate that *yucatanicus* is not simply
a diminutive version of *mexicanus* (Carleton, 1973; Huckaby, 1980).
Available data do support a broadly defined affiliation of *yucatanicus*
with *mexicanus, guatemalensis, megalops,* and their allied forms, rather
than one with *furvus* and *ochraventer,* but do not permit a finer phylogenetic resolution. Huckaby (1980) followed Lawlor (1965) in
treating *yucatanicus* as monotypic, but Hall (1981) accepted Osgood's (1909) subspecific arrangement.

Peromyscus stirtoni (Fig. 19) is known from scattered localities in
El Salvador, Guatemala, Honduras, and Nicaragua (Huckaby, 1980;
Jones and Yates, 1983). Jones and Yates recorded *stirtoni* from dry
forest and adjacent brush in the interior of Nicaragua at low elevations, a description which generally fits its habitat where previously
known. Although Hooper (1968:60) opined that ". . . when better
known [*stirtoni*] likely will be found to tie in with another previously
described species," its affinity remains unclear and its species-level
recognition seems secure. Indeed, the uniqueness of its delicate
skull with supraorbital beading and its densely-haired, bicolored
tail make questionable its assignment to the *mexicanus* group.

The *mexicanus*-group species examined karyotypically to date
display almost identical banding patterns, an unexpected finding in
view of the morphological diversity encompassed by the group. In
addition to the primitive biarmed pairs, eight sampled species—
furvus, guatemalensis, gymnotis, megalops, melanurus, mexicanus (including *nudipes*), *yucatanicus, zarhynchus*—possess inversions of chromosomes 2, 3, and 9 (Rogers *et al.*, 1984; Stangl and Baker, 1984*b*;
Smith *et al.*, 1986); only *ochraventer* departs from this condition,
having a biarmed pair 6 as well (Robbins and Baker, 1981). In cladistic evaluations of banding data, the species are portrayed as ra-

diating from a single node, usually in association with *pectoralis* and sometimes with *melanophrys* and *perfulvus*, which apparently experienced the same inversion events (Rogers *et al.*, 1984; Stangl and Baker, 1984*b*). These data provide few insights to relationships among species within the group but do not contradict their union. The limited electrophoretic information published on the *mexicanus* group portends greater discriminatory power at the species level, but too few species have been sampled to allow meaningful taxonomic deductions (Schmidly *et al.*, 1985). In view of recent electrophoretic investigations on the *leucopus* and *boylii* groups, such studies must be focused carefully, both in terms of their geographic sampling and the taxonomic questions asked.

Subgenus: Osgoodomys

Osgood (1909)	Hooper (1968)	Hall (1981)
Not recognized, *banderanus* part of *mexicanus* Group	*banderanus*	*banderanus* (2)

For a form that Osgood (1909) considered just another species of the *mexicanus* group and one that might be found to intergrade with *mexicanus*, our perception of the morphological and phyletic distinctiveness of *banderanus* has changed vastly in recent years. Actually, Osgood's suspicion of intergradation was misinformed because he confused the Oaxacan coastal form *angelensis* as a subspecies of *banderanus*. As predicted by Hooper (1968) and documented by Musser (1969), populations known as *angelensis* are not conspecific with true *banderanus* and more properly belong with the *mexicanus* complex. Musser (1969) also noted the misallocation of *coatlanensis* and *sloeops*, two forms Goodwin (1955, 1956) described from Oaxaca as subspecies of *banderanus* but which also represent forms of *mexicanus*. These specific realignments have reduced the range of *banderanus* in western Mexico (Fig. 22), where it now is known to occur only from southernmost Nayarit to southern Guerrero along the coastal plain and into arid interior valleys and basins (Musser, 1969; Carleton *et al.*, 1982).

Hooper (1968:64) anticipated the isolated phylogenetic position of *banderanus* by observing that it ". . . is a relict species which may well date back in its own phyletic branch to an early stage in the evolution of the genus." His observation was shaped by the marked dissimilarity of the glans penis of *banderanus* to most other kinds of

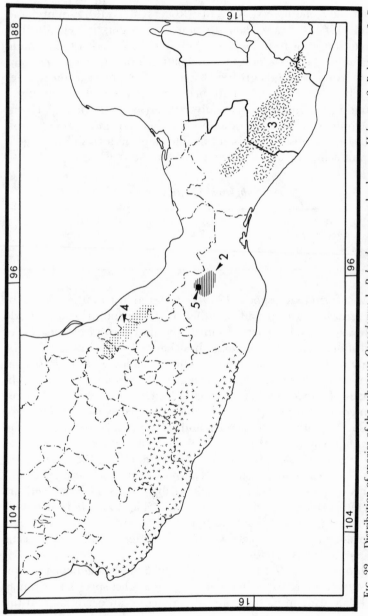

Fig. 22.—Distribution of species of the subgenus *Ogoodomys*: 1, *P. banderanus*; and subgenus *Habromys*: 2, *P. lepturus*; 3, *P. lophurus*; 4, *P. simulatus*; 5, *P. chinanteco*.

Peromyscus (Hooper, 1958; Hooper and Musser, 1964*b*). Hooper and Musser (1964*b*) accordingly created the subgenus *Osgoodomys* in recognition of this divergence. Subsequent anatomical data for *banderanus* have reinforced the view of its extraordinary differentiation and probable early separation from the ancestral stock leading to typical *Peromyscus* (Linzey and Layne, 1969; Carleton, 1980).

Karyological banding data provide little resolution of the hierarchical placement of *Osgoodomys* within the peromyscine radiation because *banderanus* possesses a chromosomal complement like that considered primitive for the genus (Rogers *et al.*, 1984). In cladograms generated from euchromatic rearrangements, *Osgoodomys* is therefore arranged as one line emerging from a basal polychotomy together with other species (such as *crinitus* and certain *boylii*) that retain the ancestral karyotype (Rogers *et al.*, 1984; Stangl and Baker, 1984*b*). As noted by the aforementioned authors, these data do not necessarily support or deny the subgeneric (or Carleton's, 1980, generic) isolation of *banderanus* but are compatible with the notion of its earlier cladistic origin.

The few biochemical investigations that have included *banderanus* do not, however, uniformly indicate a relict species of very distant affinity with other peromyscines. As discussed above, immunological comparisons of serum albumins indicated that *perfulvus* was more highly differentiated from three other species of *Peromyscus* than was *Osgoodomys*, although the two clustered together based on average albumin distances (Fuller *et al.*, 1984). One-way distances derived from a broader taxonomic sampling revealed that *Osgoodomys* is roughly equidistant in its albumin structure to species of *Haplomylomys* and *Peromyscus* (*s. s.*). Based on an electrophoretic assay of 18 structural loci in 27 species of *Peromyscus* (*s. l.*), Schmidly *et al.* (1985) found that *banderanus* clustered at a low similarity level with species of the *mexicanus* group and that *Megadontomys thomasi* was the most distinctive followed by *banderanus* and *hooperi*.

Subgenus: Habromys

Osgood (1909)	Hooper (1968)	Hall (1981)
Not recognized,	*simulatus*	*simulatus*
species part of	*lophurus*	*chinanteco*
lepturus Group (part)	*lepturus*	*lophurus*
simulatus	*ixtlani*	*lepturus* (2)
lophurus		
lepturus		

Hooper and Musser (1964*b*) diagnosed the subgenus *Habromys* in acknowledgment of the special features of the phallus observed in these mice (Hooper, 1958). Osgood (1909) clearly recognized a close affinity among *lepturus*, *lophurus*, and *simulatus*, but he enigmatically grouped them with species of the *mexicanus* complex in the subgenus *Peromyscus*. Thus, the uniqueness of *lepturus* and its relatives has been revealed comparatively late in their taxonomic history. Information gathered from other anatomical sources— particularly the striking modifications of male accessory glands (Linzey and Layne, 1969), lack of an acrosomal hook on the spermatozoa (Linzey and Layne, 1974), and atypical number of thoracolumbar vertebrae (Carleton, 1980)—have only served to underscore their phyletic isolation and lend further weight to Hooper's (1968:65) estimation that: "They appear to be relict taxa, which jointly may date to an early stage in the evolution of the genus." Nevertheless, most of our morphological knowledge of *Habromys* issues from the two species *lepturus* and *lophurus*; the diagnostic characters of the taxon remain to be verified for *simulatus* and *chinanteco*.

Species of *Habromys* occupy cloud forest habitats at intermediate to high elevations in the highlands of southern Mexico southward to Guatemala and El Salvador (Fig. 22). The most recently described species, *chinanteco*, is sympatric with *lepturus* in cloud forests of predominantly oak mixed with pine in the Sierra de Juarez, Oaxaca (Robertson and Musser, 1976). Robertson and Musser identified the rare form *simulatus* as the probable closest relative of *chinanteco*. In the same paper, the authors reported a third record of *simulatus*, appreciably extending its northern limits along the Sierra Madre Oriental in Veracruz. Hooper (1968) questioned the separation of *ixtlani* and *lepturus*, forms described from nearby mountain ranges in Oaxaca, and Musser (1969) concluded that the two were montane isolates of one species, although their differences persuaded him to retain *ixtlani* as a subspecies. Specimens of *lophurus* have been collected south of the Isthmus of Tehuantepec, from widely scattered montane localities in southern Chiapas, Guatemala, and northern El Salvador. Robertson and Musser (1976) noted appreciable morphological diversity among their samples of *lophurus*, especially between the Chiapas series and those from Guatemala and El Salvador. Larger sample sizes covering a broader geographic area are required to evaluate the extent of divergence among these widely separated populations.

To date, only *Habromys lepturus* has been analysed for its chromosomal banding patterns (Rogers *et al.*, 1984). In addition to the usual complement of presumably primitive biarmed elements, it possesses biarms at pairs 2, 3, 5, 6, 7, 9, 10, and 15, their short arms apparently derived by pericentric inversions. The large number of rearrangements affiliates *lepturus* with many species of the subgenus *Peromyscus*, but in a large unresolved clade that also comprises *Podomys* and *Neotomodon*. In their cladograms based on G-banding, Rogers *et al.* (1984) placed *lepturus* as a sister species of *Neotomodon alstoni* based on a shared inversion of chromosome 7. In contrast, Stangl and Baker (1984*b*) aligned *lepturus* with *truei* based on an alteration of chromosome 15 and failed to mention the occurrence of the pericentric inversion of chromosome 7. Nevertheless, they attributed the apparent relationship of *lepturus* and *truei* to a probable convergent event and suggested that (pg. 651) ". . . *Habromys* shared an early common ancestor with *Podomys* and *Neotomodon* . . .," as earlier proposed by Carleton (1980).

Subgenus: Podomys

Osgood (1909)	Hooper (1968)	Hall (1981)
floridanus	*floridanus*	*floridanus*

Students of *Peromyscus* have interpreted the limited geographic range of the Florida mouse, confined mainly to peninsular Florida (Fig. 23), as a relictual distribution (Osgood, 1909; Hooper, 1968; Smith *et al.*, 1973). Although subspecies have not been designated over such a small geographic area (Osgood, 1909; Hall, 1981), appreciable morphological and size variation occurs among the populations and is deserving of further study. Smith *et al.* (1973) commented that *floridanus* was among the least variable species in the genus among those (12) examined electrophoretically: only 17 of 41 structural loci proved to be polymorphic, and the average genetic similarity among their nine population samples was uniformly high (.96–.98) and exhibited no apparent geographic pattern. Their study, however, did not survey some of the more divergent populations, such as the large-bodied forms along the Atlantic coast or the smaller individuals from northern Florida.

That Osgood (1909) viewed *floridanus* as highly distinct is underscored by his diagnosis of the monotypic subgenus *Podomys* in his revision. Osgood cited the reduction in size and number (loss of the

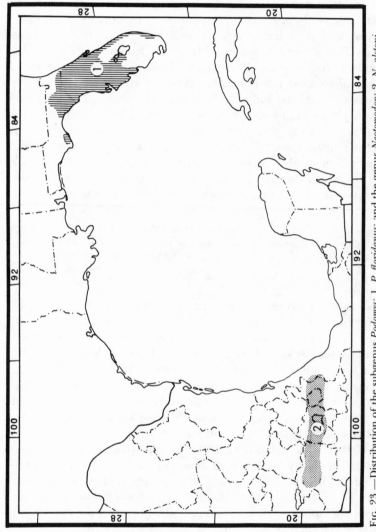

Fig. 23.—Distribution of the subgenus *Podomys*: 1, *P. floridanus*; and the genus *Neotomodon*: 2, *N. alstoni*.

hypothenar pad) of plantar tubercles as the essential distinguishing trait of his new taxon, but also remarked on the tubercular hypsodonty of *floridanus*, which he considered as intermediate between *Peromyscus* and *Onychomys*. Although a reduced sixth plantar pad has been demonstrated in some populations of *floridanus* (Layne, 1970), studies of the male reproductive tract have only served to bolster the impression of its phyletic remoteness from typical *Peromyscus* (Hooper, 1958; Hooper and Musser, 1964*b*; Linzey and Layne, 1969; Carleton, 1980).

Clues to the relationship of *floridanus* have implicated possible kinship groups in Middle America and particularly the highlands of southern Mexico (Johnson and Layne, 1961; Hooper, 1968). Phallic characteristics of *floridanus* suggest affinity to species of *Habromys, Osgoodomys banderanus,* or *Neotomodon alstoni* (Hooper, 1958; Hooper and Musser, 1964*b*), and features of the male accessory glands most resemble *Habromys* (Linzey and Layne, 1969). In his quantitative phylogenetic analysis of neotomine-peromyscine rodents, Carleton (1980) demonstrated a consistent association of *Podomys, Habromys,* and *Neotomodon*, based primarily on shared modifications of the male accessory reproductive glands and glans penis, although membership in this clade sometimes included *Osgoodomys* or *Onychomys*, or both, depending upon the algorithm and combination of OTUs employed.

Although euchromatic banding data do not refute critically an hypothesized relationship of *Podomys* to *Habromys* and *Neotomodon*, neither do they provide clear-cut support for this interpretation. Yates *et al.* (1979) arrayed *Podomys* as one lineage of an unresolved tetrachotomy that also included *gossypinus*, three species of the *maniculatus* group, and *Neotomodon*; this descendent assemblage was defined by pericentric inversions of chromosomes 2, 3, and 6. Robbins and Baker (1981) associated *Podomys* as one of seven lineages in an unresolved polychotomy consisting of *Neotomodon, pectoralis, ochraventer, truei,* the *leucopus* group, and the *maniculatus* group on the grounds of pericentric rearrangements in chromosomes 3 and 9. Rogers (1983) too aligned *Podomys* with *Neotomodon* and certain species of *Peromyscus* (*s.s.*) but on the basis of derived rearrangements in chromosomes 3 and 6. None of these studies included *Habromys*. The addition of *Habromys lepturus* to the growing karyological data base resulted in a proposed sister-group relationship of that form and *Neotomodon*, but the cladistic posture of *Podomys* remained enigmatic as one stem of an unresolved polychotomy that also subtended members of the *mexicanus* group, *pectoralis, truei,*

Neotomodon-Habromys, the *leucopus* group, and the *maniculatus* group (Rogers *et al.*, 1984). A pericentric inversion in chromosome 3 was identified as a synapomorphy defining this clade. In contrast, Stangl and Baker (1984*b*: fig. 3) portrayed *Habromys* as sharing a common ancestor with *truei* but admitted that this linkage is probably due to a convergent inversion in chromosome 15. *Podomys* appeared in their cladogram as one descendant in a seven-stemmed unresolved polychotomy, defined by pericentric changes in 3 and 9, whose taxonomic composition otherwise resembled that identified by Rogers *et al.* (1984).

There have been few investigations of allelic variation that have addressed the question of the phylogenetic affinity of *Podomys flori-danus*. In their study of the biochemical systematics of *Haplomylomys*, Avise *et al.* (1974*b*:233) also considered *floridanus* in their dendrogram of genetic similarities and found that ". . . the degree of genic similarity of *floridanus* with other species may be near the lower limit for the genus." In addition to *Podomys*, their survey encompassed eight species of *Haplomylomys*, one of the *boylii* group, two of the *maniculatus* group, and two of the *leucopus* group. Genetic distance and cladistic analyses of 14 protein loci suggested a cognate relationship of *Podomys* and *Neotomodon* in a study using several genera of sigmodontine rodents (Patton *et al.*, 1981). In the dendrogram (UPGMA) generated from a distance matrix, *Podomys* and *Neotomodon* formed a pair-group that united with one containing *Onychomys* and three species (*leucopus*, *maniculatus*, *polionotus*) of the subgenus *Peromyscus*. Cladistic evaluation of this same data set retained *Podomys* and *Neotomodon* as a sister-group but disclosed a closer phyletic relationship to species of the subgenus *Peromyscus* rather than to *Onychomys*. Other species of *Peromyscus* (*s. l.*) were not included in their biochemical survey.

Subgenus: Megadontomys

Osgood (1909)	Hooper (1968)	Hall (1981)
thomasi	*thomasi*	*thomasi* (3)
nelsoni		
flavidus		

These large deer mice inhabit cool, wet cloud forests as isolated populations (Fig. 24) in the highlands of Guerrero (*thomasi*), Oaxaca (*cryophilus*), and Hidalgo, Puebla, and Veracruz (*nelsoni*) (Musser, 1964; Heaney and Birney, 1977; Werbitsky and Kilpatrick,

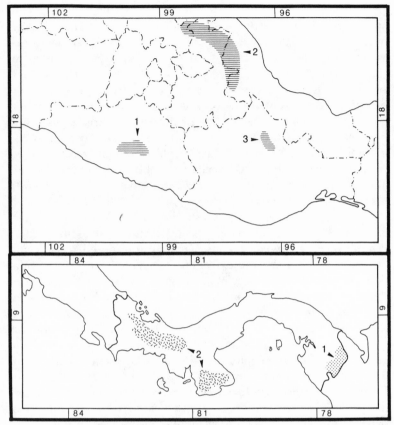

Fig. 24.—*Top*—Distribution of species of the subgenus *Megadontomys*: 1, *P. thomasi*; 2, *P. nelsoni*; 3, *P. cryophilus*. *Bottom*—Distribution of species of the subgenus *Isthmomys*: 1, *P. pirrensis*; 2, *P. flavidus*.

1987). At the same time that Musser (1964) described the new sub-species *thomasi cryophilus*, he realigned *nelsoni* as a third geographic race of *thomasi*, a pragmatic interpretation in light of their allopatric ranges and the limited sample of *nelsoni* then available. Nevertheless, some of the qualitative cranial traits distinguishing these forms (for example, development of supraorbital ridging, depth of zygomatic notch, shape of mesopterygoid fossa) are impressive in a genus where many species are recognized, often with difficulty, on the basis of subtle size and shape attributes.

Werbitsky and Kilpatrick (1987) discovered that levels of genetic similarity among the three described taxa of *Megadontomys* are low

(Nei's I = .741–.818) compared to that reported for most con-specific populations of *Peromyscus*. Although they retained the taxa as subspecies of *thomasi*, as Musser (1964) had arranged them, such values of genetic resemblance correspond more closely to forms that Zimmerman *et al.* (1978) characterized as semispecies. Werbitsky and Kilpatrick (1987) attributed the strong genetic differentiation among, and low heterozygosity within, these taxa to their lengthy isolation (estimated time of divergence three to four million years ago) as relictual populations of a formerly widespread montane species. Their cladistic and phenetic analyses of the electrophoretic data suggested a closer relationship between *nelsoni* and *thomasi* as compared to *cryophilus*, a conclusion consistent with some of Musser's (1964) statements.

Musser's (1964) conviction on this account was clearly not firm. Elsewhere in the same paper, he drew attention to the nearer resemblance of *nelsoni* and *cryophilus* or considered *thomasi* and *cryophilus* to represent the extremes in morphological differentiation with *nelsoni* somewhere intermediate. My own examinations of the type material of Merriam's *thomasi* and *nelsoni* lead me to support the former interpretation. Examples of *nelsoni* agree with those of *cryophilus* in the development of an incipient zygomatic notch, lack of pronounced supraorbital ridging, construction of the otic capsule, smaller pinnae, and darker pelage tone; in other features, the three seem to be equally divergent from one another. Furthermore, the hypothesis of a *nelsoni-cryophilus* clade seems more plausible from a zoogeographic standpoint. Distributional patterns of other inhabitants of montane wet forest in this region (such as *aztecus, boylii,* and the *mexicanus* complex) seem to be linked via the Sierra Madre del Sur of Guerrero and Oaxaca, through the mountains of middle and northern Oaxaca, to the eastern flanks of the southern Sierra Madre Oriental. The sequence of phyletic diversification required by a *thomasi-nelsoni* clade is more difficult to reconcile biogeographically. Alternatively, the separation of the three may have occurred more or less simultaneously.

In view of the discrete character differences and the low coefficients of genetic similarity so far demonstrated, I favor the recognition of *thomasi, nelsoni,* and *cryophilus* as separate species. The taxonomic decision prompted by these allopatric morphs represents yet another instance in which the static formality of our Linnaean classification is poised against the dynamism of the evolutionary process. As such, the issue may ultimately prove unresolvable. Yet, I believe that still finer resolution of their level of relationship and degree of divergence can emerge, for example through amplification of geographic sampling, more rigorous treatment of morphometric and qualitative character variation (especially of the glans penis), and application of other molecular techniques. To bury these taxa as subspecies of *thomasi* obscures their appreciable differentiation and masks the need for this further investigation.

Other than the eventual removal of *flavidus* to the new subgenus *Isthmomys* (Hooper and Musser, 1964*b*), the differentiation of *Megadontomys* from typical *Peromyscus* has been sustained in a variety of morphological studies (Hooper, 1958; Hooper and Musser, 1964*b*; Linzey and Layne, 1969; Carleton, 1980). Musser (1964:18) interpreted the earlier body of systematic information as indicating that ". . . *thomasi* is morphologically and probably phyletically peripheral to the main cluster of species in *Peromyscus* . . . ," and Hooper (1968) expressed this peripheral nature by retaining *thomasi* alone in the subgenus *Megadontomys*.

How phyletically peripheral *thomasi* actually is and how best to represent its position in our classification are two themes that have dominated research on *Megadontomys* subsequent to Hooper's (1968) review. Carleton (1980) proposed that *Megadontomys* diverged early relative to the origin of other *Peromyscus* (*s. s.*) (and *Reithrodontomys* and *Onychomys*), sharing a common ancestry with *Isthmomys*, and elevated it to generic level. Recent karyological studies have not endorsed this treatment (Rogers, 1983; Rogers *et al.*, 1984; and Stangl and Baker, 1984*b*). The autosomal complement of *thomasi* contains five biarmed pairs, the three (1, 22, 23) proposed as plesiomorphic for the genus *Peromyscus* (*s. l.*) and two (2, 9) formed by pericentric rearrangements (Rogers, 1983). Euchromatic banding homologies thus led Rogers (1983) to postulate the derivation of *Megadontomys* at the base of a clade including *Neotomodon, Podomys,* and some species of the subgenus *Peromyscus*, a grouping defined by synapomorphic pericentric inversions either in chromosome 2 or in chromosome 9, although recognizing (p. 620) ". . . that inclusion of *thomasi* within *Peromyscus* also means that either pair 2 or 9 must have been rearranged at least twice. . . ." In a broader synthesis of G-banding data for 26 species of *Peromyscus* (*s.l.*), Rogers *et al.* (1984) reaffirmed the traditional association of *Megadontomys* in the genus and its probable early divergence, but the suspected homoplasy in rearrangements of chromosomes 2 and 9 required consideration of three equally plausible phylogenetic origins: 1) from a stock of *Haplomylomys*; 2) from the lineage that eventually proliferated into species of the nominate subgenus; or 3) from a primitive stock separate from the progenitors of the subgenera *Haplomylomys* or *Peromyscus*. In an analysis of chromosomal banding data for 30 species representing all seven subgenera, Stangl and Baker (1984*b*) aligned *Megadontomys* as one stem of an unresolved trichotomy, an arrangement that basically conveys the first two phylogenetic alternatives offered by Rogers *et al.* (1984). Their analysis disclosed no karyotypic traits supporting a sister-group relationship for *Isthmomys* and *Megadontomys*.

Little information addressing the differentiation and kinship of *thomasi* is available from molecular approaches. In a phenetic assessment of genetic variation in 18 structural loci, *Megadontomys thomasi* was amalgamated at the lowest similarity level and formed a pair-group to all other species represented, including 21 species of the subgenus *Peromyscus*, three of *Haplomylomys*, and *Osgoodomys banderanus* (Schmidly *et al.*, 1985).

Subgenus: Isthmomys

Osgood (1909)	Hooper (1968)	Hall (1981)
Not recognized,	*flavidus*	*flavidus*
flavidus part of	*pirrensis*	*pirrensis*
(*Megadontomys*)		

Isthmomys flavidus and *pirrensis* are terrestrial rats that occur in evergreen forests at intermediate elevations in western Panama and at higher elevations in eastern Panama, respectively (Handley, 1966). Their allopatric distributions prompted Hooper (1968) to suggest that the two have the appearance of geographic races (Fig. 24), but this possibility has yet to be rigorously assessed. Aside from the generally smaller size of *flavidus*, the two differ qualitatively in the occurrence of the subsquamosal foramen, which is usually absent in *flavidus* and present but tiny in *pirrensis*.

Hooper and Musser (1964*b*) diagnosed *Isthmomys* as a subgenus to reflect the conspicuous penile differences between *flavidus* and *pirrensis* and other *Peromyscus* (*s. l.*). Previous works (for example, Miller and Kellogg, 1955; Hall and Kelson, 1959) followed Osgood (1909) in allocating these species to the subgenus *Megadontomys* along with the two other largest forms, *thomasi* and *nelsoni*, known in the genus. The two currently recognized species of *Isthmomys* also diverge from typical *Peromyscus* in their complement of male accessory reproductive glands (Linzey and Layne, 1969), gastric morphology (Carleton, 1973), as well as certain skeletal features, such as the lack of an entepicondylar foramen, articulation of the first rib, fenestration above the otic capsule, and ossification of the mesopterygoid fossa (Carleton, 1980). On the basis of such morphological criteria, Carleton (1980) favored recognition of *Isthmomys* as a genus and arrayed it in a clade with *Megadontomys*. The case for their sister-group relationship, however, is weak, for six of the eight derived character states uniting them are character reversals and the other two are nonunique transitions.

Samples of *Isthmomys* have been examined recently for chromosomal banding patterns and electrophoretic variation. Like many other peromyscines, *pirrensis* has a diploid count of 48 (pairs 1, 16, 17, 20, 22, and 23 biarmed), but Stangl and Baker (1984*b*) identified three chromosomal modifications unique to *pirrensis* among the 30 species of *Peromyscus* (*s. l.*) surveyed. In their cladistical analysis of the accumulated G- and C-banded data, they derived *Isthmomys* from a basal polychotomy subtending seven branches, which include those leading to *Onychomys*, *Osgoodomys*, *P. crinitus*, *P. boylii*, two species of *Haplomylomys*, and the remainder of *Peromyscus* (*s. l.*). The biarmed condition in chromosome 23 of *pirrensis* differs from that proposed as primitive for *Peromyscus* (*s. l.*) by Greenbaum and Baker (1978), in that the long arm is entirely euchromatic and the short arm entirely heterochromatic (Stangl and Baker, 1984*b*). The authors remarked that this condition entails either an inversion of number 23 to an acrocentric condition and then acquisition of a heterochromatic short arm, or forces a reconsideration of the hypothesized primitive state as being biarmed for only pairs 1 and 22. In the latter case, *Isthmomys* would necessarily originate prior to the *Onychomys-Peromyscus* (*s. l.*) clade. Preliminary electrophoretic analyses similarly indicate an earlier divergence of *Isthmomys* relative to the separation of *Onychomys* and other *Peromyscus* (*s. l.*) (J. C. Patton, cited as a personal communication in Stangl and Baker, 1984*b*). In a revisionary study wherein he argued pointedly for the retention of a generic scope that included *Ochrotomys* and *Baiomys*, it is noteworthy that *flavidus* is the only species whose placement in *Peromyscus* Osgood (1909:222) viewed as provisional.

RELATIVES OF *PEROMYSCUS* AND THE GENERIC SCOPE OF MUROIDEA

Osgood (1909) and Hooper (1968) both chronicled the nomenclatural history of species eventually consigned to the genus *Peromyscus*, and I shall not repeat the details of their reviews.

To the degree that the cumbersome genera of the middle to late 1800s may be expected to reflect phylogenetic relationship, species of *Peromyscus*, by their membership in *Hesperomys*, were broadly allied with *Onychomys, Oryzomys, Phyllotis, Thomasomys*, and other forms then ranked as subgenera but now conventionally recognized as genera. In fact, Waterhouse's (1839) genus *Hesperomys* is practically equivalent to the subfamily Sigmodontinae, as that taxon is currently defined (Carleton and Musser, 1984). In the late 1800s, *Onychomys* and South American taxa were removed from *Hesperomys* with the result that Gloger's *Peromyscus* became restricted to North and Central American species.

As a result of Osgood's revision (1909), the contents and infrageneric categories of *Peromyscus* became, for the most part, ingrained in North American mammalian systematics. Osgood employed six subgenera: the nominate taxon, *Megadontomys, Haplomylomys, Baiomys, Podomys*, and *Ochrotomys*, the latter two described as new. *Megadontomys* and *Baiomys* had been employed variously as genera by other authors, but Osgood strongly argued for the maintenance of his generic scope and the subgeneric categories recognized. Miller (1912) soon returned *Baiomys* to a genus, a decision ratified by subsequent systematic works (for example, Hooper and Musser, 1964*b*).

At this stage of our phylogenetic awareness, *Peromyscus* was arrayed in the subfamily Cricetinae with other North American genera such as *Onychomys, Reithrodontomys*, and *Baiomys*, as well as with taxa of South American affinities such as *Nyctomys, Oryzomys* and *Sigmodon* (Miller, 1912). A suggestion of an evolutionary division within North American cricetines was conveyed by Miller's retention of Merriam's (1894) subfamily Neotominae, which was then more restricted in contents than the "neotomine" lineage of Hooper and Musser (1964*a*, 1964*b*). Hooper and Musser's group included *Tylomys* and *Ototylomys* and apparently represented a tribal level category. Later classifications abandoned Merriam's Neotominae and so returned *Peromyscus* to a broad and amorphous phyletic association with numerous North and South American genera (Miller, 1924; Ellerman, 1941; Simpson, 1945; Miller and Kellogg, 1955; Hall and Kelson, 1959).

Discussion of the relatives of *Peromyscus* was almost an irrelevant exercise prior to 1958. In that year, Hooper published the first of a series of anatomical surveys of the phallus in *Peromyscus* and other New World cricetids. These anatomical reviews culminated in 1964 with two epochal papers coauthored with Guy G. Musser: one on the classification of *Peromyscus*, which is the precursor of Hooper's 1968 contribution; and one of a broader scope bearing not only on the higher order relationships of New World cricetines but also on the classification of Muroidea. In the *Peromyscus* paper, Hooper and Musser (1964*a*) substantially reordered the generic subdivisions, convincingly documented the more distant affinity of *Ochrotomys* and its generic status, and named three new subgenera (*Habromys, Osgoodomys, Isthmomys*) that contained species whose more distant phyletic separation had been largely unrecognized since the time of Osgood. In the companion paper, they (1964*b*) documented the occurrence of simple and complex phalli, an anatomical distinction fundamental to the notion of a phyletic breach between neotomine-peromyscines and South American cricetines, and they clearly depicted patterns of descent among the twelve genera of simple penile neotomine-peromyscines, as well as among the major assemblages of muroids. Truly, we are still laboring to synthesize the ramifications of these two studies.

The phylogeny of neotomine-peromyscines developed by Hooper and Musser (1964*b*: fig 9) portrayed two major lines of descent, the members of each sometimes formally grouped as the Neotomini and Peromyscini. The peromyscines consisted of seven genera of North and Central American distribution, including *Peromyscus, Reithrodontomys, Neotomodon, Ochrotomys, Onychomys, Baiomys,* and *Scotinomys,* in order of increasingly distant cladistic relationship to *Peromyscus.* Curiously, they cladistically depicted *Nelsonia* as a basal member of the peromyscine lineage but in the text (p. 54) assigned it to the neotomines. In any case, their evolutionary tree clearly expressed a sister-group relationship between *Peromyscus* and *Reithrodontomys* and constituted the first explicit statement of relationship portraying *Peromyscus* within the kinship web of New World cricetines (= Sigmodontinae), the phylogenetic hypothesis which motivated my study of 1980.

The phylogenetic scheme that I presented (Carleton, 1980) departed somewhat from that of Hooper and Musser's (1964*a*, 1964*b*). Actually, it agreed more closely with one of Hooper's (1960) preliminary arrangements of the genera with simple phalli. However, with respect to the kinship and generic boundaries of *Peromyscus,* my phylogeny accorded a closer relationship of certain *Peromyscus* (the nominate subgenus and *Haplomylomys*) to *Reithrodontomys* than to other forms of *Peromyscus,* which Hooper and Musser had arranged as subgenera (*Podomys, Habromys, Osgoodomys, Megadontomys, Isthmomys*). On this basis, as well as the degree of their morphological divergence in comparison to generic boundaries among other muroid rodents, I recommended the elevation of the latter five taxa to genera. The other forms that I included as peromyscines— *Neotomodon*, *Onychomys*, and *Ochrotomys*—correspond to the conclusions of Hooper and Musser, although my interpretation of their cladistic position within the peromyscine complex differs from theirs in detail. The phylogenetic hypothesis that I advanced in 1980 has garnered little support from cladistic analyses of chromosomal banding data. In particular, the predicted cognate relationship of *Reithrodontomys* and *Peromyscus* (*s. s.*) seems to have drawn the most disagreement (Rogers, 1983; Rogers *et al.*, 1984; Stangl and Baker, 1984*b*).

In view of the repeated lack of corroboration from the euchromatic banding data, I am compelled to abandon this relationship as reflective of the true phylogeny. Yet in doing so, I entertain a lingering reservation that persists for the following reason. Within *Reithrodontomys*, there exist species (for example, *R. montanus* and *R. megalotis*) with chromosomal rearrangements so extensive that identification of homologous chromosomes or chromosomal arms proves difficult or impossible, given the current state of banding resolution (Robbins and Baker, 1980; Hood *et al.*, 1984). Such extensive G-band repatterning during chromosomal evolution among closely related species has been christened "karyotypic megaevolution" (Baker and Bickham, 1980; Haiduk and Baker, 1982). The banding sequences in *Reithrodontomys* with higher diploid numbers, such as *R. fulvescens* and *R. mexicanus*, can be compared more completely with those of other peromyscine genera like *Peromyscus*, *Onychomys*, and *Baiomys*. Yet the inability to discern any or few euchromatic homologies of the highly derived forms of *Reithrodontomys* has not compelled us to consider them as unrelated at the generic level because other bodies of evidence support a monophyletic interpretation of *Reithrodontomys* (Hooper, 1952*a*; Hooper and Musser, 1964*b*; Carleton, 1980). Indeed, without a credible taxonomic framework, examples of karyotypic megaevolution would otherwise be difficult to identify. In a similar manner, other kinds of information persuaded Hooper and Musser, and Carleton, to hypothesize a near relationship between *Reithrodontomys* and *Peromyscus* (of either scope). If such episodes of karyotypic megaevolution confound discernment of character homologies and consequent genealogical estimates on a local part of the tree—that is, within the generic level—how are we to detect such transformations in the earlier branching sequences, over a vaster sweep of geological time? This question is similar to the reservation offered by Warner (1983) regarding karyotypic reconstructions of glossophagine bat phylogeny (but see Haiduk and Baker, 1984, for rejoinder).

Our generic constructs in Muroidea have not been seriously challenged since Ellerman's (1941) magnum opus. This work was more a state-of-the-art summarization of earlier systematic opinions than a critical evaluation of character variation designed to reexamine generic limits. For North American rodents, many of our generic usages issue from the classical revisionary work of the late 1800s and early 1900s, their identities acquiring a certain historical inertia through repetition in large and generally accepted faunal catalogs and broad systematic treatises, such as Simpson (1945), Miller and Kellogg (1955), and Hall and Kelson (1959). There were exceptions, for example, Burt and Barkalow's (1942) redefinition of generic lines of woodrats using new information drawn from bacular morphology. We have recently entered a new phase of examination of muroid phylogeny, in which workers are trying to assimilate new kinds of systematic information and methodological interpretations of that information, and to apply more critical, consistent, and defensible criteria for recognition of muroid genera and their higher level classification. Signs of this revived attention include Chaline's *et al.* (1977) integration of fossil evidence into a revised classification of Muroidea, Pavlinov's (1982) studies on gerbilline phylogeny, Gardner and Patton's (1976) karyotypic overview of oryzomyine genera, and the dialogue concerning the interrelationship and taxonomic rank of *Peromyscus* and other neotomine-peromyscines (Yates *et al.*, 1979; Carleton, 1980; Rogers, 1983; Stangl and Baker, 1984*b*). And at the American Museum of Natural History, a colleague of ours has been writing on the Giant Rat of Sumatra and its kin, in the course of which he has revolutionized perceptions of the generic scope and interrelationships of Murinae, compared to that set forth by Ellerman and others.

Such on-going research activities herald a period of flux in the classification of Muroidea and the definition of muroid genera in the near future. Yet the ultimate benefit of this belated systematic introspection will be a sounder, more explicit generic framework founded on discrete character variation, interpretations of evolutionary polarities, and synapomorphic patterns. Such research activities will achieve a marked improvement over the undefinable, nebulous state of many muroid genera, such as *Rattus*, *Akodon*, and *Oryzomys*, handed down from Ellerman's era. Furthermore, attention to these research goals will simultaneously advance efforts to carry forth species-level revisions so urgently required for many muroid rodent faunas, to work out patterns of descent among the major assemblages of Muroidea, and to encourage the use of a well-documented muroid taxonomy in biogeographic interpretations. The dynamic state of current muroid systematics aside, the question of the taxonomic rank to be accorded the five "subgenera" of *Peromyscus* remains.

Osgood (1909:24–26) devoted protracted discussion to his views on the recognition of genera versus subgenera. Specifically, he advised that "The use of subgenera provides a means of adjusting the differences usually existing between the general zoologist and the specialist. The generic name answers all the purposes of the general zoologist while the specialist may use as many subgenera as he desires and meet all the requirements of discriminating classification. This also operates to conciliate the amateur, whose outcries against the continual changing of names by specialists will thereby be lessened." His opinions in this regard express a certain sentiment still voiced today, but lack currency when viewed from the perspective of present-day phylogenetic systematics and the nature of criteria considered important in developing classifications and determining taxonomic ranks. The pragmatism central to his argument is more clearly appreciated in light of his association with the Biological Survey, then an applied scientific arm of the Department of Agriculture. Many taxonomic publications from this era of the Biological Survey's history carried a prefatory "letter of transmittal" to the Secretary of Agriculture and addressed, at least in a nominal fashion, the economic status or agricultural effects of the forms being studied. Unfortunately our growing awareness of the complexity of muroid rodent evolution has taken our taxonomy far beyond the practical classificatory objectives that Osgood considered germane to the recognition of subgenera and genera.

In light of the apparent incongruence of morphological and karyotypic interpretations of neotomine-peromyscine evolutionary history, I continue to favor generic status for the five taxa that have been conventionally included in *Peromyscus*. As mammalian systematists, we can do little to ensure the morphological comparability or degree of evolutionary divergence circumscribed by a genus of birds as compared to a genus of mammals; however, we can establish some comparable and consistent criteria for what constitutes a genus of North American sigmodontine versus a South American sigmodontine or a gerbilline or an arvicoline. As previously noted (Carleton, 1980), the patristic and phenetic distances delimiting the subgenera of *Peromyscus* in question fully match, or exceed, those of some currently recognized genera of South American sigmo-dontines. This approach, though less than one ultimately would desire, is an operational one that acknowledges the generally weak and inconsistent support of critical early branching sequences, using either data set, which, to date, has disallowed a clearer deter-mination of taxonomic ranks. The uniquely derived morphological or chromosomal features of forms such as *Isthmomys* are readily identifiable, but synapomorphic character states that would con-fidently predict its close relationship to some other peromyscine are not so apparent. Until firmer resolution and corroboration are forthcoming, the patristic separation among these sigmodontine phyletic entities, as revealed by discrete character state variation, constitutes an objective and defensible basis upon which to argue their status.

The recommendation of generic status for these forms recog-nizes that traits of the male reproductive tract account substantially for their morphological divergence (Hooper and Musser, 1964*a*, 1964*b*; Linzey and Layne, 1969) and form an integral part of their generic diagnoses (see Taxonomic Summary below). This is not wholly the case, for *Osgoodomys* exhibits a derived carotid circulatory pattern and pronounced supraorbital ridges, *Isthmomys* lacks entepi-condylar and subsquamosal foramina, *Megadontomys* possesses a sat-ellite root on the upper first molar and a complicated caecum, and *Habromys* has an extra lumbar vertebra among peromyscines exam-ined (Carleton, 1980). Still the striking modifications of the phallus and accessory reproductive glands of these taxa contribute appre-ciably to their larger distance coefficients compared to other *Pero-myscus*. Although diagnoses of muroid genera have been tradi-tionally tied to features of the external form and skull, there is no a priori reason to suspect that characters of the reproductive system are poorer indicators of phylogenetic relationship than other as-pects of the rodent phenotype.

Over the past two centuries, there has been a succession of char-acter complexes and organ systems applied to muroid systematics, and each has, in turn, obliged us to revise and adjust our estimates of relationship and classifications based thereon. Pelage and exter-nal traits enjoyed a historical preeminence in our generic construc-tion. That many North American rodents were first associated with *Mus, Apodemus,* and *Arvicola,* forms familiar to the early European naturalists, underscores the inattention paid to the dental and cra-nial characteristics that we now regularly consult. In view of this emphasis, Waterhouse's (1839) discovery that New World rodents possess a biserial arrangement of dental cusps, a condition central to his diagnosis of *Hesperomys*, must have been an exhilarating apprehension and one that elevated the systematic value of such cranial features. The marriage of the two character suites involving external form and cranium yielded the standard skin-and-skull preparation that has formed the empirical basis for much of mammalian taxonomy and that corresponds to the relatively minor portion of the organism usually referenced under the rubric of "classical morphology." The subject matter of rodent morphology is far richer and more phylogenetically informative than suggested by this conventional perception, as anyone who has studied the plates of Tullberg's (1899) treatise on rodent classification readily appreciates. Contributions like Tullberg's in part inspired Hooper (1958) to undertake his anatomical survey of the phallus of *Peromyscus* and other New World cricetines. The fresh sources of morphological variety unfolding from his investigations introduced yet another suite of systematic characters to be weighed against previous classifications and to be incorporated, where judged appropriate, into our muroid taxonomy.

The cladistic value of a character largely depends on our ability to infer homologies among the organisms involved in a phylogenetic study, on our confidence to adduce character polarities, and on the character's degree of variation within the forms being compared relative to its variation between them. Regarding the last point, character states of the reproductive tract have proven more evolutionarily conservative than certain hallowed ones of the dentition (Carleton, 1980). The proliferation of karyological and electrophoretic studies in recent years presents still other bodies of information to be appraised and synthesized. The ultimate systematic value of these data will doubtlessly hinge on their ability to meet the same three criteria enumerated above.

PROPOSED CLASSIFICATION

The basis for my revised classification of *Peromyscus* and related peromyscines (Table 4) is covered primarily in the foregoing Systematic Review. However, some additional comments and explanation regarding certain aspects of the classification are warranted.

I have recognized 53 species of *Peromyscus* arranged in 13 species groups. For reasons presented above, the five peripheral subgenera of Hooper and Musser (1964a) and Hooper (1968) are accorded separate generic status. These six genera of Sigmodontinae, together with *Neotomodon* and *Onychomys*, comprise the peromyscine assemblage, which I would informally consider as tribal in rank. Peromyscini in this sense is more taxonomically restricted compared to the peromyscines of Hooper and Musser (1964a, 1964b), which also included *Reithrodontomys*, *Ochrotomys*, *Baiomys* and *Scotinomys*, or to the peromyscines of Carleton (1980), which also included *Reithrodontomys* and *Ochrotomys*. Where these other genera fit within the radiation of North American sigmodontines is beyond the scope of the present paper. Regarding *Ochrotomys*, I would only note that its cladistic separation is probably earlier than I represented it in my final cladogram (1980: fig. 42), an interpretation supported by some of my results (for example, Wagner trees in figs. 28 and 29) and perhaps by karyotypic data (Engstrom and Bickham, 1982). On these grounds, *Ochrotomys* may eventually merit its own tribe, but future evaluation of its relationships should also consider that Rogers and Heske (1984) identified two chromosomal rearrangements shared by *Ochrotomys* and *Scotinomys*.

I have not listed the various species groups under subgenera. This is not because I believe that subgenera are inappropriate categories for a systematist to use or that there are not major lines of descent within *Peromyscus* that could merit subgeneric recognition. Rather I feel that lack of firm documentation of kinship among the species groups, coupled with persistent doubts about the monophyly of some, prevent critical, unambiguous diagnoses of subgenera at this time. As mentioned above, several lines of evidence collectively suggest the weakness of our current subgeneric alignments, such as the uncertain accomodation of *P. hooperi* (Lee and Schmidly, 1977) and the systematic implications of some karyotypic and electrophoretic results (Avise *et al.*, 1979; Rogers, 1983; Rogers *et al.*, 1984; Stangl and Baker, 1984b; Schmidly *et al.*, 1985). Those desiring to maintain the conventional arrangement may simply associate the *californicus* and *eremicus* groups under *Haplomylomys* and refer the remaining 11 species groups to the subgenus *Peromyscus*.

An earlier forecast of difficulties with the subgenera finally adopted, particularly the apportionment of species among *Haplomylomys* and *Peromyscus*, is found in Hooper's (1958) pioneering study on the glans penis. In this paper, Hooper initially associated *hylocetes* (= *aztecus*) with *eremicus* and *californicus* as a "division" apart from the *maniculatus* division and five others, most of which were later treated as subgenera (Hooper and Musser, 1964a). A broad vase-shaped glans with distal flare, lack of a pronounced protractile tip, and absence of dorsal and ventral lappets comprise the phallic traits that persuaded Hooper to group them in his *eremicus* division. Subsequently, Hooper and Musser (1964a:4) questioned the significance of these resemblances, partially attributing them to the inadequate state of preservation of the original material. Instead, they drew attention to penile similarities of *hylocetes* and *oaxacensis*, which had been historically linked with the *boylii* group of the subgenus *Peromyscus* (Osgood, 1909), and so classified them as such. Nevertheless, Hooper's (1958) description of the phallus of *hylocetes* is fundamentally accurate and partly constitutes the basis for my separation of *aztecus* and its allies from the *boylii* group. These phallic traits are believed to represent plesiomorphic states (Lawlor, 1971b; Carleton, 1980) and therefore do not provide critical evidence of kinship to *Haplomylomys*, yet neither do they, by themselves, support a phyletic unity with members of the subgenus *Peromyscus*. The evolution of the narrow elongate glans penis with protractile tip, lappets, and cartilaginous tip that characterizes many species of *Peromyscus* may have occurred independently, as suggested by the rudimentary development of these features in species such as *guardia* (Lawlor, 1971b) and *hooperi* (Lee and Schmidly, 1977; Schmidly *et al.*, 1985). Reevaluation of this character variation and assumptions of polarity within *Peromyscus* is required. Until relationships among the species groups can be delineated with greater assurance, designation of subgeneric boundaries seems moot.

Workers on *Peromyscus* have employed the species-group category with slightly different meanings and underlying assumptions regarding evolutionary or speciational processes. Osgood (1909) did not define clearly his usage of this supraspecific grouping, but in his earlier (1900:12) revision of *Perognathus*, he mentioned the arrangement of species into groups ". . . in order to show the affinities of the species . . .," but whether he was referring to morphological, evolutionary, or ecological affinities is unclear. Hooper (1968:28) applied the category in a loosely defined and informal sense to identify ". . . an assemblage of similar species within a subgenus." Hooper's meaning of "similar" was candidly identified as a morphological criterion in operation, though he presumed the similarity to reflect descent from a single phylogenetic origin. Dice (1933) and his colleagues attached a genetic criterion, adduced from interbreeding trials with stocks of *Peromyscus*, to the definition of a taxonomic species group: forms were judged to be from the same group if they were completely or partially interfertile in the laboratory and never mated successfully outside of their own group. Blair (1943a) acknowledged the same breeding studies in developing his distinction between incipient species and cenospecies, the latter equivalent to the species group, as a means of understanding stages of speciation. Whether judged primarily from morphological or genetic viewpoints, systematists seemed to agree that species groups are "natural biological units" originating from a common ancestral stock.

As an aside, it is interesting to note that the interspecific fertility criterion advanced to genetically delimit species groups of *Peromyscus* and other rodents is analogous to the argument Van Gelder (1977) developed for determining the boundary of the genus in primates, carnivores, and ungulates. The equivalence of the muroid species group to the genus of certain other mammalian orders, as suggested by this yardstick, supplies some reassurance that the proliferation of recognized muroid genera, particularly in the past decade (see Carleton, 1984), has not been as excessive as we might suppose. Perhaps more importantly, the comparison highlights the prevalence of sexual selection among these different mammal groups and the historical emphasis systematists often placed on epigamic characters.

Regrettably, the innovative program of Dice and his colleagues to document the hybridization potential among the numerous species and species groups of *Peromyscus* was never completed. Therefore, the degree to which morphological similarity may predict interspecific fertility and aid the recognition of species groups of *Peromyscus* has never been fully assessed, and as a consequence, our species groups have come to be highly uneven in morphological diversity. For instance, compare the contents of North American species groups like *leucopus* and *maniculatus* to those more poorly known in Central America like the *boylii* and *mexicanus* groups. Moreover, breeding results were inconsistently applied for certain species: *californicus* has not been segregated from *eremicus* and its kin as a different group in our formal classifications, although Dice (1933) and Blair (1943*a*, 1950) had earlier referred them to separate groups based on their failure to interbreed.

The goal to verify the monophyly of our species groups is now receiving renewed attention from a cladistic perspective that stresses identification of synapomorphic features suggesting descent from a common ancestor. Whether the nature of variation in chromosomes, proteins, crania, or other characters among some 50 species of *Peromyscus* will allow the identification of discrete character states amenable to cladistic analysis and resolution of hierarchical patterns remains to be seen, but I am not overly optimistic. Conceivably, such research may lead to the amalgamation of certain species groups in the future. For now, my usage of species groups basically follows Hooper's (1968) informal rationale and presumes monophyly, but I do not intend any connotation of superspecies, cenospecies, or speciation stage by their employment. I have avoided the proliferation of single-taxon species groups, except where there has been repeated corroboration of their differentiation and more distant kinship, as for *californicus, crinitus*, and *hooperi*. Species such as *slevini, polius*, and *stirtoni* that might potentially merit single-member species group status or whose species group placement is otherwise uncertain are indicated in Table 4.

TAXONOMIC SUMMARY

This taxonomic summary is provided to consolidate the nomenclatural foundation and diagnostic features that have been advanced for the various genus-group taxa of peromyscine rodents. The following authors have described more fully or otherwise discussed most traits cited herein: Osgood (1909); Hooper (1957, 1958); Hershkovitz (1962); Hooper and Musser (1964a); Linzey and Layne (1969, 1974); Carleton (1973, 1980); Reig (1977); Rogers *et al.* (1984); Stangl and Baker (1984b); Bradley and Schmidly (1986); and Voss (1988). Nevertheless, I have supplemented external, cranial, and dental comparisons based on my own examinations of specimens contained in USNM.

The tribal-level recognition of peromyscines is framed as a characterization because I regard this rank as an informal working hypothesis. A diagnosis is given for each peromyscine genus; the nucleus of these defining traits should still apply even if the weight of subsequent systematic research favors their ranking as subgenera. The character states listed under the tribal characterization are given as the hypothesized most-derived condition that subtends all eight peromyscine genera. The existence of additional apomorphic states for a given character is highlighted by an asterisk; the nature of these derived states is identified further in the generic diagnoses. Invariant conditions and those believed to be plesiomorphic at the tribal level are generally not included as part of the generic diagnoses.

Characterization of peromyscine rodents.—Small to medium-size muroid rodents; pinnae relatively large and pliant, thinly to densely furred; cephalic vibrissae arranged as mystacial whorl plus superciliary, interramal, and genal* tufts; three pairs of mammae*, one pair postaxial, one abdominal, and one inguinal; manus with stubby pollex bearing a nail and four clawed digits; plantar surface of manus with five tubercles (thenar, hypothenar, and three interdigitals); all five digits of pes with claws, digit V slightly shorter in length than digits II – IV*; plantar surface of pes entirely naked or sparsely furred on heel* and bearing six fleshy tubercles (thenar, hypothenar, and four interdigitals)*.

Zygomatic notches absent to rudimentary* and plates relatively narrow; zygomatic arches delicate, their midportions depressed to level of orbital floor; jugal thin but always present and separating zygomatic extensions of maxillary and squamosal; infraorbital foramen narrower ventrally; superficial masseter originates from small, mound-shaped rugosity at base of inferior zygomatic root*; supraorbital borders rounded and hourglass-shaped*; braincase smooth without temporal ridges*; interparietal wide, covering most of caudal border of parietals, but short anteriorly-posteriorly; subsquamosal fenestra present and subequal in size to postglenoid foramen, the two vacuities separated by the narrow hamular process of the squamosal*; middle lacerate fissure separated from postglenoid foramen by a stout flange of the petrosal bone that projects anteroventrally to overlap the squamosal; incisive foramina long, slitlike, penetrating between the first molars to the level of their anterocones*; bony palate flat, extending to end of toothrows; posterior palatine foramina small, round to ovate, located in the maxillopalatine suture at the level of the middle to front of M2s; posterolateral palatal pits absent; roof of mesopterygoid fossa incomplete, pierced by spacious sphenopalatine vacuities*; lateral pterygoid fossae flat, not excavated; strut of alisphenoid bone separates masticatory-buccinator foramen from foramen ovale accessorius; stapedial and sphenofrontal foramina present, posterior opening of alisphenoid canal large, and shallow groove present on inner surface of alisphenoid and squamosal bones (primitive carotid circulatory pattern)*; otic capsules small, ectotympanic bullae not covering posteromedial ventral surface of petrosal; medial flange of bony eustachian tube encloses carotid canal; pars flaccida well-developed; malleus with distinct orbicularis apophysis and manubrium oriented perpendicular to cephalic peduncle; mastoid not inflated; basihyal with entoglossal process, thyrohyal greater than or equal to length of basihyal; coronoid process of dentary short and recurved, set lower than condyloid process, sigmoid notch moderately expressed*.

Incisors weakly to strongly opisthodont, lacking sulci; molars 3/3, uppers with three roots* and lowers with two; third molars slightly shorter than second molars*, about 30 percent of toothrow length; molar crowns brachyodont, cuspidate, and crested*; primary cusps arranged biserially and alternately, alternation more pronounced in lowers than uppers; mesolophs(ids), anterolophs(ids), ectolophids, and other accessory enamel structures present (pentalophodont)*; valleys between cusps broad* and generally open to margins; M1 oval-shaped with narrow, undivided anterocone*; metacone of M1–2 attaches posteromedially to posteroloph defining a short posteroflexus*; paraflexus, metaflexus, and hypoflexus of M3 small but distinguishable*; m3 with anterolabial cingulum and protoflexid*, posteroflexid present opening to lingual margin, entoconid small but distinct and offset from hypoconid.

Axial skeleton with modal counts of 13 ribs*, 19 thoracicolumbar vertebrae*, and three or four sacral vertebrae; caudal vertebrae variable, modal counts range from 19 to 38; second thoracic vertebra with pronounced neural spine; entepicondylar foramen present*; trochlear process of calcaneum broad and set near articular facet*.

Spacious internal cheek pouches absent; tongue with single circumvallate papilla; soft palate marked by distinct anterior longitudinal ridge, three complete and four incomplete transverse palatal rugae*; stomach bilocular with glandular mucosa markedly reduced (discoglandular)*; gall bladder present; large intestine simple with few or no colonic loops*; caecum moderately long and simple internally*.

Phallus short and awl-shaped* with baculum less than length of glans body*; terminal crater, lateral bacular digits, dorsal papilla, and urethral process* absent; body of glans invested with spines* and lacking corrugation*; urinary meatus positioned terminally such that protrusible tip and distal lappets are undefined*; assortment of muroid accessory reproductive glands (preputials, anterior, ventral and dorsal prostates, bulbourethrals, ampullaries, and vesiculars) complete* and unmodified*; spermatozoa with oval head and single acrosomal hook*.

Chromosomal complement having 24 pairs (2N = 48); pairs 1, 22, and 23 biarmed (AN = 52) with euchromatic short arms, the remaining pairs acrocentric, and heterochromatin restricted to the centromere*.

Comments.—The tribal characterization given above is obviously a polythetic construction, for many of the traits enumerated either have evolved in parallel within other lineages of Muroidea or are believed to represent the ancestral condition for the Superfamily (see Carleton and Musser, 1984). The joint possession of a bilocular stomach with reduced glandular mucosa and perhaps the chromosomal formula are the most persuasive synapomorphies supporting this grouping of genera as monophyletic. Still, the paucity of uniquely derived features leaves the matter of tribal recognition as inconclusive.

Genus *Peromyscus* Gloger

Peromyscus Gloger, 1841:95 (type species, *Peromyscus arboreus* Gloger, 1841 [= *Mus sylvaticus noveboracensis* Fischer, 1829])
Peromyscus (*Haplomylomys*) Osgood, 1904:53 (type species, *Hesperomys eremicus* Baird, 1858:479)

Diagnosis.—Peromyscine rodents distinguished by a combination of traits: M3 suboval with hypoflexus indistinct or absent, paraflexus and usually metaflexus distinct; posteroflexid of m3 present as small circular fossetid, rarely open to lingual margin; phallus with urinary meatus subterminal, such that at least a rudimentary protrusible tip is developed, and baculum longer than length of glans penis; accessory glands complete and unmodified except for usual absence of preputials; carotid circulatory pattern primitive.

Comments.—The unambiguous discrimination of this group on the basis of derived traits is unsatisfactory as it embraces substantial diversity. Dental complexity ranges from complex, with frequent occurrence of a mesoloph and ectolophid (most groups), to simple, with such enamel crests usually absent (*californicus*, *eremicus*, and *crinitus* groups). The upper first molar may be ovate with a single, narrow anterocone (*crinitus* and *eremicus* groups) or rectangular with the anterocone broad and subdivided (*aztecus* and *furvus* groups). The interorbital shape is usually smooth and rounded in most forms, but some species of the *mexicanus* and *megalops* groups bear a slight shelf or low ridge along the supraorbital borders. The glans penis is relatively short and vase-shaped with a short protrusible tip in a few species groups (*californicus*, *eremicus*, and some *aztecus*), but it is narrowly elongate with a well-defined protrusible tip in most. Many of the latter groups also contain species with dorsal or ventral lappets, or both, along the distal margin of the glans body. Regular fluting of the glans surface is confined to members of the *aztecus* group, although weakly expressed fluting occurs in certain species of the *boylii* group; the epidermal surface of the glans penis in most species is smooth. Most species groups lack distinct preputials, yet these reproductive glands are large in the *califor-*

nicus and *eremicus* groups. The stomach of most species conforms to the bilocular-discoglandular plan, but species of the *megalops* and *mexicanus* groups possess a partially or fully formed glandular pouch. Three pairs of mammae typify most species groups but loss of the anterior pair has apparently transpired independently in some (for example, *eremicus*, *crinitus*, and *mexicanus* groups). The number of biarmed chromosomal pairs varies from none (AN = 46, some *P. levipes*) to 21 (AN = 88, some *P. sitkensis*), a range that reflects both pericentric changes (especially of pairs 2, 3, 6, and 9) in many species as well as numerous heterochromatic short-arm acquisitions in groups such as *eremicus* and *maniculatus*.

The lack of strong concordance among these character-state differences hinders the hierarchical association of these species groups and the credible delimitation of subgenera.

Contents.—53 species are recognized and arranged in 13 species groups (see Table 4).

Genus *Habromys* Hooper and Musser

Peromyscus (Habromys) Hooper and Musser, 1964a:12 (type species, *Peromyscus lepturus* Merriam, 1898)

Diagnosis.—Peromyscine rodents having a modal number of 13 ribs and 20 thoracicolumbar vertebrae; zygomatic notch weakly developed; digit V of pes relatively long, subequal in length to II–IV; heel furred to thenar pad; phallus long and narrow, urinary meatus terminal, baculum longer than length of glans penis, spines absent; preputials and lateral ventral prostates absent, medial ventral prostates a finely-branched tubular mass, vesiculars reduced to minute diverticula from cephalic urethra; spermatozoa lack acrosomal hook; chromosomal pairs 1, 2, 3, 5, 6, 7, 9, 10, 15, 22, and 23 biarmed (AN = 68) with short arms euchromatic.

Contents.—Four species have been described: *lepturus, lophurus, simulatus,* and *chinanteco.*

Genus *Podomys* Osgood

Peromyscus (Podomys) Osgood, 1909:226 (Type species, *Hesperomys floridanus* Chapman, 1889)

Diagnosis.—Peromyscine rodents distinguished by presence of small supraorbital shelf; zygomatic notch moderately incised; molars mostly lacking accessory enamelled ridges, or present as short spurs not reaching the edges; metacone of M1–2 joins posteroloph near its labial end, posteroflexus absent or obsolete after little wear; M3 small and round, flexi indistinct or absent; m3 without anterolabial cingulum and protoflexid, posteroflexid isolated as fossetid; plantar pads of pes reduced in size, hypothenar minute or absent; heel furred to thenar pad; phallus long and narrow, urinary meatus terminal, baculum longer than length of glans penis, spines small and limited to proximal half of glans; preputials and lateral ventral prostates absent, bulbourethrals outsized, ampullaries large and coiled, vesiculars short and saclike; chromosomal pairs 1, 2, 3, 6, 9, 11, 22, and 23 biarmed (AN = 62) with short arms euchromatic.

Contents.—Only the one species *floridanus* is known.

Genus *Neotomodon* Merriam

Neotomodon Merriam, 1898:127 (type species, *Neotomodon alstoni* Merriam, 1898)

Diagnosis.—Peromyscines with strongly incised zygomatic notches; incisive foramina very long, usually reaching to the protocone of the M1s; molars hypsodont, their occlusal surface terraced to nearly planar; accessory enamel crests and styles usually absent or incomplete and fused to adjoining cusps; valleys between cusps narrow, protoflexus of M1 a shallow crease, that of M2 absent; metacone of M1–2 attaches to labial end of posteroloph, posteroflexus absent; M3s small and circular, flexi coalesced and indistinct, single-rooted; anterolabial cingulum of m3 usually absent; heel of pes furred to thenar pad; soft palate with two complete and five incomplete transverse palatal ridges; phallus moderately long with urinary meatus terminal and baculum shorter than length of glans penis; preputials and lateral ventral prostates absent, ampullaries large and coiled, vesiculars reduced to minute diverticula from cephalic urethra; chromosomal pairs 1, 2, 3, 6, 7, 9, 22, and 23 biarmed (AN = 62) with short arms euchromatic.

Contents.—Only the one species *alstoni* is known.

Genus *Megadontomys* Merriam

Peromyscus (Megadontomys) Merriam, 1898:115 (type species, *Peromyscus thomasi* Merriam, 1898)

Diagnosis.—Peromyscine rodents with weakly expressed supraorbital shelf or ridge; sphenopalatine vacuities short and narrow;

molars robust and higher crowned; M1 rectangular with broad anterocone, labial and lingual conules defined by shallow anteromedian flexus and anteromedian fossetus; M1 with labial accessory root set medially; anteroloph of M1 a strong ridge, the anteroflexus long; superficial masseter originates from elevated masseteric tubercle; large intestine with three or four colonic loops and caecum long and complex; phallus relatively wide and long, urinary meatus subterminal and enfolded by hood of soft tissue, baculum longer than length of glans penis, glans surface with pronounced furrows and large spines, urethral flap present; preputials absent, complement of accessory glands otherwise resembles ancestral condition; chromosomal pairs 1, 2, 9, 22, and 23 biarmed (AN = 56) with euchromatic short arms.

Contents.—Three species are recognized: *thomasi, nelsoni,* and *cryophilus.*

Genus *Onychomys* Baird

Onychomys Baird, 1858:458 (type species, *Hypudaeus leucogaster* Wied-Neuwied, 1841)

Diagnosis.—Peromyscine rodents distinguished by their elevated, conical molar cusps; accessory lophs(ids) absent, styles(ids) varibly present; M1 strongly ovate with narrow, undivided anterocone; posteroflexus of M1–2 rudimentary and obscured after little wear; protoflexus of M1 broad, that of M2 absent; M3 greatly reduced, flexi indistinct, roots coalesced to two or one; m3 greatly reduced, hypoconid and entoconid positioned opposite one another or sometimes fused as narrow heel, posteroflexid absent; anterolabial cingulum of m3 weakly expressed, the protoflexid a slight crease; coronoid process of dentary scythe-shaped, as long as condyloid process, together forming a deep, half-oval sigmoid notch; trochlear process set more distally on calcaneum; pinnae smaller, thickly covered with hair; plantar surface of pes densely furred from heel to first interdigital pad, thenar and hypothenar pads absent; digit V of pes notably shorter than II–IV; tail short (only 17 to 22 caudal vertebrae), less than length of head and body; soft palate with two complete and four incomplete transverse palatal ridges; stomach bilocular, incisura angularis moderately deep, glandular mucosa infolded as diverticulum (pouch) on greater curvature; caecum a short, simple sac; phallus short with terminal urinary meatus and baculum shorter than length of glans penis; anterior prostates, lateral ventral prostates, ampullaries, and vesiculars absent; chromosomal pairs 1, 9, 19, 22, and 23 biarmed with euchromatic short arms, AN relatively high (72–92) due to numerous heterochromatic short-arm additions.

Contents.—Three species are currently recognized: *leucogaster, arenicola,* and *torridus.*

Genus *Osgoodomys* Hooper and Musser

Peromyscus (*Osgoodomys*) Hooper and Musser 1964a:12 (type species, *Peromyscus banderanus* Allen, 1897)

Diagnosis.—Peromyscine rodents having well-developed supraorbital shelves with reflected lateral edges (beaded) and incipient temporal ridges, interorbital region thus appearing constricted; zygomatic notches weakly developed; subsquamosal fenestra small, postglenoid foramen narrow, hamular process of squamosal broad; modified carotid circulatory pattern (sphenofrontal foramen and squamosal-alisphenoid groove absent, stapedial foramen and posterior opening of the alisphenoid canal minute); roof of mesopterygoid fossa wholly ossified (sphenopalatine vacuities absent); genal vibrissae absent; postaxial pair of mammae absent; M1 strongly ovate with narrow anterocone and distinct anteromedian flexus defining twin conules; posteroflexid of m3 closed off as fossetid; phallus small with terminal urinary meatus and baculum shorter than glans penis; lateral ventral prostates, anterior prostates, and vesiculars absent, bulbourethrals outsized, ampullaries filamentous; chromosomal pairs 1, 22, and 23 biarmed (AN = 52).

Contents.—Only the one species *banderanus* is known.

Genus *Isthmomys* Hooper and Musser

Peromyscus (*Isthmomys*) Hooper and Musser, 1964a:12 (type species *Megadontomys flavidus* Bangs, 1902)

Diagnosis.—Peromyscine rodents having beaded supraorbital edges and weakly developed temporal ridges; subsquamosal fenestra absent or minute, postglenoid foramen small and narrow; sphenopalatine vacuities tiny or absent; M1 strongly ovate, narrow anterocone deeply bifurcate; posteroflexid of m3 isolated as fossetid; entepicondylar foramen absent; stomach with edges of discoglandular zone partially infolded (incipient pouch); phallus long and broad distally, baculum longer than length of glans penis, spines long and heavy, urinary meatus opens almost terminally through large mound of nonspinous, soft tissue; preputials and lateral ventral prostates absent, cephalic border of vesiculars with lobulate folds; postaxial pair of mammae absent; chromosomal pairs 1, 16, 17, 20, 22, and 23 metacentric (AN = 58) with short arms of pairs 1, 16, and 22 euchromatic and those of 17, 20, and 23 heterochromatic, interstitial C-band on long arm of pair 1.

Contents.—Only two species, *flavidus* and *pirrensis*, are known.

Areas for Future Research

There is a pressing need to improve the alpha taxonomy of *Peromyscus* and other peromyscine rodents, both at the specific and infraspecific levels. Such research is of critical importance if the congruence of morphologic, karyotypic, and genic data sets is to be used as a basis for conclusions on the independence of their evolutionary rates and for inference of relationships. The relevancy of this research to the understanding of geographic variation, speciation, and biogeographic patterns is likewise self-evident. Such a recommendation may appear anomalous for a genus of native rodent whose biology is among the most exhaustively studied and best known in the world. However, it should be reiterated that our nomenclature for many of the subspecies and species currently recognized dates from early revisions undertaken with different concepts of geographic variation, species lability, and taxonomic conventions for analyzing and conveying those kinds of data. The kind of revisionary work required has largely developed since the early 1970s and as yet has involved only small segments of the genus.

Systematic research of the past two decades underscores the weakness of inferring biological species status solely from bifurcations in phenograms based on multivariate analyses of cranial dimensions, from levels of genetic similarity revealed in assays of protein variation, or from discovery of chromosomal differences presumed to signal reproductive incompatibility. Recourse to diverse kinds of information appears essential and characterizes our soundest taxonomic deductions. Although this caveat seems obvious, it bears mentioning in light of the perhaps naive expectation that the new analytical techniques and quantitative rigor ushered in by technological advancements would serve alone as a critical yardstick for delimiting species.

In this regard, I find it noteworthy that the documentation of sympatry remains a most persuasive argument and brings an element of criticality otherwise lacking from interpretation of a data set by itself, be it morphometric, allelic, or karyotypic. The educational process occurring in our continuing studies of *Peromyscus* systematics is one wherein greater taxonomic weight is being accorded to subtle differences in karyotypes, in electromorph occurrence, and in craniodental features, partly by virtue of their concordance but also by the demonstration of their persistence in sympatry or parapatry through finer geographic sampling. Yet for many peromyscine taxa, their allopatric distributions, whether on islands or on mountaintops, preclude the test of sympatry. In such instances, we seem destined to rely on some form of Mayrian postulate for assessing the taxonomic status of allopatric populations, for example, like the arguments Schmidly (1973*a*) reasonably marshalled in deciding the status of *Peromyscus comanche*.

Cladistic analyses, both classical Hennigian approaches and quantitative phyletic methods, have experienced the first shakedown in their application to questions of higher level relationships of *Peromyscus* and its relatives. Future studies should routinely include character-state by OTU data matrices. The omission of original data matrices is sometimes confusing, particularly when trying to compare cladistic interpretations of chromosomal banding data across studies and to evaluate the treatment of certain characters. Moreover, the data matrix should indicate *only those characters* actually employed to structure the tree. Similarly, there is an advantage in using explicit, quantitative phylogenetic methods to construct the karyotypic cladograms. The number of species that have now been banded, the total number of characters used, and the apparent number of homoplastic rearrangements are too numerous to work through confidently by hand—there may be equally or more parsimonious interpretations available with the data sets. Finally, objective methods exist for comparing the goodness of fit of data on alternative phylogenetic trees.

Although the degree of concordance of karyotypic or electrophoretic data sets to the phylogeny of Hooper and Musser (1964*a*, 1964*b*) vis-à-vis that of Carleton (1980) is of academic interest, it remains secondary to the more fundamental need to refine character analyses used as a basis for phylogenetic hypotheses. In order to attribute incongruences of independently derived phylogenies to differential rates of evolution or occurrences of homoplasy, we must minimize the extent to which incorrect interpretations of character homologies and polarities account for incongruity. Rogers *et al.* (1984 : 463–64) emphasized this point in regard to the current and future utility of euchromatic banding data. The same holds true for future studies drawing upon morphological data. As I earlier indicated (1980 : 127), more critical interpretations of character homologies and character-state trees are essential for several of the morphological characters used in muroid taxonomy, especially those of the reproductive system.

Many aspects of our systematic understanding of *Peromyscus* and its relatives would profit from revival of the fertile school of scientific inquiry pioneered by Sumner, Blossom, Dice, Blair, and their associates from the mid-1920s through the 1940s. The results of their investigations are contained in numerous papers, but synopses are found in Sumner (1932), Dice and Blossom (1937), Blair (1950), and Dice (1968). The topics these authors collectively pursued are still relevant to systematic and evolutionary theory today: patterns and causes of intraspecific variation, both geographic and non-geographic; studies of clines, intergradation, and hybrid zones and their relationship to environmental gradients; adaptive significance of coat color and body size and the selective role played by predation, ecology, and climate; evaluation of selection, isolation, population structure, and gene flow and the occurrence of geographic differentiation and incipient speciation; development of reproductive isolation and the process of speciation; and the significance of conventional taxonomic characters as indicators of genealogical relationship at the subspecific, specific, and species-group levels. The information produced by their efforts still comprises much of the empirical framework that supports our ideas on these subjects in systematic mammalogy. Their approach was in essence one of experimental systematics and helped to catalyze the New Systematics. Their work is all the more impressive when one considers their dependence on the cumbersome laboratory breeding protocols and arduous statistical analyses that were then employed (before computers) to apprehend underlying genetic mechanisms and patterns. In certain respects, Sumner and his successors formulated their questions three to four decades before the emergence of the systematic tools and technology needed to supply more critical answers.

Nevertheless, those questions remain central to an increased understanding of the evolutionary biology of *Peromyscus* today and deserve to be reinfused with the perspectives provided by electrophoresis, karyology, gene sequencing, morphometrics, and other modern systematic techniques now on the horizon. Evidence of movement in this direction is already suggested by some of the work of Selander, Smith, and colleagues on the genetics of *Peromyscus* populations, by the continuing interest in understanding the contact zone of the cytotypes of *P. leucopus* (Stangl and Baker, 1984*a*; Stangl, 1986; Nelson *et al.*, 1987), and by the studies of meiotic behavior in populations of *P. maniculatus* polymorphic for certain pericentric inversions (Greenbaum and Reed, 1984; Greenbaum *et al.*, 1986). By focusing our attention at the level of individual and population relationships (tokogenetic of Hennig, 1966), both in the field and in the laboratory, we also stand to gain greater insight into the nature of variation, manner of inheritance, and developmental canalization of characters that we have been attempting to use in our cladistic studies and thus will be better able to assess their reliability as indicators of phylogenetic relationship.

ACKNOWLEDGMENTS

I greatly appreciate the time and attention devoted by L. R. Heaney, G. G. Musser, and R. S. Voss to reviewing this manuscript. Their advice and substantive comments improved its final content. I also thank M. Hafner, G. G. Musser, and P. Myers for loaning specimens of *Peromyscus* from their respective instititions. Tim-Bob Miller and David Schmidt ably assisted with the drafting and labeling of the distributional maps. Carole Young of Texas Tech University Press graciously accomodated my request to incorporate the most recent literature on *Peromyscus* systematics; I appreciate her indulgence of my concerns on that matter. Last in order but first in my appreciation, I wish to thank Gordon Kirkland and James Layne for inviting me to participate and providing the opportunity to reflect again on Peromystematics, and for the many hidden hours and headaches each has invested to bring such a project to fruition.

LITERATURE CITED

ALLARD, M. W., S. J. GUNN, AND I. F. GREENBAUM. 1987. Mensural discrimination of chromosomally characterized *Peromyscus oreas* and *Peromyscus maniculatus*. J. Mamm., 68:402–406.

ALLEN, J. A. 1897. Further notes on mammals collected in Mexico by Dr. Audley C. Buller, with descriptions of new species. Bull. Amer. Mus. Nat. Hist., 9:47–58.

ALVAREZ, T. 1961. Taxonomic status of some mice of the *Peromyscus boylii* group in eastern Mexico, with description of a new subspecies. Univ. Kansas Publ., Mus. Nat. Hist., 14:111–120.

ANDERSON. S. 1972. Mammals of Chihuahua, taxonomy and distribution. Bull. Amer. Mus. Nat. Hist., 148:149–410.

AQUADRO, C. A., AND C. W. KILPATRICK. 1981. Morphological and biochemical variation and differentiation in insular and mainland deer mice (*Peromyscus maniculatus*). Pp. 214–230, *in* Mammalian population genetics (M. H. Smith and J. Joule, eds.). Univ. Georgia Press, Athens, 380 pp.

AQUADRO, C. A., AND J. C. PATTON. 1980. Salivary amylase variation in *Peromyscus*: use in species identification. J. Mamm., 61:703–707.

ARRIGHI, F. E., A. D. STOCK, AND S. PATHAK. 1976. Chromosomes of *Peromyscus* (Rodentia, Cricetidae). V. Evidence of pericentric inversions. Chromosomes Today, 5:323–329.

ASHLEY, M., AND C. WILLS. 1987. Analysis of mitochondrial DNA polymorphisms among Channel Island deer mice. Evolution, 41:854–863.

ATCHLEY, W. R., J. J. RUTLEDGE, AND D. E. COWLEY. 1981. Genetic components of size and shape. II. Multivariate covariance patterns in the rat and mouse skull. Evolution, 35:1037–1055.

AVISE, J. C. 1974. Systematic value of electrophoretic data. Syst. Zool., 23: 465–481.

AVISE, J. C., M. H. SMITH, AND R. K. SELANDER. 1974*a*. Biochemical polymorphism and systematics in the genus *Peromyscus*. VI. The *boylii* species group. J. Mamm., 55:751–763.

———. 1979. Biochemical polymorphism and systematics in the genus *Peromyscus*. VII. Geographic differentiation in members of the *truei* and *maniculatus* species groups. J. Mamm., 60:177–192.

AVISE, J. C., M. H. SMITH, R. K. SELANDER, T. E. LAWLOR, AND P. R. RAMSEY. 1974*b*. Biochemical polymorphism and systematics in the genus *Peromyscus*. V. Insular and mainland species of the subgenus *Haplomylomys*. Syst. Zool., 23:226–238.

BADER, R. S. 1965. Heritability of dental characters in the house mouse. Evolution, 19:378–384.

BAILEY, V. 1902. Synopsis of the North American species of *Sigmodon*. Proc. Biol. Soc. Washington, 15:101–116.

BAIRD, S. F. 1855. Characteristics of some new species'of North American Mammalia, collected chiefly in connection with the U. S. surveys of a railroad route to the Pacific. Proc. Acad. Nat. Sci., Philadelphia, 7(8):333–337.

BAKER, R. H. 1951. Mammals from Tamaulipas, Mexico. Univ. Kansas Publ., Mus. Nat. Hist., 5:207–218.

———. 1952. Geographic range of *Peromyscus melanophrys*, with description of a new subspecies. Univ. Kansas Publ., Mus. Nat. Hist., 5:251–258.

———. 1968. Habitats and Distribution. Pp. 98–126, *in*: Biology of *Peromyscus* (Rodentia) (J. A. King, ed.). Spec. Publ., Amer. Soc. Mamm., 2:1–593.

BAKER, R. H., AND J. K. GREER. 1962. Mammals of the Mexican state of Durango. Publ. Michigan State Univ. Mus., Biol. Ser., 2:25–154.

BAKER, R. J., AND R. K. BARNETT. 1981. Karyotypic orthoselection for additions of heterochromatin in grasshopper mice (*Onychomys*: Cricetidae). Southwestern Nat., 26:125–131.

BAKER, R. J., R. K. BARNETT, AND I. F. GREENBAUM. 1979. Chromosomal evolution in grasshopper mice (*Onychomys*, Cricetidae). J. Mamm., 60:297–306.

BAKER, R. J., AND J. W. BICKHAM. 1980. Karyotypic evolution in bats: evidence of extensive and conservative chromosomal evolution in closely related taxa. Syst. Zool., 29:239–253.

BAKER, R. J., B. F. KOOP, AND M. W. HAIDUK. 1983*a*. Resolving systematic relationships with G-bands: a study of five genera of South American cricetine rodents. Syst. Zool., 32:403–416.

BAKER, R. J., L. W. ROBBINS, F. B. STANGL, JR., AND E. C. BIRNEY. 1983*b*. Chromosomal pevidence for a major subdivision in *Peromyscus leucopus*. J. Mamm., 64:356–359.

BANKS, R. C. 1967. The *Peromyscus guardia-interparietalis* complex. J. Mamm., 48:210–218.

BLAIR, W. F. 1943a. Criteria for species and their subdivisions from the point of view of genetics. New York Acad. Sci., 44:179–188.

———. 1943b. Biological and morphological distinctness of a previously undescribed species of the *Peromyscus truei* group from Texas. Contrib. Lab. Vert. Biol., Univ. Michigan, 24:1–8.

———. 1950. Ecological factors in the speciation of *Peromyscus*. Evolution, 4:253–275.

BOLES, D. J. 1984. Chromosomal evolution of populations of *Peromyscus boylii* from western Durango, Mexico. Unpubl. M.S. thesis, Univ. Vermont, 48 pp.

BOWEN, W. W. 1968. Variation and evolution of Gulf coast populations of beach mice, *Peromyscus polionotus*. Bull. Florida State Mus., 12:1–91.

BOWERS, J. H. 1974. Genetic compatibility of *Peromyscus maniculatus* and *Peromyscus melanotis*, as indicated by breeding studies and morphometrics. J. Mamm., 55:720–737.

BOWERS, J. H., R. J. BAKER, AND M. H. SMITH. 1973. Chromosomal, electrophoretic, and breeding studies of selected populations of deer mice (*Peromyscus maniculatus*) and black-eared mice (*P. melanotis*). Evolution, 27:378–386.

BRADLEY, R. D., AND J. ENSINK. 1987. Karyotypes of five cricetid rodents from Honduras. Texas J. Sci., 39:171–176.

BRADLEY, R. D., AND D. J. SCHMIDLY. 1987. The glans penes and bacula in Latin American taxa of the *Peromyscus boylii* group. J. Mamm., 68:595–616.

BRADSHAW, W. N. 1968. Progeny from experimental mating tests with mice of the *Peromyscus leucopus* group. J. Mamm., 49:475–480.

BRADSHAW, W. N., AND T. C. HSU. 1972. Chromosomes of *Peromyscus* (Rodentia, Cricetidae). III. Polymorphism in *Peromyscus maniculatus*. Cytogenetics, 11:436–451.

BRAND, L. R., AND R. E. RYCKMAN. 1969. Biosystematics of *Peromyscus eremicus*, *P. guardia*, and *P. interparietalis*. J. Mamm., 50:501–513.

BROWN, J. H., AND C. F. WESLER. 1968. Serum albumin polymorphisms in natural and laboratory populations of *Peromyscus*. J. Mamm., 49:420–426.

BROWN, W. L. 1957. Centrifugal speciation. Quart. Rev. Biol., 32:247–277.

BROWNE, R. A. 1977. Genetic variation in island and mainland populations of *Peromyscus leucopus*. Amer. Midland Nat., 97:1–9.

BROWNELL, E. 1983. DNA/DNA hybridization studies of muroid rodents: symmetry and rates of molecular evolution. Evolution, 37:1034–1051.

BURT, W. H. 1932. Description of heretofore unknown mammals from islands in the Gulf of California, Mexico. Trans. San Diego Soc. Nat. Hist., 7:161–182.

———. 1934. Subgeneric allocation of the white-footed mouse, *Peromyscus slevini*, from the Gulf of California, Mexico. J. Mamm., 15:159–160.

BURT, W. H., AND F. S. BARKALOW. 1942. A comparative study of the bacula of wood rats (subfamily Neotominae). J. Mamm., 23:287–297.

BUTH, D. G. 1984. The application of electrophoretic data in systematic studies. Ann. Rev. Ecol. Syst., 15:501–522.

CAIRE, W., AND E. G. ZIMMERMAN. 1975. Chromosomal and morphological variation and circular overlap in the deer mouse, *Peromyscus maniculatus*, in Texas and Oklahoma. Syst. Zool., 24:89–95.

CALHOUN, S. W., I. F. GREENBAUM, AND K. P. FUXA.   1988.   Biochemical and karyo-typic variation in *Peromyscus maniculatus* from western North America. J. Mamm., 69:34–45.

CARLETON, M. D.   1973.   A survey of gross stomach morphology in New World Cricetinae (Rodentia, Muroidea), with comments on functional interpreta-tions. Misc. Pub. Mus. Zool., Univ. Michigan, 146:1–43.

————.   1977.   Interrelationships of populations of the *Peromyscus boylii* species group (Rodentia, Muridae) in western Mexico. Occas. Papers Mus. Zool., Univ. Michigan, 675:1–47.

————.   1979.   Taxonomic status and relationships of *Peromyscus boylii* from El Sal-vador. J. Mamm., 60:280–296.

————.   1980.   Phylogenetic relationships in neotomine-peromyscine rodents (Muroidea) and a reappraisal of the dichotomy within New World Cri-cetinae. Misc. Publ. Mus. Zool., Univ. Michigan, 157:1–146.

————.   1984.   Introduction to Rodents. Pp. 255–265, *in* Orders and families of Recent mammals of the world (S. Anderson and J. K. Jones, eds.). Wiley Press, New York, 686 pp.

CARLETON, M. D., AND D. G. HUCKABY.   1975.   A new species of *Peromyscus* from Guatemala. J. Mamm., 56:444–451.

CARLETON, M. D., AND G. G. MUSSER.   1984.   Muroid Rodents. Pp. 289–379, *in* Orders and families of Recent mammals of the world (S. Anderson and J. K. Jones, eds.). Wiley Press, New York, 686 pp.

CARLETON, M. D., D. E. WILSON, A. L. GARDNER, AND M. A. BOGAN.   1982.   Dis-tribution and systematics of *Peromyscus* (Mammalia: Rodentia) of Nayarit, Mexico. Smithsonian Contrib. Zool., 352:1–46.

CHALINE, J., P. MEIN, AND F. PETTER.   1977.   Les grandes lignes d'une classifcation evolutive des Muroidea. Mammalia, 41:245–252.

CHAPPELL, M. A., J. P. HAYES, AND L. R. G. SNYDER.   1988.   Hemoglobin polymor-phisms in deer mice (*Peromyscus maniculatus*): Physiology of beta-globin vari-ants and alpha-globin recombinants. Evolution, 42:681–688.

CHOATE, J. R.   1973.   Identification and recent distribution of white-footed mice (*Peromyscus*) in New England. J. Mamm., 54:41–49.

CHOATE, J. R., R. C. DOWLER, AND J. E. KRAUSE.   1979.   Mensural discrimination between *Peromyscus leucopus* and *P. maniculatus* (Rodentia) in Kansas. South-western Nat., 24:249–258.

CLARK, D. L.   1966.   Fertility of a *Peromyscus maniculatus* X *Peromyscus melanotis* cross. J. Mamm., 47:340.

COMMITTEE FOR STANDARDIZATION OF CHROMOSOMES OF *PEROMYSCUS*.   1977.   Standardized karyotype of deer mice, *Peromyscus* (Rodentia). Cytogenet. Cell Genet., 19:38–43.

CORBET, G. B., AND J. E. HILL.   1980.   A world list of mammalian species. British Museum (Nat. Hist.), London, 226 pp.

DAVIS, K. M., S. A. SMITH, AND I. F. GREENBAUM.   1986.   Evolutionary implications of chromosomal polymorphisms in *Peromyscus bolyii* from southwestern Mexico. Evolution, 40:645–649.

DICE, L. R.   1933.   Fertility relationships between some of the species and sub-species of mice in the genus *Peromyscus*. J. Mamm., 14:298–305.

————.   1937.   Fertility relations in the *Peromyscus leucopus* group of mice. Contrib. Lab. Vert. Genetics, Univ. Michigan, 4:1–3.

————. 1940a. Intergradation between two species of deer-mouse (*Peromyscus maniculatus*) across North Dakota. Contrib. Lab. Vert. Biol., Univ. Michigan, 13:1–14.

————. 1940b. Relationships between the wood-mouse and cotton-mouse in eastern Virginia. J. Mamm., 21:14–23.

————. 1968. Speciation. Pp. 75–97, *in* Biology of *Peromyscus* (Rodentia) (J. A. King, ed.). Spec. Publ., Amer. Soc. Mamm., 2:1–593.

DICE, L. R., AND P. M. BLOSSOM. 1937. Studies of mammalian ecology in southwestern North America with special attention to the colors of desert mammals. Publ. Carnegie Inst. Washington, 458:1–129.

DICKEY, D. R. 1928. Five new mammals of the genus *Peromyscus* from El Salvador. Proc. Biol. Soc. Washington, 41:1–6.

DIERSING, V. E. 1976. An analysis of *Peromyscus difficilis* from the Mexican-United States boundary area. Proc. Biol. Soc. Washington, 89:451–466.

ELDER, F. B. 1980. Tandem fusion, centric fusion, and chromosomal evolution in the cotton rats, genus *Sigmodon*. Cytogenet. Cell Genet., 26:199–210.

ELDER, F. B., AND M. R. LEE. 1985. The chromosomes of *Sigmodon ochrognathus* and *S. fulviventer* suggest a realignment of *Sigmodon* species groups. J. Mamm., 66:511–518.

ELLERMAN, J. R. 1941. The families and genera of living rodents. Vol. II. British Museum (Nat. Hist.), London, 690 pp.

ENGSTROM, M. D., AND J. W. BICKHAM. 1982. Chromosome banding and phylogenetics of the golden mouse, *Ochrotomys nuttalli*. Genetica, 59:119–126.

————. 1983. Karyotype of *Nelsonia neotomodon*, with notes on the primitive karyotype of peromyscine rodents. J. Mamm., 64:685–688.

ENGSTROM, M. D., D. J. SCHMIDLY, AND P. K. FOX. 1982. Nongeographic variation and discrimination of species within the *Peromyscus leucopus* species group (Mammalia: Cricetinae) in eastern Texas. Texas J. Sci., 34:149–162.

FISHER, R. A. 1930. The genetical theory of natural selection. Clarendon Press, Oxford, 291 pp.

FULLER, F., M. R. LEE, AND L. R. MAXSON. 1984. Albumin evolution in *Peromyscus* and *Sigmodon*. J. Mamm., 65:466–473.

GARDNER, A. L., AND J. L. PATTON. 1976. Karyotypic variation in oryzomyine rodents (Cricetinae) with comments on chromosomal evolution in the Neotropical cricetine complex. Occas. Papers Mus. Zool., Louisiana St. Univ., 49:1–48.

GLAZIER, D. S. 1980. Ecological shifts and the evolution of geographically restricted species of North American *Peromyscus* (mice). J. Biogeog., 7:63–83.

GOODWIN, G. G. 1941. A new *Peromyscus* from western Honduras. Amer. Mus. Novitates, 1121:1.

————. 1955. Two new white-footed mice from Oaxaca, Mexico. Amer. Mus. Novitates, 1732:1–5.

————. 1956. Seven new mammals from Mexico. Amer. Mus. Novitates, 1791:1–10.

————. 1964. A new species and a new subspecies of *Peromyscus* from Oaxaca, Mexico. Amer. Mus. Novitates, 2183:1–8.

GREENBAUM, I. F., AND R. J. BAKER. 1978. Determination of the primitive karyotype of *Peromyscus*. J. Mamm., 59:820–834.

GREENBAUM, I. F., R. J. BAKER, AND J. H. BOWERS. 1978a. Chromosomal homology and divergence between sibling species of deer mice: *Peromyscus maniculatus* and *P. melanotis* (Rodentia, Cricetidae). Evolution, 32:334–341.

GREENBAUM, I. F., R. J. BAKER, AND P. R. RAMSEY. 1978*b*. Chromosomal evolution and its implications concerning the mode of speciation in three species of deer mice of the genus *Peromyscus*. Evolution, 32:646–654.

GREENBAUM, I. F., D. W. HALE, AND K. P. FUXA. 1986. Synaptic adaptation in deer mice: a cellular mechanism for karyotypic orthoselection. Evolution, 40:208–213.

GREENBAUM, I. F., AND M. J. REED. 1984. Evidence for heterosynaptic pairing of the inverted segment in pericentric inversion heterozygotes of the deer mouse (*Peromyscus maniculatus*). Cytogenet. Cell Genet., 38:106–111.

GUNN, S. J., AND I. F. GREENBAUM. 1986. Systematic implications of karyotypic and morphologic variation in mainland *Peromyscus* from the Pacific northwest. J. Mamm., 67:294–304.

HAIDUK, M. W., AND R. J. BAKER. 1982. Cladistical analysis of G-banded chromosomes of nectar-feeding bats (Glossophaginae: Phyllostomidae). Syst. Zool., 31:252–265.

———. 1984. Scientific method, opinion, phylogenetic reconstruction, and nectar-feeding bats: a response to Griffiths and Warner. Syst. Zool., 33:343–350.

HALL, E. R. 1971. Variation in the blackish deer mouse, *Peromyscus furvus*. An. Inst. Biol., Mexico, Ser. Zool., 1:149–154.

———. 1981. The mammals of North America. Second ed. John Wiley and Sons, New York, 2:601-1181 + 90.

HALL, E. R., AND W. W. DALQUEST. 1963. The mammals of Veracruz. Univ. Kansas Publ., Mus. Nat. Hist., 14:165–362.

HALL, E. R., AND D. F. HOFFMEISTER. 1942. Geographic variation in the canyon mouse, *Peromyscus crinitus*. J. Mamm., 23:51–65.

HALL, E. R., AND K. R. KELSON. 1959. The mammals of North America. Ronald Press, New York, 2:547–1083 + 79.

HANDLEY, C. O., JR. 1966. Checklist of the mammals of Panama. Pp. 753–795, *in* Ectoparasites of Panama (R. L. Wenzel and V. J. Tipton, eds.) Field Mus. Nat. Hist., Chicago, xii + 861 pp.

HEANEY, L. R., AND E. C. BIRNEY. 1977. Distribution and natural history notes on some mammals from Puebla. Southwestern Nat., 21:543–545.

HENNIG, W. 1966. Phylogenetic systematics. Univ. Illinois Press, Urbana, 263 pp.

HERSHKOVITZ, P. 1962. Evolution of Neotropical cricetine rodents (Muridae) with special reference to the phyllotine group. Fieldiana Zool., 46:1–524.

HOFFMEISTER, D. F. 1951. A taxonomic and evolutionary study of the pinon mouse, *Peromyscus truei*. Illinois Biol. Monogr., 21:1–104.

HOFFMEISTER, D. F., AND L. DE LA TORRE. 1961. Geographic variation in the mouse *Peromyscus difficilis*. J. Mamm., 42:1–13.

HOFFMEISTER, D. F., AND V. E. DIERSING. 1973. The taxonomic status of *Peromyscus merriami goldmani* Osgood, 1904. Southwestern Nat., 18:354–357.

HOFFMEISTER, D. F., AND M. R. LEE. 1963. The status of the sibling species *Peromyscus merriami*, and *Peromyscus eremicus*. J. Mamm., 44:201–213.

HONACKI, J. H., K. E. KINMAN, AND J. W. KOEPPL. 1982. Mammalian species of the world. Assoc. Syst. Coll., Lawrence, Kansas, 694 pp.

HOUSEAL, T. W., I. F. GREENBAUM, D. J. SCHMIDLY, S. A. SMITH, AND K. M. DAVIS.

1987. Karyotypic variation in *Peromyscus boylii* from Mexico. J. Mamm., 68:281–296.

HOOD, C. S., L. W. ROBBINS, R. J. BAKER, AND H. S. SHELLHAMMER. 1984. Chromosomal studies and evolutionary relationships of an endangered species, *Reithrodontomys raviventris*. J. Mamm., 65:655–667.

HOOPER, E. T. 1942. An effect on the *Peromyscus maniculatus* rassenkreis of land utilization in Michigan. J. Mamm., 23:193–196.

———. 1947. Notes on Mexican mammals. J. Mamm., 28:40–57.

———. 1952a. A systematic review of harvest mice (Genus *Reithrodontomys*) of Latin America. Misc. Publ. Mus. Zool., Univ. Michigan, 77:1–255.

———. 1952b. Notes on mice of the species *Peromyscus boylii* and *P. pectoralis*. J. Mamm., 33:371–378.

———. 1955. Notes on mammals of western Mexico. Occas. Papers Mus. Zool., Univ. Michigan, 565:1–26.

———. 1957. Dental patterns in mice of the genus *Peromyscus*. Misc. Publ. Mus. Zool., Univ. Michigan, 99:1–59.

———. 1958. The male phallus in mice of the genus *Peromyscus*. Misc. Publ. Mus. Zool., Univ. Michigan, 105:1–24.

———. 1960. The glans penis in *Neotoma* (Rodentia) and allied genera. Occas. Papers Mus. Zool., Univ. Michigan, 618:1–21.

———. 1961. Notes on mammals from western and southern Mexico. J. Mamm., 42:120–122.

———. 1968. Classification. Pp. 27–74, *in* Biology of *Peromyscus* (Rodentia) (J. A. King, ed.). Spec. Publ., Amer. Soc. Mamm., 2:1–593.

HOOPER, E. T., AND G. G. MUSSER. 1964a. Notes on classification of the rodent genus *Peromyscus*. Occas. Papers Mus. Zool., Univ. Michigan, 635:1–13.

———. 1964b. The glans penis in Neotropical cricetines (Family Muridae) with comments on classification of muroid rodents. Misc. Publ. Mus. Zool., Univ. Michigan, 123:1–57.

HSU, T. C., AND F. E. ARRIGHI. 1966. Chromosomal evolution in the genus *Peromyscus* (Cricetidae, Rodentia). Cytogenet., 5:355–359.

———. 1968. Chromosomes of *Peromyscus* (Rodentia, Cricetidae). I. Evolutionary trends in 20 species. Cytogenet.,7:417–446.

HUCKABY, D. G. 1973. Biosystematics of the *Peromyscus mexicanus* group (Rodentia). Unpubl. Ph.D. dissert., Univ. Michigan, Ann Arbor, 144 pp.

———. 1980. Species limits in the *Peromyscus mexicanus* group (Mammalia: Rodentia: Muroidea). Contrib. Sci., Los Angeles Co. Mus. Nat. Hist., 326:1–24.

JENNINGS, M. R. 1987. A biography of Dr. Charles Elisha Boyle, with notes on his 19th century natural history collection from California. The Wasmann J. Biol., 45:59–68.

JENSEN, J. N., AND D. I. RASMUSSEN. 1971. Serum albumins in natural populations of *Peromyscus*. J. Mamm., 52:508–514.

JOHNSON, G. L., AND R. L. PACKARD. 1974. Electrophoretic analysis of Peromyscus comanche Blair, with comments on its systematic status. Occas. Papers Mus., Texas Tech. Univ., 24:1–16.

JOHNSON, P. T., AND J. N. LAYNE. 1961. A new species of *Polygenis* Jordon from Florida, with remarks on its host relationships and zoogeographic significance. Ent. Soc. Washington, 63:115–123.

JONES, J. K., AND T. L. YATES. 1983. Review of the white-footed mice, genus Peromyscus, of Nicaragua. Occas. Papers Mus., Texas Tech. Univ., 82: 1–15.

KILPATRICK, C.W. 1971. Distribution of the brush mouse, *Peromyscus boylii*, and the encinal mouse, *Peromyscus pectoralis*, in north-central Texas. Southwestern Nat., 16:211–213.

———. 1984. Molecular evolution of the Texas mouse, *Peromyscus attwateri*. Pp. 87–96, *in* Festschrift for Walter W. Dalquest in honor of his sixty-sixth birthday (N. Horner, ed.). Midwestern State Univ., Wichita Falls, Texas, 163 pp.

KILPATRICK, C. W., AND W. CAIRE. 1973. First record of the encinal mouse, *Peromyscus pectoralis*, for Oklahoma, and additional records for north-central Texas, Southwestern Nat., 18:351.

KILPATRICK, C. W., AND E. G. ZIMMERMAN. 1975. Genetic variation and systematics of four species of mice of the *Peromyscus boylii* species group. Syst. Zool., 24:143–162.

———. 1976. Biochemical variation and systematics of *Peromyscus pectoralis*. J. Mamm., 57:506–522.

KING, J. A. (ed.) 1968. Biology of *Peromyscus* (Rodentia). Spec. Publ., Amer. Soc. Mamm., 2:1–593.

KOH, H. S., AND R. L. PETERSON. 1983. Systematic studies of deer mice, *Peromyscus maniculatus* Wagner (Cricetidae, Rodentia): analysis of age and secondary sexual variation in morphometric characters. Canadian J. Zool., 61: 2618–2628.

KOOP, B. F., R. J. BAKER, AND H. H. GENOWAYS. 1983. Numerous chromosomal polymorphisms in a natural population of rice rats (*Oryzomys*, Cricetidae). Cytogenet. Cell Genet., 35:131–135.

KOOP, B. F., R. J. BAKER, M. W. HAIDUK, AND M. D. ENGSTROM. 1984. Cladistical analysis of primitive G-band sequences for the karyotype of the ancestor of the Cricetidae complex of rodents. Genetica, 64:199–208.

KOOP, B. F., R. J. BAKER, AND J. T. MASCARELLO. 1985. Cladistical analysis of chromosomal evolution within the genus Neotoma. Occas. Papers Mus., Texas Tech Univ., 96:1–9.

KUHN, T. S. 1970. The structure of scientific revolutions. Second ed. Univ. Chicago Press, Chicago, 210 pp.

LANSMAN, R. A., J. C. AVISE, C. F. AQUADRO, J. F. SHAPIRA, AND S. W. DANIEL. 1983. Extensive genetic variation in mitochondrial DNA's among geographic populations of the deer mouse, *Peromyscus maniculatus*. Evolution, 37:1–16.

LAWLOR, T. E. 1965. The Yucatan deer mouse, *Peromyscus yucatanicus*. Univ. Kansas Publ., Mus. Nat. Hist., 16:421–438.

———. 1971a. Distribution and relationships of six species of *Peromyscus* in Baja California and Sonora, Mexico. Occas. Papers Mus. Zool., Univ. Michigan, 661:1–22.

———. 1971b. Evolution of *Peromyscus* on northern islands in the Gulf of California, Mexico. Trans. San Diego Soc. Nat. Hist., 16:91–124.

———. 1974. Chromosomal evolution in *Peromyscus*. Evolution, 28:689–692.

———. 1983. The mammals. Pp. 265–289, *in* Island biogeography in the Sea of Cortez (T. J. Case and M. L. Cody, eds.). Univ. California Press, Berkeley, 508 pp.

LAYNE, J. N. 1970. Climbing behavior of *Peromyscus floridanus* and *Peromyscus gossypinus*. J. Mamm., 51:580–591.

LEAMY, L. J. 1977. Genetic and environmental correlations of morphometric traits in randombred house mice. Evolution, 31:357–369.

LEE, M. R., AND F. B. ELDER. 1977. Karyotypes of eight species of Mexican rodents (Muridae). J. Mamm., 58:479–487.

LEE, M. R., AND D. J. SCHMIDLY. 1977. A new species of *Peromyscus* (Rodentia: Muridae) from Coahuila, Mexico. J. Mamm., 58:263–268.

LEE, M. R., D. J. SCHMIDLY, AND C. C. HUHEEY. 1972. Chromosomal variation in certain populations of *Peromyscus boylii* and its systematic implications. J. Mamm., 53:697–707.

LIDICKER, W. Z. 1962. The nature of subspecies boundaries in a desert rodent and its implications for subspecies taxonomy. Syst. Zool., 11:160–171.

LINZEY, A. V., AND J. N. LAYNE. 1969. Comparative morphology of the male reproductive tract in the rodent genus *Peromyscus* (Muridae). Amer. Mus. Novitates, 2355:1–47.

———. 1974. Comparative morphology of spermatozoa of the rodent genus *Peromyscus* (Muridae). Amer. Mus. Novitates, 2532:1–20.

LOFSVOLD, D. 1986. Quantitative genetics of morphologoical differentiation in *Peromyscus*. I. Tests of the homogeneity of genetic covariance structure among species and subspecies. Evolution, 40:559–573.

———. 1988. Quantitative genetics of morphological differentiation in *Peromyscus*. II. Analysis of selection and drift. Evolution, 42:54–67.

MACEY, M., AND L. K. DIXON. 1987. Chromosomal variation in *Peromyscus maniculatus* populations along an elevational gradient. Evolution, 41:676–678.

MAILLAIRD, J. 1924. A new mouse (*Peromyscus slevini*) from the Gulf of California. Proc. California Acad. Sci., Ser. 4, 12:1219–1222.

MASCARELLO, J. T., A. D. STOCK, AND S. PATHAK. 1974. Conservatism in the arrangement of genetic material in rodents. J. Mamm., 55:695–704.

MAYR, E. 1963. Animal species and evolution. Harvard Univ. Press, Cambridge, 797 pp.

MCCARLEY, W. H. 1954. Natural hybridization in the *Peromyscus leucopus* species group of mice. Evolution, 8:314–323.

MCDANIEL, V., R. TUMLISON, AND P. McLARTY. 1983. Mensural discrimination of the skulls of Arkansas *Peromyscus*. Ark. Acad. Sci. Proc., 37:50–53.

MERRIAM, C. H. 1894. A new subfamily of murine rodents—the Neotominae—with descriptions of a new genus and species and a synopsis of the known forms. Proc. Acad. Nat. Sci., Philadelphia, 46:225–252.

———. 1898a. Mammals of Tres Marias Islands, off western Mexico. Proc. Biol. Soc. Washington, 12:13–19.

———. 1898b. Descriptions of twenty new species and a new subgenus of *Peromyscus* from Mexico and Guatemala. Proc. Biol. Soc. Washington, 12:115–125.

MILLER, G. S. 1912. List of North American land mammals in the United State National Museum, 1911. Bull. U.S. Natl. Mus., 79:1–455.

———. 1924. List of North American Recent mammals. Bull. U.S. Natl. Mus., 128:1–673.

MILLER, G. S., AND R. KELLOGG. 1955. List of North American Recent mammals. Bull. U.S. Natl. Mus., 205:1–954.

MILLER, G. S., AND J. A. REHN. 1901. Systematic results of the study of North American land mammals to the close of the year 1900. Proc. Boston Soc. Nat. Hist., 30:1–352.

MODI, W. S., AND M. R. LEE. 1984. Systematic implication of chromosomal banding analyses of populations of *Peromyscus truei* (Rodentia, Muridae). Proc. Biol. Soc. Washington, 97:716–723.

MURIE, A. 1933. The ecological relationships of two subspecies of *Peromyscus* in the Glacier Park region, Montana. Occas. Papers Mus. Zool., Univ. Michigan, 270:1–17.

MUSSER, G. G. 1964. Notes on geographic distribution, habitat, and taxonomy of some Mexican mammals. Occas. Papers Mus. Zool., Univ. Michigan, 636:1–22.

———. 1969. Notes on *Peromyscus* (Muridae) of Mexico and Central America. Amer. Mus. Novitates, 2357:1–23.

———. 1971. *Peromyscus allophylus* Osgood: a synonym of *Peromyscus gymnotis* Thomas (Rodentia, Muridae). Amer. Mus. Novitates, 2453:1–10.

NELSON, E. W. 1921. Lower California and its natural resources. Mem. Nat. Acad. Sci., 26:1–194.

NELSON, K., R. J. BAKER, AND R. L. HONEYCUTT. 1987. Mitochondrial DNA and protein differentiation between hybridizing cytotypes of the white-footed mouse, *Peromyscus leucopus*. Evolution, 41:864–872.

NEVO, E. 1979. Adaptive convergence in subterranean mammals. Ann. Rev. Ecol. Syst., 10:269–308.

OSGOOD, W. H. 1900. Revision of the pocket mice of the genus *Perognathus*. N. Amer. Fauna, 18:1–72.

———. 1904a. *Haplomylomys*, a new subgenus of *Peromyscus*. Proc. Biol. Soc. Washington, 17:53–54.

———. 1904b. Thirty new mice of the genus *Peromyscus* from Mexico and Guatemala. Proc. Biol. Soc. Washington, 19:55–77.

———. 1909. Revision of the mice of the American genus *Peromyscus*. N. Amer. Fauna, 28:1–285.

———. 1945. Two new rodents from Mexico. J. Mamm., 26:299–301.

PARADISO, J. L. 1969. Mammals of Maryland. N. Amer. fauna, 66:1–193.

PATHAK, S., T. C. HSU, AND F. E. ARRIGHI. 1973. Chromosomes of *Peromyscus*. IV. The role of heterochromatin in karyotypic evolution. Cytogenet. Cell Genet., 12:315–326.

PATTON, J. C., R. J. BAKER, AND J. C. AVISE. 1981. Phenetic and cladistic analyses of biochemical evolution in peromyscine rodents. Pp. 288–308, *in* Mammalian population genetics (M. H. Smith and J. Joule, eds.). Univ. Georgia Press, Athens, 380 pp.

PATTON, J. L., AND S. W. SHERWOOD. 1983. Chromosome evolution and speciation in rodents. Ann. Rev. Ecol. Syst., 14:139–148.

PAVLINOV, I. 1982. Phylogeny and classification of the subfamily Gerbillinae. Bull. Moscow Nat. Hist. Soc., 87:19–31.

PENGILLY, D., G. H. JARRELL, AND S. D. MacDONALD. 1983. Banded karyotypes of *Peromyscus sitkensis* from Baranof Island, Alaska. J. Mamm., 64:682–685.

PRICE, P. K., AND M. L. KENNEDY. 1980. Genic relationships in the white-footed mouse, *Peromyscus leucopus*, and the cotton mouse, *Peromyscus gossypinus*. Amer. Midland Nat., 103:73–82.

RAMIREZ-PULIDO, J., A. MARTINEZ, AND G. URBANO. 1977. Mammiferos de la Costa Grande de Guererro, Mexico. An. Inst. Biol., Mexico, 48:243–292.

RASMUSSEN, D. I. 1970. Biochemical polymorphisms and genetic structure in populations of *Peromyscus*. Symp. Zool. Soc. London, 26:335–349.

REIG, O. A. 1977. A proposed unified nomenclature for the enamelled components of the molar teeth of the Cricetidae (Rodentia). J. Zool., London, 181:227–241.

RENNERT, P. D., AND C. W. KILPATRICK. 1986. Biochemical systematics of populations of *Peromyscus boylii*. I. Populations from east-central Mexico with low fundamental numbers. J. Mamm., 67:481–488.

———. 1987. Biochemical systematics of *Peromyscus boylii*. II. Chromosomally variable populations from eastern and southern Mexico. J. Mamm., 68: 799–811.

ROBBINS, L. W., AND R. J. BAKER. 1980. G- and C-band studies on the primitive karyotype for *Reithrodontomys*. J. Mamm., 61:708–714.

———. 1981. An assessment of the nature of rearrangements in eighteen species of *Peromyscus* (Rodentia:Cricetidae). Cytogenet. Cell Genet., 31:194–202.

ROBBINS, L. W., M. P. MOULTON, AND R. J. BAKER. 1983. Extent of geographic range and magnitude of chromosomal evolution. J. Biogeog., 10:533–541.

ROBBINS, L. W., M. H. SMITH, M. C. WOOTEN, AND R. K. SELANDER. 1985. Biochemical polymorphism and its relationship to chromosomal and morphological variation in *Peromyscus leucopus* and *Peromyscus gossypinus*. J. Mamm., 66:498–510.

ROBERTSON, P. B., AND G. G. MUSSER. 1976. A new species of *Peromyscus* (Rodentia: Cricetidae), and a new specimen of *P. simulatus* from southern Mexico, with comments on their ecology. Occas. Papers Mus. Nat. Hist., Univ. Kansas, 47:1–8.

ROGERS, D.S. 1983. Phylogenetic affinities of *Peromyscus* (*Megadontomys*) *thomasi*: evidence from differentially stained chromosomes. J. Mamm., 64:617–623.

ROGERS, D. S., I. F. GREENBAUM, S. J. GUNN, AND M. D. ENGSTROM. 1984. Cytosystematic value of chromosomal inversion data in the genus *Peromyscus*. J. Mamm., 65:457–465.

ROGERS, D. S., AND E. J. HESKE. 1984. Chromosomal evolution of the brown mice, genus *Scotinomys* (Rodentia: Cricetidae). Genetica, 63:221–228.

SCHMIDLY, D. J. 1972. Geographic variation in the white-ankled mouse, *Peromyscus pectoralis*. Southwestern Nat., 17:113–138.

———. 1973a. The systematic status of *Peromyscus comanche*. Southwestern Nat., 18:269–278.

———. 1973b. Geographic variation and taxonomy of *Peromyscus boylii* from Mexico and southern United States. J. Mamm., 54:111–130.

SCHMIDLY, D. J., R. D. BRADLEY, AND P. S. CATO. 1988. Morphometric differentiation and taxonomy of three chromosomally characterized groups of *Peromyscus boylii* from east-central Mexico. J. Mamm., 69:462–480.

SCHMIDLY, D. J., M. R. LEE, W. S. MODI, AND E. G. ZIMMERMAN. 1985. Systematics and notes on the biology of Peromyscus hooperi. Occas. Papers Mus., Texas Tech Univ., 97:1–40.

SCHMIDLY, D. J., AND G. L. SCHROETER. 1974. Karyotypic variation of *Peromyscus boylii* (Rodentia, Cricetidae) from Mexico and corresponding taxonomic implications. Syst. Zool., 23:333–342.

SCHNELL, G. D., AND R. K. SELANDER. 1981. Environmental and morphological correlates of genetic variation in mammals. Pp. 60–99, *in* Mammalian population genetics (M. H. Smith and J. Joule, eds.). Univ. Georgia Press, Athens, 380 pp.

SELANDER, R. K., AND W. E. JOHNSON. 1973. Genetic variation among vertebrate species. Ann. Rev. Ecol. Syst., 4:75–91.

SELANDER, R. K., M. H. SMITH, S. Y. YANG, W. E. JOHNSON, AND J. B. GENTRY. 1971. Biochemical polymorphism and systematics in the genus *Peromyscus*. I. Variation in the old-field mouse (*Peromyscus polionotus*). Studies in Genetics VI, Univ. Texas Publ., 7103:49–90.

SEVERINGHAUS, W. D., AND D. F. HOFFMEISTER. 1978. Qualitative cranial characters distinguishing *Sigmodon hispidus* and *Sigmodon arizonae* and the distribution of these two species in northern Mexico. J. Mamm., 59:868–870.

SHEPPE, W., JR. 1961. Systematic and ecological relations of *Peromyscus oreas* and *P. maniculatus*. Proc. Amer. Phil. Soc., 105:421–446.

SIMPSON, G. G. 1945. The principles of classification and a classification of mammals. Bull. Amer. Mus. Nat. Hist., 85:1–350.

SMITH, M.F. 1979. Geographic variation in genic and morphological characters in *Peromyscus californicus*. J. Mamm., 60:705–722.

SMITH, M. H., AND J. JOULE(eds.). 1981. Mammalian population genetics. Univ. Georgia Press, Athens, 380 pp.

SMITH, M. H., R. K. SELANDER, AND W. E. JOHNSON. 1973. Biochemical polymorphism and systematics in the genus *Peromyscus*. III. Variation in the Florida deermouse (*Peromyscus floridanus*), a Pleistocene relic. J. Mamm., 54:1–13.

SMITH, S. A., R. D. BRADLEY, AND I. F. GREENBAUM. 1986. Karyotypic conservatism in the *Peromyscus mexicanus* group. J. Mamm., 67:584–586.

SNYDER, L. R. G., J. P. HAYES, AND M. A. CHAPPELL. 1988. Alpha-chain hemoglobin polymorphisms are correlated with altitude in the deer mouse, *Peromyscus maniculatus*. Evolution, 42:689–697.

STANGL, F. B., JR. 1986. Aspects of a contact zone between two chromosomal races of *Peromyscus leucopus* (Rodentia: Cricetidae). J. Mamm., 67:465–473.

STANGL, F. B., AND R. J. BAKER. 1984a. A chromosomal subdivision in *Peromyscus leucopus*: implications for the subspecies concept as applied to mammals. Pp. 139–145, *in* Festschrift for Walter W. Dalquest in Honor of His Sixty-sixth Birthday (N. Horner, ed.). Dept. Biology, Midwestern State Univ., 163 pp.

———. 1984b. Evolutionary relationships in *Peromyscus*: congruence in chromosomal, genic, and classical data sets. J. Mamm., 65:643–654.

STRANEY, D. O. 1981. The stream of heredity: genetics in the study of phylogeny. Pp. 100–138, *in* Mammalian population genetics (M. H. Smith and J. Joule, eds.). Univ. Georgia Press, Athens, 380 pp.

SUMNER, F. B. 1932. Genetic, distributional, and evolutionary studies of the subspecies of deer mice (*Peromyscus*). Bibliographia Genet., 9:1–106.

THOMAS, B. 1973. Evolutionary implications of karyotypic variation in some insular *Peromyscus* from British Columbia, Canada. Cytologia, 38:485–495.

THOMAS, O. 1903. On three new forms of *Peromyscus* obtained by Dr. Hans Gadow, F. R. S., and Mrs. Gadow in Mexico. Ann. Mag. Nat. Hist., Ser. 7, 11:484–487.

THROCKMORTON, L. H. 1977. *Drosophila* systematics and biochemical evolution. Ann. Rev. Ecol. Syst., 8:235–254.

TOLLIVER, D. K., J. R. CHOATE, D. W. KAUFMAN, AND G. A. KAUFMAN. 1987. Microgeographic variation of morphometric and electrophoretic characters in *Peromyscus leucopus*. Amer. Midland Nat., 117:420–427.

TROUESSART, E. L. 1904–1905. Catalogus mammaliantam viventum quam fossilium. R. Friedlander and Sohn, Berolini, 929 pp.

TRUE, F. W. 1885. A provisional list of the mammals of North and Central America and the West Indian Islands. Proc. U.S. Natl. Mus., 7:587–611.

TULLBERG, T. 1899. Uber das system der Nagethiere, eine phylogenetische studie. Nova Acta R. Soc. Scient. Upsala, ser. 3, 18:1–514.

VAN GELDER, R. G. 1977. Mammalian hybrids and generic limits. Amer. Mus. Novitates, 2635:1–25.

VOSS, R. S. 1988. Systematics and ecology of ichthyomyine rodents (Muroidea): patterns of morphological evolution in a small adaptive radiation. Bull. Amer. Mus. Nat. Hist., 188:259–493.

WARNER, R. M. 1983. Karyotypic megaevolution and phylogenetic analysis: New World nectar-feeding bats revisited. Syst. Zool., 32:279–282.

WATERHOUSE, C. R. 1839. The zoology of the voyage of H.M.S. Beagle. Part 2. Mammalia. 97 pp.

WERBITSKY, D., AND C. W. KILPATRICK. 1987. Genetic variation and genetic differentiation among allopatric populations of *Megadontomys*. J. Mamm., 68:305–312.

WHITE, M. J. D. 1978. Modes of speciation. W. H. Freeman and Co., San Francisco, 455 pp.

WILSON, E. O., AND W. L. BROWN, JR. 1953. The subspecies concept. Syst. Zool., 3:97–111.

WOLFE, J. L., AND J. N. LAYNE. 1968. Variation in dental structures of the Florida mouse, *Peromyscus floridanus*. Amer. Mus. Novitates., 2351:1–7.

YATES, T. L., R. J. BAKER, AND R. K. BARNETT. 1979. Phylogenetic analysis of karyological variation in three genera of peromyscine rodents. Syst. Zool., 28:40–48.

ZIMMERMAN, E. G. 1970. Karyology, systematics and chromosomal evolution in the rodent genus, *Sigmodon*. Publ. Mus., Michigan St. Univ., Biol. Ser., 4:385–454.

———. 1974. Chromsomes of the Mexican Plateau mouse, *Peromyscus melanophrys*, and a new sex-determining mechanism in mammals. Canadian J. Genet. Cytol., 16:797–804.

ZIMMERMAN, E. G., B. J. HART, AND C. W. KILPATRICK. 1975. Biochemical genetics of the *boylii* and *truei* groups of the genus *Peromyscus* (Rodentia). Comp. Biochem. Physiol., 52B:541–545.

ZIMMERMAN, E. G., C. W. KILPATRICK, AND B. J. HART. 1978. The genetics of speciation in the rodent genus *Peromyscus*. Evolution, 32:565–579.

ADAPTIVE PHYSIOLOGY

RICHARD E. MacMILLEN AND THEODORE GARLAND, JR.

Abstract.—Herein we examine and analyze the available data for the genus *Peromyscus* that relate to energy metabolism, water metabolism, and the interrelationships between these variables, approaching our study from an allometric perspective. The genus is nearly ubiquitous in North America, and there appears to be a well-defined geographic gradation in body mass, with the largest species in the tropical south and smaller species at higher latitudes. Allometric comparisons of basal metabolism, evaporative water loss, and water regulatory efficiency with other rodents indicate that *Peromyscus* typically are intermediate between dietary specialist granivores and herbivores; this is in keeping with their intermediate dietary position of omnivory. Thus, *Peromyscus* species may be viewed as physiological generalists. Their frequent success in surviving physically stressful environmental conditions, we propose, is made possible through the intermittent use of torpor until the stress is moderated. We propose further that torpor is a generalized syndrome of small (less than 40 grams) species of *Peromyscus*, and that selection has favored smaller size in those species inhabiting more stressful (more northerly) regions, where torpor provides the added margin for survival.

In 1983, R. W. Hill and R. E. MacMillen published companion papers on energy regulation (Hill, 1983) and water regulation (MacMillen, 1983*a*) in *Peromyscus* in an attempt to compensate for the absence of a chapter devoted to physiology in King's (1968) book, *Biology of* Peromyscus *(Rodentia)*. As these papers effectively brought the topic of adaptive physiology of *Peromyscus* up to the present and because the purpose of this volume is to update information accrued on the biology of *Peromyscus* since 1968, we were faced with the dilemma of merely reiterating a topic already updated and addressed in detail, or of attempting a new and different approach based on some of the same information. We elected the latter and will confine our comments to consideration of two metabolic variables that are of major regulatory importance and hence should be subject to environmentally related selective pressures. These are: 1) basal (or standard) metabolic rate (BMR), a major component of energy metabolism; and 2) evaporative water loss (EWL), the major avenue of water loss of most terrestrial mammals and therefore a major component of water metabolism. We will also consider in detail the information that is available on the interrelationships between energy and water metabolism.

With regard to water regulatory capabilities, MacMillen (1983*a*) concluded that *Peromyscus* were intermediate in nearly all aspects of their water economies compared to other similar-sized rodents and that this intermediacy reflected their intermediate dietary position, omnivory. Because dietary choices dictate qualitatively and quantitatively both water intake and energy intake, we hypothesize that omnivorous *Peromyscus* should be intermediate in energy metabolism when compared to similar-sized rodents having more specialized diets (for example, granivorous, herbivorous, carnivorous). This notion that regulatory processes and diet should be interrelated in mammals is not a new one and is examined and supported in detail by McNab (1980). In addition, because so many features of the lives of animals are related to their sizes (that is, they are allometric; for example, Peters, 1983; Calder, 1984), we will stress the relationships between 1) the above-mentioned metabolic variables, 2) diet, and 3) body size (as expressed through mass) in *Peromyscus* compared to other rodents. Throughout this paper we will adhere to the more traditional taxonomic treatment of Rodentia as presented by Hall (1981), rather than that of Honacki *et al.* (1982). Except for analyses already available in the literature (and as cited below) all of the raw data employed in our analyses are provided along with their sources in Tables 1 and 2.

GEOGRAPHIC DISTRIBUTION OF BODY SIZE

"The genus *Peromyscus* is one of the most widespread and geographically variable of North American rodents . . ." occurring in almost every habitat from arctic America in the north to the tropics of extreme southern Panama, and from the Pacific to the Atlantic coasts (Hall, 1981). Within this distribution two discernible patterns of variation in body mass occur with latitude: 1) an overall negative relationship with larger species in the tropical south and small species at higher latitudes, the converse of Bergmann's rule (Fig. 1 and Table 1); and 2) with the possible exception of *Peromyscus californicus*, body mass within the genus is independent of latitude between about 30° and 50° N latitudes, as is particularly exemplified by populations of *P. maniculatus* that nearly span that range (Fig. 1 and Table 1). We must admit, however, that our body mass data are limited to those species and populations for which physiological data exist and we largely ignore *Peromyscus* from latitudes lower than 30° N. However, inspection of standard specimen measurements from Hall (1981) seems to confirm our contention that *Peromyscus* species are larger in size in tropical and subtropical regions of Central and North America, decreasing in size as one proceeds northward. More data on body mass (a seldom-published, yet the

TABLE 1.—*Basal metabolic rates, body masses, climatic data, and latitude for* Peromyscus *populations.*

Subspecies or population	Body mass (g)	BMR (ml O$_2$/ g.h)	Precipitation (cm)	Mean July temp. (°C)	Desert index (°C/cm)	Latitude (°N)	Source
P. boylii	23.2	2.34	63.0	26.2	0.42	39.8	Mazen and Rudd, 1980
P. californicus							
parasiticus	49.6	1.17	58.7	16.4	0.28	37.9	McNab and Morrison, 1963
insignis	45.5	1.03	64.4	25.2	0.39	34.2	McNab and Morrison, 1963
Coastal	41.3	1.37	33.8	19.3	0.57	33.5	Mazen and Rudd, 1980
P. crinitus							
stephensi	15.9	1.58	12.3	30.1	2.45	36.2	McNab and Morrison, 1963
pergracilis	20.9	1.48	14.8	25.9	1.75	40.3	McNab and Morrison, 1963
crinitus	13.6	1.33					Kenagy and Vleck, 1982
?	23.1[1]	0.92				30.0	McNab, 1968
P. eremicus							
Nevada	21.5	1.48	12.3	30.1	2.45	36.2	McNab and Morrison, 1963
New Mexico	19.9	1.60	23.4	26.3	1.12	31.8	McNab and Morrison, 1963
eremicus	24.2	1.45				33.3	Murie, 1961
fraterculus	17.4	1.56	47.1	23.2	0.49	34.1	MacMillen, 1965
Desert	18.4	1.32	12.1	31.3	2.59	34.2	Hulbert *et al.*, 1985
Intermediate	19.1	1.41	34.8	26.7	0.77	33.8	Hulbert *et al.*, 1985
P. floridanus	30.8	1.68	121.4	27.7	0.23	29.1	Glenn, 1970
P. gossypinus	21.5	1.72	124.7	27.1	0.22	29.6	Glenn, 1970
P. leucopus							
leucopus	21.2	1.66				42.5	Deavers and Hudson, 1981
noveboracensis	26.0	2.2					Hart, 1953
P. maniculatus							
gambeli	19.1	2.04	58.7	16.4	0.28	37.9	McNab and Morrison, 1963
sonoriensis	24.2	1.67	30.2	23.6	0.78	38.0	McNab and Morrison, 1963
gambeli	20.5	2.55	58.7	16.4	0.28	37.9	Murie, 1961
sonoriensis	20.8	2.28				37.3	Murie, 1961
gracilis	17.0	1.8				44.4	Brower and Cade, 1966
nebrascensis	19.0	2.1	46.7	16.8	0.36	45.2	Hayward, 1965
austerus	19.5	2.0	106.8	17.4	0.16	49.2	Hayward, 1965
sonoriensis	20.0	2.0	14.0	23.2	1.66	37.7	Hayward, 1965
artemisiae	23.0	2.0	29.9	21.1	0.71	49.2	Hayward, 1965
oreas	25.0	1.8	145.2	12.2	0.08	49.2	Hayward, 1965
cooledgei	20.8	1.82	7.4	31.1	4.20	28.9	Abbott, 1974
sonoriensis	19.6	1.88	13.3	28.8	2.17	34.6	Abbott, 1974
gambelii	18.5	2.13	32.7	22.1	0.68	33.7	Abbott, 1974
rubidus	19.7	2.26	98.3	15.6	0.16	39.5	Abbott, 1974
austerus	17.3	2.36	62.6	14.0	0.22	48.0	Abbott, 1974
?	16.7	1.74					Kenagy and Vleck, 1982
Chico, CA	17.2	2.56	63.0	26.2	0.42	39.8	Mazen and Rudd, 1980
P. megalops	66.2	1.37				17.6	Musser and Shoemaker, 1965
P. pirrensis	138.0	0.88				9.1	Hill, 1975
P. polionotus	12.0	1.79	135.5	27.3	0.20	29.2	Glenn, 1970
P. sitkensis	28.0	1.65	126.1	13.9	0.11	53.3	Hayward, 1965
P. thomasi	110.8	1.12				17.6	Musser and Shoemaker, 1965
P. truei							
truei	33.2	1.53	30.2	23.6	0.78	38.0	McNab and Morrison, 1963
gilberti	33.3	1.88	58.7	16.4	0.28	37.9	McNab and Morrison, 1963

[1] Judging from the published body weight, this population was probably misidentified as *P. crinitus;* it was not included in the *P. crinitus* species average value.

TABLE 2.—*Basal metabolic rates and body masses of rodent species.*

Species	Body mass (g)	BMR (ml O$_2$/g.h)	Source
	CRICETINAE (28 species)		
Akodon azarae	23.5	1.70	Hayssen and Lacy, 1985
Baiomys taylori	7.3	1.95	Hudson, 1965
Neotoma albigula [1]	161.8	0.735	
High desert	173.0	0.74	Brown and Lee, 1969
Low desert	150.6	0.73	Brown and Lee, 1969
Neotoma cinerea [1]	262.6	0.728	
Coastal	299.0	0.78	Brown and Lee, 1969
Colorado highland	288.9	0.70	Brown and Lee, 1969
California highland	261.0	0.63	Brown and Lee, 1969
High desert	201.3	0.80	Brown and Lee, 1969
Neotoma fuscipes	186.7	0.79	Brown and Lee, 1969
Neotoma lepida [1]	116.4	0.767	
Coastal	138.5	0.72	Brown and Lee, 1969
Intermediate desert	110.3	0.79	Brown and Lee, 1969
Low desert	100.4	0.79	Brown and Lee, 1969
Ochrotomys nuttali	19.5	1.39	Hayssen and Lacy, 1985
Onychomys torridus	19.1	1.55	Whitford and Conley, 1971
Peromyscus boylii	23.2	2.34	Mazen and Rudd, 1980
Peromyscus californicus [1]	45.5	1.19	McNab and Morrison, 1963; Hulbert *et al.*, 1985
Peromyscus crinitus [1]	16.8	1.463	McNab and Morrison, 1963; Kenagy and Vleck, 1982
Peromyscus eremicus [1]	20.1	1.47	Murie, 1961; McNab and Morrison, 1963; Hulbert *et al.*, 1985; MacMillen, 1965
Peromyscus (Podomys) floridanus	30.8	1.68	Glenn, 1970
Peromyscus gossypinus	21.5	1.72	Glenn, 1970
Peromyscus leucopus [1]	23.6	1.93	Hart, 1953; Deavers and Hudson, 1981
Peromyscus maniculatus [1]	19.9	2.058	Murie, 1961; McNab and Morrison, 1963; Hayward, 1965; Brower and Cade, 1966; Abbott, 1974; Mazen and Rudd, 1980; MacMillen, 1965
Peromyscus megalops	66.2	1.37	Musser and Shoemaker, 1965
Peromyscus (Isthmomys) pirrensis	138.0	0.88	Hill, 1975
Peromyscus polionotus	12.0	1.79	Glenn, 1970
Peromyscus sitkensis	28.0	1.65	Hayward, 1965
Peromyscus (Megadontomys) thomasi	110.8	1.12	Musser and Shoemaker, 1965
Peromyscus truei [1]	33.3	1.705	McNab and Morrison, 1963
Reithrodontomys megalotis [1]	8.8	2.455	
megalotis	9.0	2.50	Pearson, 1960
longicaudus	7.9	2.63	Thompson, 1985
ravus	9.5	2.235	Thompson, 1985
Sigmodon alleni	137.8	1.475	Bowers, 1971
Sigmodon fulviventer	137.8	1.505	Bowers, 1971

TABLE 2.—(Continued).

Species	Body mass (g)	BMR (ml O₂/g.h)	Source
Sigmodon hispidus	139.3	1.654	Bowers, 1971
Sigmodon leucotis	128.6	1.450	Bowers, 1971
Sigmodon ochrognathus	115.1	1.340	Bowers, 1971
	MICROTINAE (25 species)		
Apodemus agrarius	21.0	2.27	Deavers and Hudson, 1981
Apodemus flavicolis	30.0	2.28	Deavers and Hudson, 1981
Apodemus sylvaticus	22.0	2.60	Deavers and Hudson, 1981
Arvicola terrestris	97.5	1.16	Deavers and Hudson, 1981
Clethrionomys gapperi [1]	23.4	2.84	
	22.1	2.3	Deavers and Hudson, 1981
	22.9	4.7	Deavers and Hudson, 1981
	23.3	2.08	Deavers and Hudson, 1981
	24.0	3.14	Deavers and Hudson, 1981
	24.9	1.96	Deavers and Hudson, 1981
Clethrionomys glareolus [1]	18.9	2.85	
	17.4	3.2	Deavers and Hudson, 1981
	20.4	2.5	Deavers and Hudson, 1981
Clethrionomys rufocanus	27.5	2.1	Deavers and Hudson, 1981
Clethrionomys rutilus	28.0	2.75	Hayssen and Lacy, 1985
Dicrostonyx groenlandicus [1]	54.0	1.785	
	61.0	1.6	Deavers and Hudson, 1981
	47.0	1.97	Casey *et al.*, 1979
Lemmus sibiricus	64.0	2.47	Casey *et al.*, 1979
Microtus arvalis	23.9	2.5	Deavers and Hudson, 1981
Microtus californicus	44.0	1.55	Bradley, in Wunder, 1975
Microtus longicaudus [1]	33.4	2.185	
	25.3	2.67	Deavers and Hudson, 1981
	41.4	1.70	Kenagy and Vleck, 1979
Microtus mexicanus	29.0	1.63	Bradley, in Wunder, 1975
Microtus minutus	8.7	5.0	Hayssen and Lacy, 1985
Microtus montanus	30.8	2.65	Deavers and Hudson, 1981
Microtus nivalis	32.8	2.47	Hayssen and Lacy, 1985
Microtus ochrogaster [1]	46.6	1.693	
	38.5	2.16	Wunder *et al.*, 1977
	47.4	1.74	Wunder *et al.*, 1977
	54.0	1.18	Bradley, in Wunder, 1985
Microtus oeconomus	32.0	3.15	Casey *et al.*, 1979
Microtus pennsylvanicus [1]	35.3	2.21	
	39.0	1.93	Bradley, in Wunder, 1985
	35.6	2.5	Deavers and Hudson, 1981
	31.2	2.2	Deavers and Hudson, 1981
Microtus pinetorum	25.0	1.98	Bradley, in Wunder, 1985
Microtus richardsoni	51.0	1.74	Bradley, in Wunder, 1985
Microtus townsendi	52.2	1.64	Kenagy and Vleck, 1982
Ondatra zibethicus [1]	873.0	0.883	
	1100.0	0.97	Hart, 1962
	869.0	0.80	Fish, 1979
	649.0	0.88	Fish, 1982
Pitymys pinetorum [1]	24.4	2.58	
	22.5	2.6	Deavers and Hudson, 1981
	26.3	2.56	Deavers and Hudson, 1981

TABLE 2.—*(Continued).*

Species	Body mass (g)	BMR (ml O$_2$/g.h)	Source
	HETEROMYIDAE (21 species)		
Dipodomys agilis	60.6	1.050	Hayssen and Lacy, 1985
Dipodomys deserti	104.7	0.898	Hinds and MacMillen, 1985
Dipodomys merriami	35.8	1.257	Hinds and MacMillen, 1985
Dipodomys microps	55.7	1.080	Hayssen and Lacy, 1985
Dipodomys ordii	46.8	1.372	Hinds and MacMillen, 1985
Dipodomys panamintinus	64.2	1.157	Hinds and MacMillen, 1985
Dipodomys anomalus	69.3	1.450	Hayssen and Lacy, 1985
Heteromys desmarestianus	75.8	1.308	Hinds and MacMillen, 1985
Liomys irroratus	44.9	1.341	Hinds and MacMillen, 1985
Liomys salvini	42.7	1.314	Hinds and MacMillen, 1985
Microdipodops megacephalus	11.0	2.743	Hinds and MacMillen, 1985
Microdipodops pallidus	15.2	1.300	Hayssen and Lacy, 1985
Perognathus baileyi	29.1	1.187	Hinds and MacMillen, 1985
Perognathus californicus	22.0	0.970	Hayssen and Lacy, 1985
Perognathus fallax	19.6	1.370	Hinds and MacMillen, 1985
Perognathus flavus	8.3	2.085	Hinds and MacMillen, 1985
Perognathus hispidus	32.0	1.434	Hinds and MacMillen, 1985
Perognathus intermedius	14.6	1.070	Hayssen and Lacy, 1985
Perognathus longimembris	8.0	1.759	Hinds and MacMillen, 1985
Perognathus parvus	19.2	1.719	Hayssen and Lacy, 1985
Perognathus penicillatus	16.0	1.400	Hayssen and Lacy, 1985

[1]Calculated (unweighted) average value. For *Peromyscus* species, individual subspecies or populations are listed in Table 1.

most revealing size dimension) of species of the genus clearly are needed to delineate and confirm this suspected relationship between body mass and latitude in *Peromyscus.*

BASAL METABOLIC RATES

The basal metabolic rate (BMR) of an endothermic animal is the rate of metabolism of a fasting adult animal at rest in its thermal neutral zone (Bartholomew, 1982*a*). Basal rate has long been known to vary in a predictable manner with body mass (Kleiber, 1932) and rather fixed relationships exist between BMR, rate of increase in metabolism below thermal neutrality, and the lower limit of thermal neutrality, which are functions of insulative capacity and physical principles of heat exchange (Scholander *et al.*, 1950*a*, 1950*b*, 1950*c*). In addition, the daily energy expenditure of an animal in nature is definable as BMR plus additives (specific dynamic action, thermoregulation, activity, production), and thus BMR offers an

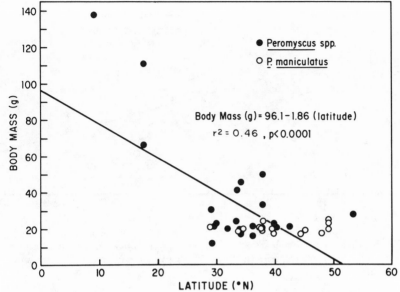

FIG. 1.—Relationship between body mass and latitude in *Peromyscus* populations (N = 39 of which 16 are *P. maniculatus*) for which physiological data are available. The line is fit to the data by least-squares regression analysis. The highly significant regression is due to the inclusion of the three largest species.

essential baseline from which to evaluate the rates of energy metabolism that actually exist under natural conditions of stress or activity (Bartholomew, 1982*a*). As such, BMR is a meaningful index of the metabolic or energetic intensity at which an endotherm lives.

In order to determine the relation of BMR of *Peromyscus* to that of other rodents, we have examined this variable in detail in the murid rodent subfamily Cricetinae to which *Peromyscus* belongs, as well as in the subfamily Microtinae and the family Heteromyidae (Table 2). In many North American habitats representatives of at least two of these three taxa co-occur, resulting in potentially competitive interactions. Figure 2 depicts the allometric relationship between BMR and body mass in these three taxonomic groups. The regression lines are from analysis of covariance, which showed that slopes did not differ significantly among taxa (P > 0.063), but all intercepts (elevations) did differ significantly (P < 0.009). Using a pooled slope of -0.266 ± 0.057 ($\pm$ 95 percent confidence interval), intercepts are 5.62 for Microtinae, 3.93 for Cricetinae, and

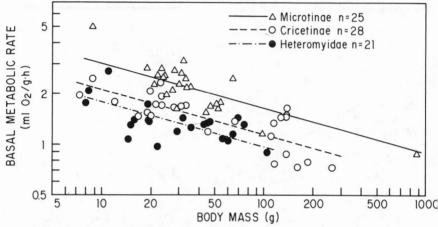

FIG. 2.—Double logarithmic plot of mass-specific basal metabolic rate regressed on body mass in three rodent taxa. Each point represents an average value from the literature for a single species (see Table 2). Regression lines are from analysis of covariance.

3.32 for Heteromyidae. Separate allometric equations for the three taxa are [BMR (ml O_2/g.h) = Body Mass $(g)^b$]:

Microtinae, $7.61 \div/\times 1.452 \ M^{-0.351 \pm 0.101}$, $r^2 = 0.69$

Cricetinae, $3.59 \div/\times 1.402 \ M^{-0.243 \pm 0.085}$, $r^2 = 0.57$

Heteromyidae, $2.74 \div/\times 1.564 \ M^{-0.210 \pm 0.129}$, $r^2 = 0.38$

Coefficients are followed by 95 percent confidence intervals. The equation for the 14 species of *Peromyscus* (n = 14, using species means from Table 2), is $4.21 \div/\times 1.645 \ M^{-0.288 \pm 0.141}$, $r^2 = 0.62$.

It is apparent (Fig. 2) that the metabolic intensity of cricetines is intermediate to that of the largely granivorous heteromyids, which is lower, and the nearly exclusively herbivorous microtines, which is higher. Although most of the cricetine BMRs are for *Peromyscus* species, which are omnivorous, also included are representatives of several other genera: *Onychomys* (carnivorous-insectivorous), *Baiomys* and *Reithrodontomys* (probably also omnivorous), and *Neotoma* and *Sigmodon* (herbivorous) (Fig. 3). Thus, either within the genus *Peromyscus* or cricetine genera collectively, most of the dietary items sought by rodents are included in the diets of cricetines. If a relationship exists between diet and BMR, the allometric relationship for BMR of cricetines should not differ from that of rodents in general, regardless of taxonomic affinity; such in fact is the case (Fig. 3). It is interesting to note, however, that of the two chiefly

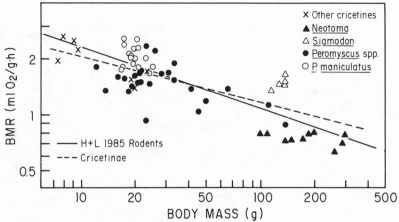

FIG. 3.—Double logarithmic plot of BMR regressed on body mass for species or populations of cricetine rodents. Solid regression line is from Hayssen and Lacy's (1985) equation for 122 species of rodents. Dashed line is the regression for 28 species of Cricetinae (see text).

herbivorous cricetine genera depicted, one (*Neotoma*) has species whose BMRs are rather low and similar to those of heteromyids, whereas the other (*Sigmodon*) apparently exhibits BMRs that are unusually high and similar to those of microtines (Figs. 2, 3). Nevertheless the BMRs of the omnivorous *Peromyscus* are intermediate to those of the dietarily more specialized rodents with whom they co-occur and potentially interact (Fig. 3).

The *Peromyscus* populations for which BMR data are available divide readily into two groups: 1) *Peromyscus maniculatus*, with significantly (P < 0.0001) elevated BMRs, and 2) other *Peromyscus* species with lower BMRs (Fig. 4, Table 1). The regression lines are from analysis of covariance, which showed that slopes did not differ significantly for *P. maniculatus* populations (n = 17) versus other *Peromyscus* populations (n = 25) but the former had significantly higher BMRs (*P* = 0.0001). Using a pooled slope of −0.208 ± 0.121 (± 95 percent confidence interval), intercepts are 3.80 for *P. maniculatus* and 2.93 for other *Peromyscus*. The equation for all *Peromyscus* populations (n = 42) is:

$$BMR = 4.38 \div / \times 1.549 \, M^{-0.301 \pm 0.135}, \, r^2 = 0.34$$

The relevance of these intergroup differences in terms of ecological energetics and diet are unclear to us, but *P. maniculatus*, the most ubiquitous species of the genus, appears to operate at a level

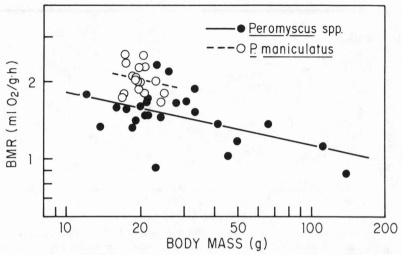

Fig. 4.—Double logarithmic plot of BMR regressed on body mass in *Peromyscus*. Regression lines are from analysis of covariance.

of metabolic intensity that exceeds that of most other species of the genus.

In addition to the strong relationship between body mass and BMR in rodents, climatic conditions, particularly aridity, have been postulated as ultimate (evolutionary) determinants of reduced BMR, especially in burrowing rodents (for example, Hinds and MacMillen, 1985; MacMillen and Lee, 1970; McNab, 1966). McNab and Morrison (1963), in particular, have contended that desert-dwelling *Peromyscus* have reduced BMRs, proposing that the extent of physiological adaptation to desert conditions in *Peromyscus* (the sum of the percentage reduction from allometric expectations of surface-specific thermal conductance and BMR) is sensitive to mean annual rainfall and mean July (summer) temperature, either separately or collectively. As a collective index of aridity McNab and Morrison (1963) proposed the use of a Desert Index [= mean July temperature (°C)/mean annual rainfall (cm)]. To determine the relative influences of body mass and geographic or climatic factors, or both on BMR, we conducted a multiple regression analysis of BMR on body mass, mean annual precipitation, mean July temperature, the McNab-Morrison Desert Index, and latitude for 31 populations representing 10 species of *Peromyscus*, from localities between 29° and 53° N latitude and with body mass between 12 and

TABLE 3.—*Alternative multiple regression equations for 31* Peromyscus *populations (data from Table 1).*

Dependent variable	Independent variables (% of variance explained) (significance level)
$\text{Log}_{10}\text{BMR} = -\text{Log}_{10}\text{Body mass (28.4)} + \text{Precipitation (3.8)}$	
(0.0016) (0.2193)	
$\text{Log}_{10}\text{BMR} = -\text{Log}_{10}\text{Body mass (28.4)} - \text{Temperature (20.8)}$	
(0.0001) (0.0021)	
$\text{Log}_{10}\text{BMR} = -\text{Log}_{10}\text{Body mass (28.4)} - \text{Desert index (11.1)}$	
(0.0004) (0.0315)	
$\text{Log}_{10}\text{BMR} = -\text{Log}_{10}\text{Body mass (28.4)} + \text{Latitude (11.2)}$	
(0.0009) (0.0305)	

50 grams (Table 1). Precipitation and temperature data were taken from climatic summaries for the weather station nearest to the collecting locality, as given in the following sources: British Columbia Dept. Agriculture (1975), U.S. Dept. Agriculture (1941), and Hastings and Humphrey (1969).

As expected, the single most important correlate of BMR in this analysis is body mass, explaining 28.4 percent of the variability (Table 3). After entering body mass into the multiple regression, an additional significant negative correlation accounting for 20.8 percent of the variability exists between mean July temperature and BMR. Also significantly correlated with BMR are the Desert Index (negative) and latitude (positive), each explaining about 11 percent of the variation. The partial correlation between precipitation and BMR is not significant. If body mass and temperature are entered first into a multiple regression, neither the Desert Index nor latitude (nor precipitation) explains a significant amount of the remaining variance in BMR.

Thus energy metabolism in *Peromyscus* as reflected in BMR appears to be sensitive to high summer temperatures, at least within the intermediate latitudinal range represented by available data. That the reduction in BMR with high midsummer temperature is not actually attributable to aridity rather than temperature per se is demonstrated by the weak and nonsignificant correlation with precipitation. The weak correlation with the Desert Index likely is attributable exclusively to the influence of the temperature component in the index. The results of our analysis are in contrast to those of McNab and Morrison (1963), and we emphasize that the adaptive significance of reduced BMR under conditions of high ambient temperature has yet to be elucidated experimentally. It should be stressed that our analysis and most other studies of relationships between metabolic characteristics and climatic factors have employed only coarse meteorological variables, which may not accurately reflect actual microclimatic conditions encountered by the animals (Gates, 1980).

EVAPORATIVE WATER LOSS

The major avenue of water loss in rodents is by evaporation from the lungs and skin, termed evaporative water loss (EWL; Chew, 1951; Schmidt-Nielsen and Schmidt-Nielsen, 1951). Measurements of EWL in *Peromyscus* are limited to only a few species and are equivalent to, or higher than, those for other rodents of similar size (MacMillen, 1983*a*). Moderate to rather high rates of EWL in *Peromyscus* are in keeping with their omnivorous diet, which includes a high intake of succulent foods (herbage, fruit, insects), so long as such foods can be found, thereby promoting both positive energy and water balance. The available data on EWL in *Peromyscus* are plotted for comparison with the more extensive data for other mammalian groups in Figure 5. Only measurements made between ambient temperatures of 5° and 25° C are used, because it is within this temperature range that EWL remains nearly constant (Hinds and MacMillen, 1985). The regression line for Eutheria relating mass-specific EWL to body mass is based on 23 species and 40 populations (including three species and five populations of *Peromyscus*), ranging in mass from 16 grams (pallid bat, *Antrozous pallidus*) to 3630 kilograms (Asian elephant, *Elephas maximus*) (from Chew, 1965, as computed by Hinds and MacMillen, 1985). The regression line for heteromyid rodents represents 13 species ranging from 8 to 105 grams (Hinds and MacMillen, 1985). Obviously heteromyids have unusually low rates of EWL, whereas *Peromyscus* species have unusually high rates of EWL, higher even than might be expected given their omnivorous diet. We must emphasize, however, that the data available for *Peromyscus* are largely for populations of *P. maniculatus*, whose BMRs are unusually elevated for *Peromyscus* (Fig. 4). Clearly many more data are required for a variety of species and sizes of *Peromyscus* before meaningful generalizations and comparisons may be made concerning mass-specific rates of evaporative water loss in this group of mammals. Because pulmonary water loss and oxygen consumption ($\dot{V}O_2$) are inter-related in the respiratory process, another commonly-used expression is the ratio of EWL to $\dot{V}O_2$, when the two variables are measured simulta-

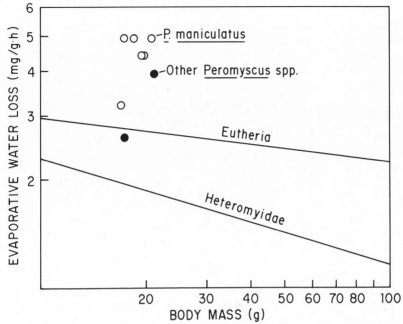

FIG. 5.—Relationship between evaporative water loss (expressed as a function of body mass) and body mass in *Peromyscus* and other mammals. The *Peromyscus* species are given in Table 4. The regression lines are fit to the data for eutherian mammals (EWL = 3.90 g $^{-0.122}$) given by Chew (1965), as computed by Hinds and MacMillen (1985), and for heteromyid rodents (EWL = 4.51 g $^{-0.295}$) given by Hinds and MacMillen (1985).

neously and across a broad range of ambient temperature (T_a). The regression line (Fig. 6) calculated by MacMillen and Grubbs (1976) for the relation between EWL to $\dot{V}O_2$ ratio and T_a for 23 species representing four families of rodents, which ranged broadly across the spectrum of dietary specialization, included only a single species of *Peromyscus* (*P. eremicus*). For comparison, a separate regression line fit to the data now available for *Peromyscus* (*P. eremicus*, *P. leucopus*, and six subspecies of *P. maniculatus*) is also shown in Figure 6. Again, *Peromyscus* have rather high rates of EWL, about 35 percent greater than those of other rodents. At moderate to high temperatures, the rates of EWL in *P. maniculatus* subspecies are conspicuously higher than those in the other two *Peromyscus* species. If there were a fixed ratio of EWL to $\dot{V}O_2$ for all rodents, the data points should cluster around the regression line for combined rodents in Figure 6. That *Peromyscus* generally fall on or above that

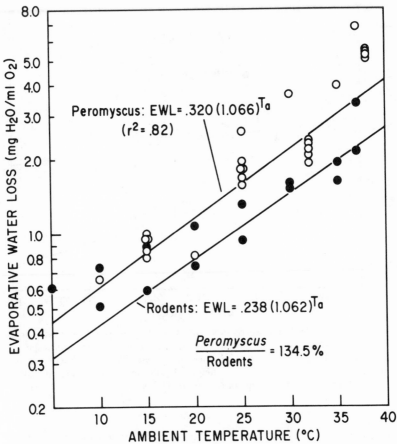

F IG. 6.—Relationship between evaporative water loss (expressed as a function of oxygen consumption) and ambient temperature in *Peromyscus* and other rodents. The regression line for *Peromyscus* is fit by the method of least squares; that for other rodents is from MacMillen and Grubbs (1976). The hollow circles represent measurements for subspecies of *P. maniculatus*, and the filled circles for other *Peromyscus* species as indicated in Table 4.

line is indicative that not only may VO_2 be elevated, as in *P. maniculatus* (Fig. 4), but that EWL must be disproportionately elevated. Because at least half of these populations of *Peromyscus* are from desert or semi-desert habitats, we conclude that among normothermic rodents *Peromyscus* are quite liberal with regard to water expenditures from evaporative routes.

Interrelationships Between Energy and Water Metabolism

Metabolic water production (MWP) represents a major component of the water requirements of at least certain granivorous rodents (Howell and Gersh, 1935). MacMillen (1972) was among the first to point out clearly in rodents that 1) at ambient temperatures (T_a) below thermal neutrality, MWP (as translated directly from measures of oxygen consumption = energy metabolism) is negatively related to T_a; 2) EWL, the chief avenue of water loss, is either positively related to or is independent of T_a; and 3) for each species a T_a exists at which MWP equals EWL, resulting in positive water balance. These relationships were further refined by MacMillen and Grubbs (1976) who analyzed data for all rodents for which simultaneous measures of $\dot{V}O_2$ (translatable into MWP) and EWL existed. They concluded that differences in T_a at which MWP equals EWL did not exist between desert and non-desert species and that every species of rodent possesses the capacity for preformed water independence (that is, exclusive reliance upon MWP) at some moderate to low T_a, depending upon the species and the composition of the energy source being oxidized. More recently, MacMillen and Hinds (1983) applied this concept to representative species of all genera of the chiefly granivorous family Heteromyidae and demonstrated that these dietary specialists on seeds achieve equality of MWP and EWL at relatively high T_as. Heteromyids, therefore, commonly meet their water needs under laboratory conditions from MWP while oxidizing carbohydrate-rich seeds, augmented by the small amount of preformed water they contain. MacMillen and Hinds (1983) defined as an index of water regulatory efficiency as that T_a at which MWP equals EWL, based upon the compositional ratio (protein:lipid:carbohydrate) of the oxidative substrate and the metabolic water yield of that composition. Species with a higher T_a at which MWP equals EWL are more efficient than those with a lower T_a at which MWP equals EWL.

If there is a strict relationship between the degree of dietary specialization and water regulatory efficiency as defined above and as suggested by MacMillen (1983a), we would anticipate that omnivores such as *Peromyscus* would be less efficient than granivorous heteromyids but more efficient than strict herbivores or insectivore-carnivores that subsist on succulent diets. By the same token we would anticipate that water regulatory efficiency in *Peromyscus* would compare closely with that of rodents in general, for the latter represent a broad range of diets (that is, equivalent to omnivory).

The EWL data from Figure 6 for *Peromyscus* can be transformed into the ratio MWP to EWL by assuming oxidation of a standard seed, for example, millet (13.5 percent protein, 5.1 percent lipid, 81.4 percent carbohydrate) from which the consumption of 1 milliliter of oxygen yields 0.62 milligrams of metabolic water (MacMillen and Hinds, 1983). Regression lines for the limited data available for *Peromyscus*, the more extensive data for heteromyids (MacMillen and Hinds, 1983), and data for 23 species in other rodent groups (MacMillen and Grubbs, 1976) are shown in Figure 7. These three groups rank in water regulatory efficiency as follows: heteromyids (T_a at MWP = EWL = 19.8°C) are greater than rodents in general (T_a at MWP = EWL = 16.1°C) which are greater than *Peromyscus* (T_a at MWP = EWL = 10.3°C). That *Peromyscus*

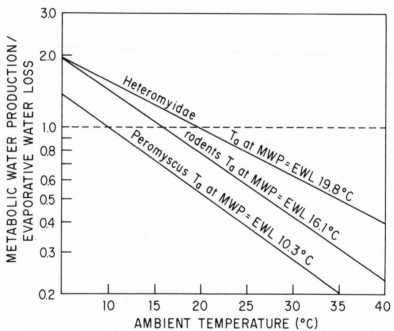

Fig. 7.—Relationship between metabolic water production (MWP)/evaporative water loss (EWL) and ambient temperature (T_a) in heteromyid rodents [MWP/EWL = 2.493(0.955)T_a; MacMillen and Hinds, 1983], rodents in general [MWP/EWL = 2.618(0.942)T_a; MacMillen and Hinds, 1976], and *Peromyscus* species as indicated in Table 4 [MWP/EWL = 1.932(0.938)T_a]. The index of water regulatory efficiency for each group is defined as that T_a at which the regression line intercepts the line of unity (dashed line) at which MWP = EWL.

appear to be less efficient than rodents in general may well be a sampling artifact, as very few dietary specialists on succulent foods were available for rodents in general, and data for the most efficient *Peromyscus, P. crinitus* (MacMillen, 1983*a*), were not available. However, the fact that *Peromyscus* appear to have relatively high rates of EWL (Fig. 5) but intermediate rates of metabolism (Fig. 3) suggests that they may have as a genus low indices of water regulatory efficiency among rodents. Clearly *Peromyscus* are not highly efficient with regard to water regulation while subsisting on a dry-seed diet. In addition, their relative inefficiency is in keeping with their typically omnivorous diet that normally provides the necessary preformed water input to compensate for rather high rates of water loss. As a cautionary note, only heteromyids (MacMillen and Hinds, 1983) have had critical analysis of the relationship between MWP and EWL, requiring simultaneous measurements of $\dot{V}O_2$ and EWL. Although there are measurements of $\dot{V}O_2$ for many rodent species, few investigators have bothered to obtain the easily-made measurements of EWL. In particular, the available data for *Peromyscus* are insufficient for broad comparisons, and simultaneous measurements of $\dot{V}O_2$ and EWL are almost completely lacking for such rodent specialists on succulent diets as *Microtus, Neotoma,* and *Onychomys.*

In heteromyid rodents, MacMillen and Hinds (1983) demonstrated not only an unusually high degree of water regulatory efficiency, but also that the index of water regulatory efficiency (T_a at MWP = EWL) scales negatively with body mass, especially among quadrupedal species. This scaling regression equation for heteromyids is T_a at MWP = EWL = $29.682g^{-0.137}$. Comparisons of water regulatory efficiency of individual *Peromyscus* populations with the heteromyid regression line relating the index of water regulatory efficiency to body mass (Fig. 8, Table 4) demonstrate that even when body mass is accounted for, *Peromyscus* are still far less efficient in water regulation than heteromyids. The relative inefficiency with regard to water regulation in *Peromyscus* is even more apparent when the individual regression lines relating the quotient of MWP and EWL to T_a in *Peromyscus* are compared to those of heteromyids (Fig. 9, Table 4). Even the most efficient *Peromyscus* (a desert population) is a poorer water regulator than the least efficient heteromyid (a tropical population). We must reiterate, however, that data on water regulatory efficiency are not available for *P. crinitus*, whose ability to tolerate water deprivation while subsist-

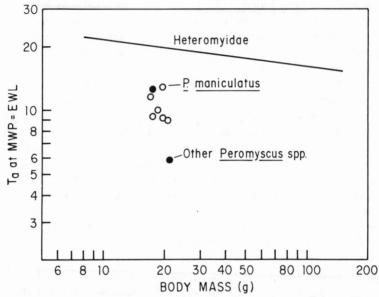

FIG. 8.—Relationship between the index of water regulatory efficiency (T_a at MWP = EWL) and body mass in heteromyid rodents (solid line; based on Mac-Millen and Hinds, 1983) and in species of *Peromyscus* listed in Table 4.

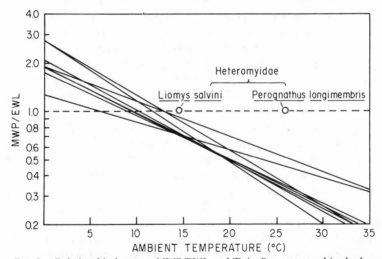

FIG. 9.—Relationship between MWP/EWL and T_a in *Peromyscus* and in the least efficient (*L. salvini*) and most efficient (*P. longimembris*) heteromyid rodents. Equations for the regression lines fit to the *Peromyscus* data are given in Table 4. The horizontal dashed line represents unity between MWP and EWL. Hollow circles represent the indices of water regulatory efficiency (T_a at MWP = EWL) for the two heteromyids.

TABLE 4.—*Body mass, index of water regulatory efficiency (T_a when MWP = EWL), and regression equations relating MWP/EWL to T_a in Peromyscus and two heteromyid species (see Figs. 8 and 9).*

Species	Body mass (g)	T_a when MWP=EWL (°C)	Regression equation	Source
CRICETIDAE				
Peromyscus eremicus	17.4	12.7	MWP/EWL=1.887(0.951)^{T_a}	MacMillen, 1965
P. leucopus	22.1	5.7	MWP/EWL=1.247(0.962)^{T_a}	Deavers and Hudson, 1981
P. maniculatus austerus	17.3	9.4	MWP/EWL=1.849(0.937)^{T_a}	Abbott, 1974
P. m. cooledgei	20.8	9.0	MWP/EWL=1.749(0.940)^{T_a}	Abbott, 1974
P. m. gambelii	18.5	10.0	MWP/EWL=2.040(0.931)^{T_a}	Abbott, 1974
P. m. gracilis	17.0	11.6	MWP/EWL=2.774(0.916)^{T_a}	Brower and Cade, 1966
P. m. rubidus	19.7	9.2	MWP/EWL=1.842(0.936)^{T_a}	Abbott, 1974
P. m. sonoriensis	19.6	12.9	MWP/EWL=2.776(0.924)^{T_a}	Abbott, 1974
HETEROMYIDAE				
Liomys salvini	42.7	14.4	MWP/EWL=2.021(0.952)^{T_a}	MacMillen and Hinds, 1983
Perognathus longimembris	8.0	25.8	MWP/EWL=4.521(0.942)^{T_a}	MacMillen and Hinds, 1983

ing on air-dry millet far exceeds that of other *Peromyscus* that have been studied and is equivalent to that of some of the more efficient heteromyids (MacMillen, 1983*a*).

In summary, in regard to water regulatory efficiency as herein defined, *Peromyscus* species appear to be rather inefficient compared to rodents in general. Their lack of well-developed physiological capacities for efficient water regulation is compensated for by an omnivorous diet which ensures adequate inputs of both energy and water. However, because of insufficient data for species that subsist on more succulent diets, it remains to be positively demonstrated that *Peromyscus* are actually intermediate among rodents in general with regard to water regulatory efficiency.

CONCLUSIONS

With respect to metabolic rate and evaporative water loss, both being physiological variables useful in gauging levels of regulatory efficiencies, *Peromyscus* are rather unspectacular among rodents, being neither extremely efficient nor extremely inefficient. This apparent physiological intermediacy is consistent with an omnivorous diet that provides substantial flexibility in meeting both energy and water needs, even under restrictive situations imposed by climate, competitive interactions, or both. It likely is this flexibility that helps explain the paradox of the ubiquity in North America of this unspecialized genus, whose species inhabit virtually every terrestrial situation regardless of degree of environmental rigor and often co-occur with other rodents (for example, *Dipodomys* and *Perognathus* spp.) whose physiological specializations are credited for their survival under demanding circumstances. In this regard *Peromyscus* are spectacular examples of the success of physiologically unspectacular animals, demonstrating that such generalists are as capable of coping with rigorous circumstances as are rodents with greater physiological specializations.

We believe the success of *Peromyscus* in inhabiting harsh environments in the absence of apparent specializations may be attributed to their omnivorous diets and their employment of torpor to escape, either seasonally or for shorter times, periods of environmental stress. At least among the smaller species of *Peromyscus* (less than 40 grams) torpor provides temporary relief from demanding circumstances related either to energy or water regulation, and its utility is discussed in some detail in Hill (1983) and MacMillen (1983*a*). Torpor has been viewed historically as a physiological specialization in the extreme, but its common occurrence among smaller mammalian and avian species of widely separated taxa (McNab, 1983) argues that it more likely is a general endothermic phenomenon associated with small body size. Torpor differs in various endothermic species both with regard to temporal patterns and depths of hypothermia (Bartholomew, 1982*b*), and the degree of complexity of this patterning we believe represents the specialization of this more general phenomenon. Torpor in *Peromyscus* appears to be of a very non-specialized nature, being basically limited to successive diurnal bouts and involving only moderate hypothermia; nevertheless, it provides effective relief from otherwise intolerable conditions (Hill, 1983; MacMillen, 1983*a*).

McNab (1983) proposed the concept of a minimal boundary curve for endothermy that relates BMR to body mass, with the notion that species falling below that curve frequently employ torpor. Among rodents, with the exception of microtines, nearly all species with masses less than 40 grams for which data exist have BMRs that fall below this minimal boundary curve. However, an equally valid and biologically reasonable interpretation is that in rodents, with the exception of microtines, a body mass threshold exists at about 40 grams, below which species may commonly employ torpor and above which torpor is far less commonly employed. In these small species having high mass-relative energy and water needs while normothermic, ecologically stressful periods may occur frequently enough to result in morbidity unless survival alternatives exist; torpor (or hypothermia) is such an alternative, reducing energy and water requirements to levels consistent with their availability during temporarily stressful periods.

The concept of a size-related threshold for torpor has been applied effectively to heteromyid rodents (MacMillen, 1983*b*; MacMillen and Hinds, 1983). Herein we apply the same concept to *Peromyscus*. We believe that larger (greater than 40 grams) *Peromyscus* species abound in subtropical and tropical regions of greater resource (food and water) availability throughout the year, where normothermic survival is accompanied by few risk-related constraints with regard to energy and water regulation. In proceeding from tropical to subtropical settings with increasing seasonality, body mass of *Peromyscus* decreases consistently with seasonally-determined levels of energy and water production, assuming normothermia and physiological generalization. At about 30° N latitude, continuous normothermia becomes locally incompatible with seasonally-stressful periods (aridity, low temperatures, or both) in spite of reductions in body mass to less than 40 grams and accompanying reductions in absolute energy and water needs. At this point torpor provides the necessary physiological relief to ensure survival. Once the size threshold for torpor has been achieved, environmental conditions leading to energy and water stresses can be ameliorated by torpor without further reduction in body mass. This view is consistent with the apparent relationship between body mass and latitude in *Peromyscus* (Fig. 1), and with the ubiquity of this biologically generalized genus throughout North America. We propose this view not as a proven fact, but as an internally consistent hypothesis whose testing should bear fruitful results.

A question arises, however, concerning the positioning of this size threshold for torpor in both *Peromyscus* and heteromyids at about 40 grams, particularly considering the frequent use of torpor by sciurids that exceed (often considerably) 40 grams in mass. We have no answer based on experimental evidence to this question, but propose that the difference lies in the capacity of sciurids to store fat, often to the point of obesity, as an energy reserve to draw upon during the torpor period. *Peromyscus* and heteromyids lack this conspicuous fat storage capacity and must rely upon lower-energy food stores in or near nest chambers. These relationships between body mass, use of torpor, and quantity, quality, and site of storage of energy reserves drawn upon during the torpor period merit further investigation.

Finally, we should emphasize that, because of their generalized biological characteristics and success in occupying a wide range of habitats, *Peromyscus* are ideal mammalian models to employ in investigations of many of the questions in physiological ecology that remain unsolved. Frequently such questions have arisen in studies of more specialized rodents whose degrees of specialization together with frequent rarity and limited distributions make them less tractable for study. Following are some of the problems that could be addressed profitably by employing *Peromyscus* as model study organisms. First, what is the relationship between basal metabolic rate, as commonly measured in laboratory studies, and daily energy expenditures (DEE) in the field, as determined using doubly-labelled water techniques? Do species with a high BMR, such as *P. maniculatus*, necessarily have a high DEE? Conversely, does a low BMR necessarily translate into low total energy requirements in nature? Similarly, what is the relationship between "basal" evaporative water loss and actual field water-turnover rates? What is the functional significance of basal metabolic rate? Does a correlation, perhaps indicating a causal or at least a permissive relationship, exist between BMR and 1) maximal rates of oxygen consumption, which would have consequences for locomotor endurance; 2) summit metabolism, which would have consequences for thermoregulatory abilities; 3) rates of evaporative water loss; 4) reproductive and growth rates; 5) ability to detoxify plant secondary compounds or invertebrate toxins encountered in the diet. We suggest that examining such correlations in relation to ecological, life history, and phylogenetic factors would be fruitful not only in increasing our understanding of the biology of *Peromyscus* but of small mammals and their adaptive physiologies in general.

ACKNOWLEDGMENTS

Many of the ideas expressed in this paper were first developed in studies of heteromyid rodents under NSF grants DEB-7620116 and DEB-7923808, and it is gratifying to observe that they apply equally well to *Peromyscus*. The ideas have been refined through countless discussions with D. S. Hinds, whom we thank. Participation in the symposium on The Biology of *Peromyscus* was made possible by a grant from the School of Biological Sciences Faculty Research Committee, University of California, Irvine. Thanks also go out to Hi Ho Sai Gai and G. I. Nebra for continuing aid. The editors contributed substantially to the clarity of expression of our ideas, for which we are grateful.

LITERATURE CITED

ABBOTT, K. D. 1974. Ecotypic and racial variation in the water and energy metabolism of *Peromyscus maniculatus* from the western United States and Baja California, Mexico. Unpubl. Ph.D. dissert., Univ. California, Irvine, 155 pp.

BARTHOLOMEW, G. A. 1982a. Energy metabolism. Pp. 46–93, *in* Animal physiology, principles and adaptations (Fourth ed.) (M. S. Gordon, ed.). Macmillan Publ. Co., New York, 635 pp.

———. 1982b. Body temperature and energy metabolism. Pp. 333–406, *in* Animal physiology, principles and adaptations (Fourth ed.) (M. S. Gordon, ed.). Macmillan Publ. Co., New York, 635 pp.

BOWERS, J. R. 1971. Resting metabolic rate in the cotton rat: *Sigmodon*. Physiol. Zool., 44:137–147.

BRITISH COLUMBIA DEPARTMENT OF AGRICULTURE. 1975. Climate of British Columbia. 85 pp.

BROWER, J. E., AND T. J. CADE. 1966. Ecology and physiology of *Napaeozapus insignis* (Miller) and other woodland mice. Ecology, 47:46–63.

BROWN, J. H., AND A. K. LEE. 1969. Bergmann's rule and climatic adaptation in woodrats (*Neotoma*). Evolution, 23:329–338.

CALDER, W. A., III. 1984. Size, function and life history. Harvard Univ. Press, Cambridge, Massachusetts, 431 pp.

CASEY, T. M., P. C. WITHERS, AND K. K. CASEY. 1979. Metabolic and respiratory responses of Arctic mammals to ambient temperature during the summer. Comp. Biochem. Physiol., 64A:331–341.

CHEW, R. M. 1951. The water exchanges of some small mammals. Ecol. Monogr., 21:215–225.

———. 1965. Water metabolism in mammals. Pp. 43–178, *in* Physiological mammalogy. Vol. 2 (W. V. Mayer and R. G. Van Gelder, eds.). Academic Press, New York, 326 pp.

DEAVERS, D. R., AND J. W. HUDSON. 1981. Temperature regulation in two rodents (*Clethrionomys gapperi* and *Peromyscus leucopus*) and a shrew (*Blarina brevicauda*) inhabiting the same environment. Physiol. Zool., 54:94–108.

FISH, F. E. 1979. Thermoregulation in the muskrat (*Ondatra zibethicus*): the use of regional heterothermia. Comp. Biochem. Physiol., 64A:391–397.

———. 1982. Aerobic energetics of surface swimming in the muskrat *Ondatra zibethicus*. Physiol. Zool., 55:180–189.

GATES, D. M. 1980. Biophysical ecology. Springer-Verlag, New York, 611 pp.

GLENN, M. E. 1970. Water relations in three species of deer mice (*Peromyscus*). Comp. Biochem. Physiol., 33:231–248.

HALL, E. R. 1981. The mammals of North America. John Wiley & Sons, New York, 1:xv + 1–600 + 90 and 2:vi + 601–1181 + 90.

HART, J. S. 1953. Energy metabolism of the white-footed mouse, *Peromyscus leucopus leucopus*, after acclimation at various environmental temperatures. Canadian J. Zool., 31:99–105.

———. 1962. Mammalian cold acclimation. Pp. 203–213, *in* Comparative physiology of temperature regulation. Vol. 2 (P. Hannon and E. Viereck, eds.). Arctic Aeromedical Laboratory, Fort Wainwright, Alaska, 278 pp.

HASTINGS, J. R., AND R. R. HUMPHREY. 1969. Climatological data and statistics for Baja California. Tech. Rept. on Meteorology and Climatology, No. 18. Univ. Arizona Inst. Atmospheric Physics, 96 pp.

HAYSSEN, V., AND R. C. LACY. 1985. Basal metabolic rates in mammals, taxonomic differences in the allometry of BMR and body mass. Comp. Biochem. Physiol., 81A:741–754.

HAYWARD, J. S. 1965. Metabolic rate and its temperature-adaptive significance in six geographic races of *Peromyscus*. Canadian J. Zool., 43:309–323.

HILL, R. W. 1975. Metabolism, thermal conductance, and body temperature in one of the largest species of *Peromyscus*, *P. pirrensis*. J. Thermal Biology, 1:109–112.

———. 1983. Thermal physiology and energetics of *Peromyscus*; ontogeny, body temperature, metabolism, insulation, and micro-climatology. J. Mamm., 64:19–37.

HINDS, D. S., AND R. E. MACMILLEN. 1985. Scaling of energy metabolism and evaporative water loss in heteromyid rodents. Physiol. Zool., 58:282–298.

HONACKI, J. H., K. E. KINMAN, AND J. W. KOEPPL (EDS.). 1982. Mammal species of the world. Allen Press and Assoc. Syst. Collections, Lawrence, Kansas, 694 pp.

HOWELL, A. B., AND I. GERSH. 1935. Conservation of water by the rodent *Dipodomys*. J. Mamm., 16:1–9.

HUDSON, J. W. 1965. Temperature regulation and torpidity in the pygmy mouse, *Baiomys taylori*. Physiol. Zool., 38:243–254.

HULBERT, A. J., D. S. HINDS, AND R. E. MACMILLEN. 1985. Minimal metabolism, summit metabolism and plasma thyroxine in rodents from different environments. Comp. Biochem. Physiol., 81A:687–693.

KENAGY, G. J., AND D. VLECK. 1982. Daily temporal organization of metabolism in small mammals: adaptation and diversity. Pp. 322–338, *in* Vertebrate circadian systems (J. Aschoff, S. Dean, and G. Groos, eds.). Springer-Verlag, Berlin, 363 pp.

KING, J. A. (ED.). 1968. Biology of *Peromyscus* (Rodentia). Spec. Publ., Amer. Soc. Mamm., 2:1–593.

KLEIBER, M. 1932. Body size and metabolism. Hilgardia, 6:315–353.

MACMILLEN, R. E. 1965. Aestivation in the cactus mouse, *Peromyscus eremicus*. Comp. Biochem. Physiol., 16:227–248.

———. 1972. Water economy of nocturnal desert rodents. Symp. Zool. Soc. Lond., 31:147–174.

———. 1983*a*. Water regulation in *Peromyscus*. J. Mamm., 64:38–47.

———. 1983*b*. Adaptive physiology of heteromyid rodents. Great Basin Nat. Memoirs, 7:65–76.

MACMILLEN, R. E., AND D. E. GRUBBS. 1976. Water metabolism in rodents. Pp. 63–69, *in* Progress in animal biometeorology. Vol. 1, Part 1 (D. H. Johnson, ed.). Swetz and Zeitlinger, Lisse, The Netherlands, 603 pp.

MACMILLEN, R. E., AND D. S. HINDS. 1983. Water regulatory efficiency in heteromyid rodents: a model and its application. Ecology, 64:152–164.

MACMILLEN, R. E., AND A. K. LEE. 1970. Energy metabolism and pulmocutaneous water loss of Australian hopping mice. Comp. Biochem. Physiol., 35:355– 369.

MAZEN, W. S., AND R. L. RUDD. 1980. Comparative energetics in two sympatric species of *Peromyscus*. J. Mamm., 61:573–574.

MCNAB, B. K. 1966. The metabolism of fossorial rodents: a study in convergence. Ecology, 47:712–733.

———. 1968. The influence of fat deposits on the basal rate of metabolism in desert homoiotherms. Comp. Biochem. Physiol., 26:337–343.

———. 1980. Food habits, energetics, and the population biology of mammals. Amer. Nat., 116:106–124.

———. 1983. Energetics, body size, and the limits to endothermy. J. Zool., 199:1–29.

MCNAB, B. K., AND P. MORRISON. 1963. Body temperature and metabolism in subspecies of *Peromyscus* from arid and mesic environments. Ecol. Monogr., 33:63–82.

MURIE, M. 1961. Metabolic characteristics of mountain, desert and coastal populations of *Peromyscus*. Ecology, 42:723–740.

MUSSER, G. G., AND V. H. SHOEMAKER. 1965. Oxygen consumption and body temperature in relation to ambient temperature in the Mexican deer mice, *Peromyscus thomasi* and *P. megalotis*. Occas. Papers Mus. Zool., Univ. Mich., 643:1–15.

PEARSON, O. P. 1960. The oxygen consumption and bioenergetics of harvest mice. Physiol. Zool., 33:152–160.

PETERS, R. H. 1983. The ecological implications of body size. Cambridge Univ. Press, Cambridge, 324 pp.

SCHMIDT-NIELSEN, B., AND K. SCHMIDT-NIELSEN. 1951. A complete account of the water metabolism in kangaroo rats and an experimental verification. J. Cell. Comp. Physiol., 38:165–182.

SCHOLANDER, P. F., R. HOCK, V. WALTERS, AND L. IRVING.   1950*a*.   Adaptations to cold in arctic and tropical mammals and birds in relation to body temperature, insulation and basal metabolic rate.   Biol. Bull., 99:259–271.

SCHOLANDER, P. F., R. HOCK, V. WALTERS, F. JOHNSON, AND L. IRVING.   1950*b*. Heat regulation in some arctic and tropical mammals and birds.   Biol. Bull., 99:236–258.

SCHOLANDER, P. F., V. WALTERS, R. HOCK, AND L. IRVING.   1950*c*.   Body insulation of some arctic and tropical mammals and birds.   Biol. Bull., 99:225–236.

THOMPSON, S. D.   1985.   Subspecific differences in metabolism, thermoregulation, and torpor in the western harvest mouse *Reithrodontomys megalotis*. Physiol. Zool., 58:430–444.

U.S. DEPARTMENT OF AGRICULTURE.   1941.   Climate and Man.   U.S. Govt. Printing Office, 1248 pp.

WHITFORD, W. G., AND M. I. CONLEY.   1971.   Oxygen consumption and water metabolism in a carnivorous mouse.   Comp. Biochem. Physiol., 40A: 797–803.

WUNDER, B. A.   1975.   A model for estimating metabolic rate of active or resting mammals.   J. Theor. Biol., 49:345–354.

———.   1985.   Energetics and thermoregulation.   Pp. 812–844, *in* Biology of New World Microtus (R. H. Tamarin, ed.).   Spec. Publ., Amer. Soc. Mamm., 8:1–892.

WUNDER, B. A., D. S. DOBKIN, AND R. D. GETTINGER.   1977.   Shifts of thermogenesis in the prairie vole (*Microtus ochrogaster*), strategies for survival in a seasonal environment.   Oecologia, 29:11–26.

REPRODUCTION AND DEVELOPMENT

John S. Millar

Abstract.—Layne (1968) reviewed the literature on *Peromyscus* in order ". . . to obtain some indication of the extent to which growth and developmental patterns and associated aspects of reproduction reflect ecological and phylogenetic relationships." He concluded that much of the variation in growth and development could be interpreted in terms of body size and environmental relationships. This review, based on more than twice as many studies, examines prenatal growth, postnatal growth and development, litter size, weaning, frequency of reproduction, duration of the breeding season, and maturation among and within species. Layne's conclusions are generally supported. However, intraspecific trends are unclear. This may be related in part to the fact that many data have been recorded incidental to other studies, but also to the extremely conservative nature of developmental patterns. Interspecific patterns appear dominated by body size. Future advances in our understanding of reproductive and developmental patterns will depend upon more detailed studies being conducted. Laboratory studies need to be combined with field studies; specific hypotheses need to be formulated and tested. Studies of intrapopulational variation in reproductive and developmental traits are particularly important.

Layne (1968) reviewed ontogeny within the genus *Peromyscus*, with the primary objective being ". . . to obtain some indication of the extent to which growth and developmental patterns and associated aspects of reproduction reflect ecological and phylogenetic relationships. . . ." To do so, he compiled data from 83 literature sources, as well as personal observations. He found pertinent data for only 10 of 58 then-recognized species (now 49 species, Honacki *et al.*, 1982). Despite limited data, he was able to conclude that much of the variation in growth and development could be interpreted in terms of body size and environmental relationships. Large-bodied species exhibited slower growth and development than small-bodied species, although this relationship was weak; major differences in body size were associated with minor differences in development. Species occupying sylvan habitats generally had large litters; pastoral species tended to produce small litters of relatively large neonates. Some pastoral species showed prolonged development, whereas others were relatively precocial. Such trends appeared evident among subspecies as well as among species. He concluded that "the reproductive features of *Peromyscus* are too readily modified in response to environmental conditions to provide a useful criterion of taxonomic relationships."

Since 1968, a considerable number of studies have examined *Peromyscus* in the laboratory and under natural conditions. These include investigations in the tropics (Anderson, 1982; Lackey, 1976, 1978; Rickart, 1977; Robertson, 1975) as well as near the northern limits of the range of the genus (Fuller, 1969; Gilbert and Krebs, 1981; Gyug, 1979; Gyug and Millar, 1981; Krebs and Wingate, 1985; Mihok, 1979; Millar, 1981; Millar and Gyug, 1981; Van Horne, 1981). Several studies have compiled and reviewed specific aspects of reproduction and development in *Peromyscus* (Drickamer and Bernstein, 1972; Glazier, 1979; Millar 1982, 1984*a*; Millar *et al.*, 1979; Modi, 1984). However, Layne's (1968) objectives have not yet been met. We still do not fully understand patterns of variation in reproduction and development among *Peromyscus* populations and species. The objective of this review, therefore, is to reconsider the extent to which reproduction and development varies in this genus and to consider the environmental factors that have shaped this variation.

Comparisons of reproductive and developmental patterns of mammals are complicated because data available from laboratory and field studies may not be comparable and relationships among traits vary with the taxonomic level being considered (Case, 1978; Clutton-Brock and Harvey, 1984; Millar, 1984*b*; Stearns, 1983). For these reasons, the effects of captivity on reproductive and developmental traits and comparisons at different taxonomic levels will be treated separately.

LABORATORY *VERSUS* FIELD STUDIES

Peromyscus are among the easiest of small mammals to maintain in captivity, a fact that has made them popular for laboratory experimentation. Such studies have tended to assume that laboratory stocks are representative of wild populations. This may not be true given that almost all laboratory colonies are derived from limited genetic stock and usually maintained under sterile conditions, limited space, artificial light and temperature, and with foreign, but superabundant food.

Three studies have specifically compared reproduction between wild or wild-caught *Peromyscus* and their laboratory-raised descendants. Williams *et al.* (1965*b*) compared gestation period, litter size, and frequency of pregnancy of *Peromyscus polionotus* among free-living wild mice, wild-caught mice, and first generation laboratory-bred mice. First generation mice maintained a seasonal pattern of breeding similar to that of wild mice, but a smaller proportion of first generation mice than wild-caught mice bred in captivity. Neonates from wild-caught and first generation females were the same weight. However, time between subsequent litters averaged 2.5 days less for first generation females than wild-caught females, and, although not significantly different, litter size tended to be smaller among wild mice than first generation mice. Price (1967) compared reproductive performance between wild-caught *Peromyscus maniculatus bairdii* and captive stock maintained in the laboratory for approximately 25 generations. The semidomesticated mice bred more readily and had a greater rate of litter production than wild caught mice, but showed more cannibalism and desertion of litters than wild caught parents. Variability in litter size and weight at weaning was greater among semidomesticated mice than among wild-caught mice, although mean litter size and weaning weights were not different. Forrester (1975) examined the effects of domestication of *P. m. gambelii* on reproduction and found the pregnancy rate and mean litter size to increase with domestication.

These studies are consistent in indicating that laboratory-raised mice breed more readily than wild-caught mice under laboratory conditions, but do not show consistent trends with regard to rates of development or litter size. Further comparison of the effects of captivity or domestication on reproduction and development of *Peromyscus*, based primarily on incidental reports, shows few statistically significant differences (Table 1). Litter size was largest among

TABLE 1.—*The effects of captivity or domestication on reproduction and development in* Peromyscus.

Population	Characteristic	Change with captivity or domestication	Source
P. maniculatus	Litter size	Increase, more variable*	Price, 1967
P. maniculatus	Weaning weight	More variable*	Price, 1967
P. maniculatus	Cannibalism and desertion	Increase	Price, 1967
P. maniculatus	Litter size	Increase	Forrester, 1975
P. maniculatus	Litter size	Decrease	McCabe and Blanchard, 1950
P. maniculatus	Fat content of young	Increase*	Gyug and Millar, 1980
P. maniculatus	Post-weaning growth	Decrease	Millar and Innes, 1983
P. leucopus	Litter size	Decrease	Millar, 1978
P. leucopus	Fat content of breeding females	Increase	Millar, 1975
P. leucopus	Time spent with young	Increase	Hill, 1972
P. leucopus	Nestling growth	Increase	Hill, 1972
P. leucopus	Litter size	Decrease	Lackey, 1976
P. polionotus	Neonate weight	Decrease	Williams *et al.*, 1965*b*
P. polionotus	Litter size	Increase	Williams *et al.*, 1965*b*
P. floridanus	Litter size	Decrease	Layne, 1966
P. californicus	Litter size	Decrease	McCabe and Blanchard, 1950
P. truei	Litter size	Decrease	McCabe and Blanchard, 1950
P. melanocarpus	Litter size	Decrease	Rickart, 1977
P. mexicanus	Litter size	Decrease	Rickart, 1977
P. yucatanicus	Litter size	Increase	Lackey, 1976

*$P < 0.05$

laboratory mice in four cases and largest among wild or wild-stock mice in eight cases; however, none of the differences was statistically different. The Laboratory-born nestlings may grow faster and be fatter than wild nestlings due to their reduced activity and superabundant food, and postweaned captive mice may grow slower than wild mice due to crowding. In general, there is no evidence that reproduction and development of captive mice differs markedly from that of wild mice, although firm conclusions cannot be reached until more detailed studies are done. In this review, I assume that data from laboratory colonies are applicable to wild populations. Tabulated data (Appendices I through VI) include all studies found prior to February 1985.

DEVELOPMENT

Prenatal Development

Layne (1968) described prenatal development of *P. polionotus*, based primarily on unpublished graduate theses by Smith (1939) and Laffoday (1957). His description still stands as a key reference. Little has been done to investigate prenatal development in *Peromyscus*, with the exception of further documentation of gestation periods (Appendix I). These data must be viewed with some caution because gestation has been measured in a variety of ways among studies. Time of conception has been interpreted from observed matings or vaginal smears, intervals between pairing and parturition, or intervals between parturitions. Svihla (1932) documented gestation periods by pairing mice for 24 hours; his study still provides some of the best data on gestation periods in *Peromyscus*. He found that gestation periods were variable among individuals. He also noted that postpartum pregnancies have longer gestation periods than those of nonlactating mice. Layne (1968) attributed the prolongation of gestation in postpartum pregnancies to delayed implantation, although I could find no verification of this. Myers and Master (1983) found a negative relationship between postpartum weight of females and their subsequent postpartum gestation period.

Recent records of gestation in *Peromyscus* often have been recorded incidental to general studies of reproduction and development (Appendix I). Among populations of *P. maniculatus*, average gestation of nonlactating females ranges from 22.4 to 25.5 days in nonlactating females and 24.1 to 30.6 days in lactating females, a difference of 3.6 ± 0.7 days. Average gestation periods of nonlactating *P. maniculatus* show no significant relationships with size of young at birth (r = 0.60, N = 7, $P > 0.05$), or mean litter size for the population (r = 0.35, N = 7, $P > 0.05$), perhaps due to the low variation among populations. Fetal growth rates, calculated using the formula of Huggett and Widdas (1951), are consistently higher for nonlactating than for lactating *P. maniculatus*. Samples of other species are too few to permit the examination of inter-populational variation in gestation periods.

A comparison among the 15 species for which data on gestation periods are available (Appendix I) indicates three significant relationships. Gestation periods increase with increasing adult weight (r = 0.70, N = 11, $P < 0.05$ for nonlactating mice; r = 0.69, N = 11, $P < 0.05$ for lactating mice), large neonates are associated with long gestation periods (r = 0.59, N = 12, $P < 0.05$ for non-lactating mice; r = 0.62, N = 12, $P < 0.05$ for lactating mice), and large litters are associated with short gestation periods, at least among lactating mice (r = -0.41, N = 12, $P > 0.05$ for nonlactating mice; r = -0.78, N = 12, $P < 0.01$ for lactating mice). However, fetal growth rates, calculated from gestation periods and neonate weights (Huggett and Widdas, 1951), were not significantly related to adult size (r = 0.34, N = 10, $P > 0.05$ for nonlactating mice; r = 0.47, N = 10, $P > 0.05$ for lactating mice) or to number of young per litter (r = -0.50, N = 11, $P > 0.05$ for nonlactating mice; r = 0.30, N = 11, $P > 0.05$ for lactating mice).

Parturition

Layne (1968) described the preparturitional behavior of female *Peromyscus* and the behavior of newborn mice in detail. The birth of a litter may occur in less than an hour or may take many hours. In captivity, females may or may not show antagonism towards males during parturition; paternal care of newborn young has been noted in *P. maniculatus* and *P. leucopus* (Horner, 1947), *P. californicus* (Dudley, 1974a, 1974b), and *P. megalops* and *P. thomasi* (Rickart, 1977). What little is known about paternal care under natural conditions comes from nest-box studies of *P. maniculatus* and *P. leucopus*. Howard (1949) noted the presence of male *P. maniculatus* in nest boxes with females and young, although Nicholson (1941) thought that female *P. leucopus* with recent litters were antagonistic to males. Both monogamy (*P. polionotus*; Foltz, 1981) and promiscuity (*P. maniculatus*; Birdsall and Nash, 1973) are known for the genus, so that a variety of parental care patterns may be expected.

Maternal care does not appear to be extensive, at least in *P. leucopus*. Lactating *P. leucopus* in seminatural enclosures have much longer activity (foraging) bouts than nonbreeding mice (Harland and Millar, 1980); most of the nocturnal period appears to be spent away from the young (Harland and Millar, 1980; Hill, 1972).

Newborn *Peromyscus* are highly altricial, with poorly developed motor responses, although they can vocalize (Layne, 1968). The size of neonates appears to be important to their survival. Myers and Master (1983) examined survival of newborn *P. maniculatus* (average neonate wt. = 1.70 g) and found that a high proportion of neonates weighing less than 1.30 grams failed to survive 24 hours.

Neonate weights vary within and among populations within species, and among species. Within population variation in neonate weight is attributable to a number of factors. Maternal effects account for 67 percent of the total variation in birth weight of *P. polionotus* (Carmon *et al.*, 1963). Large female *P. maniculatus* have been reported to produce larger offspring than small females (Millar, 1985; Myers and Master, 1983), although the relationship between female size and neonate size is not always clear (Millar, 1982, 1983). Neonate weight was negatively related to litter size in a number of populations of *P. maniculatus* (Millar, 1979, 1982; Millar and Innes, 1983; Myers and Master, 1983) and *P. leucopus* (Lackey, 1978; Millar, 1975, 1978) as well as in *P. yucatanicus* (Lackey, 1976) and *P. melanocarpus* (Rickart, 1977). Parity had no effect on neonate weight of *P. maniculatus* when the effects of female size and litter size were controlled (Myers and Master, 1983).

At the interpopulation level, some variation in neonate weight is undoubtedly due to sampling bias. For example, multiple samples of *P. maniculatus* and *P. leucopus* from single localities (studies of *P. maniculatus* by Millar, 1979; Millar *et al.*, 1979; Svihla, 1932, 1934, 1935; and studies of *P. leucopus* by Millar, 1975, 1978 and Millar *et al.*, 1979) have different mean neonate weights that differ by as much as 0.13 grams (*P. leucopus*, London, Ontario). These differences likely reflect founder effects of laboratory colonies or neonate weights being recorded at different times of day. Differences among populations can be considerable (Appendix II). For example, mean neonate weights of *P. maniculatus* range from 1.40 grams in California (McCabe and Blanchard, 1950) to over 2.0 grams in island populations in British Columbia (Thomas, 1971). Neonate weights are positively related to adult weight in *P. maniculatus* ($r = 0.76$, $N = 23$, $P < 0.01$) and in *P. leucopus* ($r = 0.66$,

N = 10, $<P < 0.05$). They are negatively related to litter size in *P. maniculatus* (r = -0.48, N = 23, $P < 0.05$), but not in *P. leucopus* (r = 0.001, N = 10, $P > 0.05$).

Geographic trends in neonate weights are complicated by body size relationships. Among all *P. maniculatus*, neonate weight is positively related to latitude (r = 0.57, N = 24, $P < 0.01$). However, when the large-bodied mice from coastal British Columbia (Thomas, 1971) are removed from the sample, neonate weight does not vary significantly with latitude (r = -0.15, N = 19, $P > 0.05$). A multiple regression of neonate weight on latitude and longitude is also not significant (r = 0.18, N = 15, $P > 0.05$). Therefore, although neonate weight is somewhat variable among populations, it is not predictable geographically. Among species, neonate weights increase with increasing adult size (r = 0.85, N = 15, $P < 0.01$) and decrease with increasing litter size (r = -0.63, N = 15, $P < 0.05$).

Growth and Development of Nestlings

Little work has been done on growth and development of individual *Peromyscus*. Carmon et al.(1963) found that family effects influenced individual weights of *P. polionotus* until after weaning. A negative relationship between nestling size and litter size tends to hold throughout the nestling period in *P. maniculatus* (Millar, 1979, 1982; Millar and Innes, 1983; Myers and Master, 1983), *P. leucopus* (Fleming and Rauscher, 1978; Millar, 1978; Lackey, 1978), *P. yucatanicus* (Lackey, 1976), and *P. melanocarpus* (Rickart, 1977), although such relationships are not always clear (Guetzow and Judd, 1981; Hill, 1972; Lackey, 1978; Millar, 1975, 1985; Rickart, 1977). Data on growth of males and females within populations (Drickamer and Bernstein, 1972; Morrison et al., 1977; Carmon et al., 1963) indicate that sex-specific differences in growth are not consistent among populations. Size at weaning has an effect on adult size attained by *P. maniculatus* (Millar, 1983; Myers and Master, 1983).

Several studies have examined nestling growth in populations of *P. maniculatus* and *P. leucopus* (Appendix III). Arithmetic growth tends to increase after birth, decrease when the young become active in the nest, and increase again prior to weaning (Millar, 1979, 1982, 1985; Millar et al., 1979). Von Bertalanffy growth constants based on growth of nestlings, (Appendix III) range from 0.029 to 0.048 among populations of *P. maniculatus*. Growth of young *P. maniculatus* is not related to adult size (r = -0.15, N = 15, $P > 0.05$), litter size (r = -0.48, N = 14, $P > 0.05$), latitude (r = 0.28,

N = 15, $P > 0.05$), or longitude (r = 0.27, N = 15, $P > 0.05$). Samples of other species are too small to examine intraspecific variation.

Among species, growth is negatively related to adult size (r = -0.82, N = 12, $P < 0.01$) and positively related to litter size (r = 0.62, N = 12, $P < 0.05$).

Layne (1968) compiled qualitative data on morphological development of *Peromyscus*. Most studies of morphological development have examined characteristics such as elevation of the ear pinnae, development of digits, eruption of incisors, opening of the auditory meatus, and opening of eyes, which serve as quantitative measures of morphological development. Individual variation in developmental patterns has not been investigated, although Lackey (1978) found development to vary with litter size in *P. leucopus*; one population showed rapid development in large litters, whereas the other showed rapid development in small litters. King (1970) demonstrated that age of eye opening in *P. maniculatus* responds to artificial selection.

Quantitative data on developmental patterns among populations of *P. maniculatus* generally are incomplete, usually concentrating on age that eyes open as a measure of the rate of development (Appendix IV). Few studies document both growth and developmental characteristics so that the relationship between growth and developmental rates cannot be determined. However, age at which eyes open can be examined in relation to adult size, litter size, size of neonates, and growth rate. Among populations of *P. maniculatus*, young from large-bodied adults and young with large neonatal weights open their eyes at a relatively late age (r = 0.70, N = 15, $P < 0.01$ in both cases). Age at which eyes opened did not vary significantly with litter size among populations of *P. maniculatus* (r = -0.16, N = 19, $P > 0.05$).

Among species, there is no significant relationship between litter size and age at which eyes open (r = -0.45, N = 15, $P > 0.5$), but age at which eyes open is positively correlated with adult size (r = 0.56, N = 14, $P < 0.05$) and size of neonates (r = 0.62, N = 16, $P < 0.01$). Age at which eyes open is also positively related to age at which the ear pinnae are free (r = 0.75, N = 13, $P < 0.01$), age at eruption of the lower and upper incisors (r = 0.88, N = 8, $P < 0.01$; and r = 0.89, N = 6, $P < 0.05$, respectively), and age at opening of the auditory meatus (r = 0.98, N = 10, $P < 0.01$). These data generally indicate that small-bodied species develop more quickly than large-bodied species. Species with fast growth rates also opened their eyes earlier than those with slow growth rates, as indicated by the negative relationship between age at which eyes opened (Appendix IV) and the von Bertalanffy growth constants (Appendix III) (r = -0.87, N = 11, $P < 0.01$). Hence, growth and rate of development appear to be related, at least among species.

Age at Independence

The attainment of independence is a gradual process involving changes in nutrition, behavior, and dispersal of young, and a variety of indicators have been used to quantify independence. These include the earliest age at which young are known to consume solid food (Layne, 1968), age at which young leave the nest (Vestal et al., 1980), age at which females exhibit aggression towards their off-spring (Savidge, 1974), age at which the young no longer associate with the female (Howard 1949; Nicholson, 1941), age at which the young enter a trappable population (Millar et al., 1979), and age at which young can maintain body weight when isolated from the female (King et al., 1963). Clearly, age at independence will vary a great deal, depending on the criteria used to measure it.

Variation in age at independence within populations has not been studied specifically, although qualitative observations summarized by Layne (1968) indicate that nonpregnant females nurse their litters somewhat longer than pregnant females, and that litters born at the end of a breeding season associate with the female longer than those born early in the breeding season. There appears to be no evidence that age at independence varies with number of young in a litter. Weight at which young P. maniculatus enter the trappable population varies seasonally in some populations (Gyug and Millar, 1981); these differences were attributed to different rates of growth rather than different ages at independence.

Age at independence is difficult to compare among populations because methodologies have varied. However, the method of King et al. (1963) has been applied to enough populations that some comparisons can be made. This method, involving weight change of young with isolation from the female, is thought to reflect a combination of factors including the ability to consume solid food, handle the stress of an unfamiliar cage, and cope with behavioral isolation from the female and siblings (King et al., 1963). A comparison of this estimate of independence among populations of P. maniculatus indicates that age at independence ranges from 18 to

TABLE 2.—*Age of* Peromyscus *at independence, as indicated by age at which zero weight change occurs with isolation from the female.*

Location	Adult weight (g)	Litter size	Size at independence (g)	Age at independence (days)	Source
		P. maniculatus			
Pinawa, Manitoba	20	5.3	7.8	21.6	Millar *et al.*, 1979
Great Slave Lake, North West Territories	19	5.0	9.6	21.4	Millar, 1982
S.E. Alberta	20	5.1	9.3	19.2	Millar, 1985
Kananaskis, Alberta	20	5.2	9.9	24.9	Millar and Innes, 1983
Mt. Evans, Colorado	22	6.4	7.1	18.9	Halfpenny, 1980
Ward, Colorado	23	5.1	8.2	15.6	Halfpenny, 1980
Brighton, Colorado	21	4.1	8.6	18.0	Halfpenny, 1980
Michigan	—	—	7.7	18.1	King *et al.*, 1963
Michigan	—	—	11.2	23.8	King *et al.*, 1963
		P. leucopus			
London, Ontario	22	4.9	9.2	22.2	Millar *et al.*, 1979
Lower Rio Grande, Texas	20	3.5	9.4	21.3	Guetzow and Judd, 1981
		P. melanocarpus			
Oaxaca, Mexico	59	2.3	21	25.4	Rickart, 1977[1]
		P. mexicanus			
Oaxaca, Mexico	61	2.6	22	23.8	Rickart, 1977[1]

[1] Approximate age and weight

25 days (Table 2). Size at independence is positively related to age at independence ($r = 0.71$, $N = 9$, $P < 0.05$), but age at independence does not vary significantly with adult size ($r = -0.58$, $N = 7$, $P > 0.05$) or litter size ($r = 0.11$, $N = 7$, $P > 0.05$).

Samples are insufficient to compare age at independence among species, but the large-bodied *P. melanocarpus* and *P. mexicanus* have ages at weaning similar to those of the small-bodied *P. maniculatus* and *P. leucopus* (Table 2).

Postweaning Growth and Development

Postweaning growth has been less studied than nestling growth. It has, however, been studied under natural conditions. Studies of wild mice indicate that early postweaning growth of *P. maniculatus* does not vary seasonally (Gyug and Millar, 1981; Millar and Innes, 1983) or consistently among habitats (Millar and Innes, 1983, 1985). Mice born early in the breeding season attain full size during the season of their birth, whereas those born late in the season tend to cease growth after 40 days of age and enter the nonbreeding season at a relatively small size (Fuller, 1969; Gyug and Millar, 1981; Millar and Innes, 1983, 1985).

Postweaning growth rate of *P. maniculatus* in the laboratory is variable, ranging from 0.030 to 0.091 (Appendix III). Samples are too small to compare postweaning growth with adult weight, litter size, or nestling growth, but males grew faster than females in two of three cases. Weight dimorphism favors males in *P. maniculatus* and *P. leucopus* (Dewsbury *et al.*, 1980).

Among species, postweaning growth ranges from 0.022 to 0.070 (Appendix III), increases with increasing litter size (r = 0.62, N = 12, $P < 0.05$), and decreases with increasing adult size (r = -0.82, N = 12, $P < 0.01$). Among species, postweaning growth is highly correlated with nestling growth (r = 0.88, N = 12, $P < 0.01$).

REPRODUCTION

Maturation

Age at maturation has a great influence on population dynamics and the production of potential descendants within lineages. As such, patterns of maturation are exceedingly important in the study of populations. Unfortunately, maturation has usually been defined in terms of several developmental events, including perforation of the vagina, vaginal smear patterns (estrus), observation of coitus, molt patterns, and the presence of pregnancy or lactation for females, and epididymal smears, testes development, scrotum development, molt patterns, observed coitus, and known inseminations for males (Clark, 1938; Lackey, 1978; Layne, 1968; Millar, 1982; Rogers and Beauchamp, 1974). Although these events are generally related (Rogers and Beauchamp, 1974), their variability makes precise determination of age at maturation difficult. In addition, some individuals in laboratory colonies do not breed at all, making it difficult to assess maximum ages at first reproduction.

Most studies dealing with variation in maturation within populations have focused on various aspects of reproductive stimulation and inhibition under controlled conditions. Reproductive stimulation has been fairly well documented in *P. maniculatus*. The presence of females stimulates the development of male reproductive organs (Babb and Terman, 1982), as does the presence of female urine (Terman and Bradley, 1981). The presence of urine from adult males accelerates the opening of the vagina and stimulates the development of the reproductive tract (Lombardi and Whitsett, 1980; Teague and Bradley, 1978). Reproductive suppression has been recorded more frequently than reproductive stimulation in *P. maniculatus*. Mice in free-living populations at high densities have

small reproductive organs (Terman, 1969) and inhibited maturation (Terman, 1965, 1968). Sibling pairs have inhibited maturation,
particularly if they are paired at an early age (Hill, 1974). The presence of mothers or unrelated females inhibits maturation in *P.
maniculatus* (Lombardi and Whitsett, 1980; Haigh, 1983*b*), *P. leucopus* (Haigh, 1983*a*), and *P. eremicus* (Skryja, 1978). Young males
also are inhibited by the presence of adult or same-aged males
(Babb and Terman, 1982; Bediz and Whitsett, 1979; Terman,
1979). Inhibition appears to be mediated through a variety of cues,
including tactile, auditory, olfactory, and visual cues (Terman, 1980)
but can be indirectly mediated through the presence of soiled bed-
ding or urine (Lawton and Whitsett, 1979; Lombardi and Whitsett,
1980; Rogers and Beauchamp, 1976). Removal of the olfactory
bulb prevented reproductive inhibition of young male *P. maniculatus* (Lawton and Whitsett, 1979). Recovery from inhibition is
possible after removal of the cue (Terman, 1973; Haigh, 1983*b*).
These factors potentially contribute to variation in age at maturation, but their role in natural populations is unknown. Natural
populations of *Peromyscus* have been reported to show a cessation
of breeding at high densities (Catlett and Brown, 1961; Canham,
1969; Hoffmann, 1955), although seasonal breeding patterns were
not well documented in these cases. Development and maintenance
of reproductive organs is also highly sensitive to restrictions in food
intake (Blank and Desjardins, 1984; Merson and Kirkpatrick, 1981;
Merson *et al.*, 1983).

A consideration of ages at maturation among laboratory colonies
(Table 3) indicates that all species are capable of maturing at less
than three months of age, although large-bodied species tend to
mature somewhat later than small-bodied species. These ages at
maturation should be considered as the minimum possible, which
may or may not be applicable to individuals in the wild. Some studies on wild populations indicate that ages at first breeding are similar to the potential indicated in the laboratory. Howard (1949) recorded a conception at 32 days of age, and Millar *et al.* (1979)
recorded a median age at first conception among spring-born young
of about 60 days in *P. maniculatus*. Young *P. leucopus* breed at about
2.5 months of age in the wild (Burt, 1940) and a median age at first
conception of 72 days has been recorded in this species (Millar
et al., 1979). *Peromyscus polionotus* may breed at 37 days of age
(Dapson, 1979) and *P. mexicanus* have been reported to breed at
110 days of age (Anderson, 1982). These ages are similar to those
recorded under laboratory conditions (Table 3). However, some

TABLE 3.—*Age at maturity in captive female* Peromyscus.

Species	Measure of maturity	Age (days)	Source
P. maniculatus	Mean age at first estrus	49	Clark, 1938
P. maniculatus	Frequent coitus	49	McCabe and Blanchard, 1950
P. maniculatus	Minimum age at first conception	90	Millar, 1982
P. maniculatus	Median age at first perforate	32	Lombardi and Whitsett, 1980
P. maniculatus	Minimum age at first conception	35	Millar, 1985
P. maniculatus	Mean age at first breeding	84	Haigh, 1983b
P. leucopus	Mean age at first estrus	46	Clark, 1938
P. leucopus	Mean age at first perforate	38	Lackey, 1978
P. leucopus	Mean age at first perforate	44	Lackey, 1978
P. leucopus	Mean age at first perforate	39	Rogers and Beauchamp, 1974
P. leucopus	Mean age at first estrus	51	Rogers and Beauchamp, 1974
P. polionotus	Mean age at first estrus	30	Clark, 1938
P. polionotus	Minimum age at conception	35	Rand and Host, 1942
P. mexicanus	Mean age at first perforate	65	Anderson, 1982
P. mexicanus	Minimum age at first perforate	46	Rickart, 1977
P. gossypinus	Mean age at first perforate	43	Pournelle, 1952
P. gossypinus	Minimum age at first conception	73	Pournelle, 1952
P. crinitus	Approximate age at first breeding	70	Egoscue, 1964
P. boylii	Mean age at first estrus	51	Clark, 1938
P. megalops	Mean age at first perforate	48	Layne, 1968
P. eremicus	Mean age at first estrus	39	Clark, 1938
P. eremicus	Approximate mean age at first conception	80	Skryja, 1978
P. floridanus	Mean age at first perforate	35	Layne, 1966
P. yucatanicus	Mean age at first perforate	51	Lackey, 1976
P. melanocarpus	Mean age at first perforate	79	Rickart, 1977

laboratory estimates do not reflect the actual age at maturation under natural conditions. For example, northern *P. maniculatus* do not mature under natural conditions until they are approaching one year of age (Gyug and Millar, 1981; Millar, 1982), whereas laboratory mice from this population mature at a much earlier age (minimum 90 days, Millar, 1982). In contrast, maturation in this laboratory colony appears later than in other laboratory colonies (Table 3). Other northern populations also exhibit maturation at approximately one year of age (Gilbert and Krebs, 1981; Millar and Innes, 1983, 1985; Van Horne, 1981), although breeding in the year of birth occasionally occurs (Millar and Innes, 1985).

Patterns of maturation in males are even less understood than those of females. Layne (1968) recorded some evidence that male *P. leucopus, P. maniculatus, P. gossypinus,* and *P. floridanus* first breed at a greater age than females. Similar evidence is available for *P. melanocarpus* (Rickart, 1977) and *P. mexicanus* (Anderson, 1982). However, the opposite appears true for northern *P. leucopus* and *P. maniculatus* (Millar *et al.*, 1979). In one northern population of

P. maniculatus, males commonly exhibit some gonadal development during the summer of their birth, whereas most females do not breed (Millar and Innes, 1985).

Litter Size

Litter size, as determined by embryo counts, number of young born, or number of nestlings, is one of the most easily and frequently measured reproductive traits of *Peromyscus*. Litter size is highly variable within populations. This is attributable to variation within as well as among individuals. Despite this inherent variation, some trends are evident within populations. First, mean litter size varies with parity; first litters are consistently smaller than subsequent litters (Table 4). The greatest difference appears to be between first and second litters. Second litters average 0.9 young more than first litters, but third litters average only 0.1 young more than second litters. Parity effects are complicated because litter size has been shown to increase with size of the mother in some studies (Caldwell and Gentry, 1965; Davis, 1956; Lackey, 1976; Millar, 1985; Myers and Master, 1983). Thus, both growth factors and prior reproductive experience may be involved in parity effects. Litter size has not been shown to vary with age of the female (Millar, 1982; Millar, 1985), and growth cannot be implicated in the tendency towards smaller litters of old multiparous females. Litter size is heritable (Leamy, 1981), although influenced by negative maternal effects whereby large females produce large litters of small young that in turn become small adults producing small litters of large young (Leamy, 1981; Millar, 1983).

The inherent variability in litter size makes the analysis of field data difficult because age, experience, and parity effects generally are not known. However, some trends (most not statistically significant) indicate a degree of phenotypic plasticity in litter size within populations. Primiparous females tend to have smaller litters than overwintered and multiparous females (Millar, 1978; Millar *et al.*, 1979). Seasonal changes in litter size have been reported, but they are not consistent. Some populations exhibit no seasonal trends in litter size (Brown, 1964*a*; Jameson, 1953; Krebs and Wingate, 1985; Millar and Innes, 1985). Some show trends towards larger litters early in the breeding season (Millar, 1978; Millar *et al.*, 1979), whereas others show an increasing litter size as the breeding season progresses (Borell and Ellis, 1934; Brown, 1966; Caldwell and Gentry, 1965; Dapson, 1979; Fuller, 1969; Gashwiler, 1979; Jameson,

TABLE 4.—*Litter size in relation to parity in Peromyscus.*

Species	Location	1	2	3	4	5	6	7	8	9	10	Source
P. maniculatus	Great Slave Lake, North West Territories	4.7	5.9	4.7	5.0	4.5	4.0	—	—	—	—	Millar, 1982
P. maniculatus	N. Michigan	4.3	—	5.2	—	5.2	—	4.5	—	4.3	—	Drickamer and Vestal, 1973
P. maniculatus	Ann Arbor, Michigan	4.2	—	5.4	4.7	5.2	4.9	4.9	—	4.2	—	Drickamer and Vestal, 1973
P. maniculatus	Siskiyou Co., California	3.3	4.7	4.7	4.7	4.8	4.9	5.1	5.1	5.2	4.5	Forrester, 1975
P. maniculatus	East Lansing, Michigan	4.3	5.0	5.3	5.2	—	—	—	—	—	—	Price, 1967
P. maniculatus	Ann Arbor, Michigan	4.8	6.3	6.4	6.8	6.9	7.1	6.7	6.3	—	—	Myers and Master 1983
P. maniculatus	S.E. Alberta	4.7	5.4	5.6	5.2	6.3	—	—	—	—	—	Millar, 1985
P. leucopus	Lansing, Michigan	4.1	—	5.2	—	5.1	—	5.3	—	5.0	—	Drickamer and Vestal, 1973
P. leucopus	Ann Arbor, Michigan	4.0	4.7	5.1	5.3	5.1	5.3	5.3	—	5.4	—	Lackey, 1973
P. leucopus	Campeche, Mexico	3.9	4.8	5.2	5.3	6.0	6.0	6.3	—	6.4	—	Lackey, 1973
P. polionotus	Ocala, Florida	3.0	—	3.6	—	4.2	—	3.7	—	4.2	—	Drickamer and Vestal, 1973
P. polionotus	Aiken, South Carolina	3.2	3.9	4.1	4.1	4.2	4.2	4.0	—	—	—	Williams *et al.*, 1965
P. crinitus	Bernadino Co., California	2.5	—	3.0	—	3.0	—	3.1	—	—	—	Drickamer and Vestal, 1973
P. californicus	Berkeley, California	1.8	—	1.9	—	2.0	—	2.1	—	2.2	—	Drickamer and Vestal, 1973
P. floridanus	Delray Beach, Florida	2.0	—	2.3	—	2.5	—	2.5	—	2.7	—	Drickamer and Vestal, 1973
P. eremicus	Tucson, Arizona	2.2	2.5	2.3	2.5	2.8	2.9	2.6	2.6	2.4	2.6	Davis and Davis, 1947
P. yucatanicus	Yucatan, Mexico	2.6	3.2	4.0	3.7	4.2	4.0	4.5	—	—	—	Lackey, 1976

1953; Judd *et al.*, 1984; Sadleir, 1974). Myers *et al.* (1985) reported an effect of environmental temperature during early development on litter size in *P. maniculatus*. They found large litters to be associated with high temperatures in spring and small litters to be associated with high temperatures in autumn. Small litters were also associated with heavy rainfall. Differences in litter size among years (Fuller, 1969) may reflect different seasonal responses. Litter size does not vary with seed crops among years (Gashwiler, 1979) or with habitat (Martell, 1983). In general, *Peromyscus* exhibit differences within populations, but cannot be considered to exhibit a great deal of phenotypic plasticity.

Variation in litter size among populations of *P. maniculatus* and *P. leucopus* is not related to adult size (r = -0.01, N = 29, *P* > 0.05; and r = 0.15, N = 15, *P* > 0.05, respectively). However, a number of geographic trends in litter size have been suggested for *P. maniculatus* and *P. leucopus*. Dunmire (1960), Spencer and Steinhoff (1968) and Halfpenny (1980) reported increases in litter size with increasing elevation, whereas McLaren and Kirkland (1979), Millar and Innes (1983), and Smith and McGinnis (1968) found no such relationship. Latitudinal trends have also been suggested by Long (1973) and McLaren and Kirkland (1979), but Lackey (1978) and Smith and McGinnis (1968) found no such relationship. Latitude and elevation together explain a significant amount of the variation in litter size among *P. maniculatus* populations (Smith and McGinnis, 1968), but length of the breeding season does not (Millar, 1984).

Data compiled here indicate that litter size does vary geographically in both *P. maniculatus* and *P. leucopus*. Among all records for *P. maniculatus* (Appendix V), the best descriptor of litter size relates litter size to latitude and longitude as

$$Y = -1.62 + 0.0103 \ X + 0.106 \ Z + 0.0004 \ X^2 -0.0005 \ Z^2, \quad (1)$$

where Y = mean litter size, X = latitude, and Z = longitude (r = 0.39, df = 72, *P* < 0.01). Predicted litter sizes of *P. maniculatus* appear greatest in northwestern North America (Fig. 1).

Among all records of *P. leucopus*, the best descriptor of litter size is

$$Y = 11.4 - 0.411 \ X - 0.0136 \ Z + 0.0065 \ X^2, \quad (2)$$

where Y = mean litter size, X = latitude, and Z = longitude (r = 0.77, df = 27, *P* < 0.01). As with *P. maniculatus*, the largest pre-

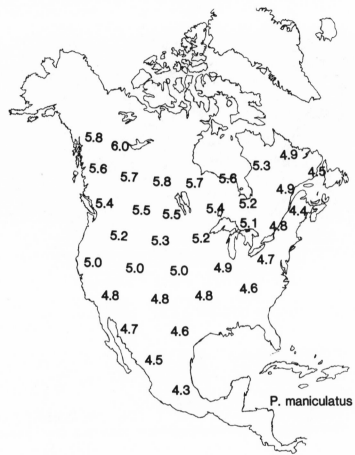

Fig. 1.—Geographic trends in predicted litter size in *P. maniculatus*. Litter sizes based on Equation 1.

dicted litter sizes of *P. leucopus* occur in northwestern North America (Fig. 2).

Comparisons that include several species indicate an increase in litter size with elevation (Smith and McGinnis, 1968) and latitude (Millar *et al.*, 1979; Smith and McGinnis, 1968), but these relationships are confounded by body size effects. For example, litter size clearly is negatively related to body size among species (r = -0.53, N = 23, P < 0.01; Appendix V) and all large-bodied species have southern distributions.

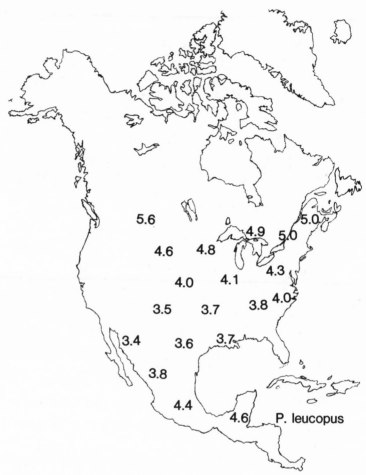

FIG. 2.—Geographic trends in predicted litter size in *P. leucopus*. Litter sizes based on Equation 2.

Energetic Costs of Reproduction

Studies on the energetic demands of reproduction have been re-stricted to the small-bodied *Peromyscus*, but some trends are appar-ent. First, breeding female *P. leucopus* and *P. maniculatus* exhibit no significant increase in fat deposition during pregnancy, and fat content does not vary significantly among inactive, pregnant, and lactating females, although fat content during lactation tends to be low (Gyug and Millar, 1980; Judd *et al.*, 1984; Millar, 1975, 1981).

This indicates that increased food ingestion is necessary for the successful production of offspring. A number of studies have monitored food intake by breeding females, particularly during lactation. These results indicate an increase in daily ingestion of about 15 percent during pregnancy and a more than doubling of food ingestion during lactation (Glazier, 1979; Millar, 1975, 1978, 1979, 1982, 1985; Millar and Innes, 1983; Stebbins, 1977). Females tend to maintain above average weights throughout lactation (Millar, 1975, 1978, 1979, 1982, 1985; Millar and Innes, 1983; Stebbins, 1977). The energetic costs of raising young increases with increasing litter size (Glazier, 1979; Millar, 1978, 1979, 1982, 1985; Millar and Innes, 1983). Available data are insufficient to compare relative energy costs among populations of *P. maniculatus*, but among all populations and species (Table 5) food ingestion also tends to increase with increasing litter size (r = 0.62, N = 10, P < 0.05). The consequences of these increased demands under natural conditions are not known.

Breeding Seasons

Peromyscus exhibit a great deal of variation in seasonal breeding patterns, ranging from aseasonal reproduction to breeding being restricted to only a few weeks per year (Appendix VI).

Seasonal breeding appears to be a response to a complex combination of cues, including photoperiod, temperature, and food availability. Photoperiodic responses vary geographically and among individuals within populations in *P. leucopus* and *P. maniculatus* (Desjardins and Lopez, 1983; Lynch *et al.*, 1981). Temperature (Millar and Gyug, 1981; Sadlier, 1974; Sadlier *et al.*, 1973) and precipitation (Bradford, 1974, 1975; Brown, 1964*a*; Robertson, 1975) have been implicated in the initiation and cessation of breeding under natural conditions and both have been implicated in the occurrence of breeding at unusual times of the year (Beidleman, 1954; Brown, 1945; Manville, 1952; Reed, 1955). Reproductive activity of *Peromyscus* is extremely sensitive to food restrictions (Blank and Desjardins, 1984; Merson and Kirkpatrick, 1981; Merson *et al.*, 1983), and food resources have been implicated in explaining winter breeding (Linduska, 1942). Duration of breeding varies with levels of natural foods (Gashwiler, 1979) and supplemental food results in extended breeding seasons under some conditions (Taitt, 1981) but not others (Gilbert and Krebs, 1981). The simplest generalization is that *Peromyscus* will breed whenever con-

TABLE 5.—*Relative energy demands of lactation among populations and species of* Peromyscus.

Species	Location	Non-breeding weight (g)	Litter size	Age at weaning (days)	Non-breeding ingestion (A)	Ingestion during lactation (B)	B/A	Source
P. maniculatus	Bar Harbor, Maine	14.5	4.3	18–20	58.5[1]	144.6[1]	2.47	Glazier, 1979
P. maniculatus	Pinawa, Manitoba	19.9	5.0	21	72.1[2]	161.3[2]	2.23	Millar, 1979
P. maniculatus	S.E. Alberta	20.1	5.0	19	66.5[2]	163.0[2]	2.45	Millar, 1985
P. maniculatus	Great Slave Lake, North-west Territories	19.2	5.0	21	60.5[2]	172.0[2]	2.84	Millar, 1982
P. maniculatus	Kananaskis, Alberta	20.3	5.0	24	85.2[2]	193.3[2]	2.27	Millar and Innes, 1983
P. leucopus	London, Ontario	21.6	5.0	21	46.2[2]	166.7[2]	3.60	Millar, 1978; Millar et al., 1979
P. leucopus	Ithaca, New York	21.0	3.9	18–20	72.4[1]	143.7[1]	1.98	Glazier, 1979
P. polionotus	Lake Placid, Florida	13.4	3.6	18	54.0[1]	94.2[1]	1.74	Glazier, 1979
P. eremicus	Portal, Arizona	21.5	2.4	18	49.3[1]	104.9[1]	2.13	Glazier, 1979
P. floridanus	Lake Placid, Florida	42.0	2.2	22–24	83.1[1]	144.3[1]	1.73	Glazier, 1979

[1] Kj/day

[2] Cumulative grams ingested during period of lactation

ditions are collectively suitable for the successful raising of off-spring, with some intrapopulational variation due to local probabilities of success (Fairbairn, 1977).

Conditions suitable for successful reproduction clearly vary a great deal geographically (Appendix VI), although the average opportunities for reproduction by individual females do not. In almost all recorded cases, individual females survive only long enough to produce two or three litters during a breeding season; this number does not vary significantly with length of the breeding season ($r = 0.36$, $N = 10$, $P > 0.05$; Table 6). The frequency with which females breed, however, increases with decreasing length of the breeding season ($r = -0.77$, $N = 7$, $P < 0.05$; Table 6). Thus, the intensity of postpartum breeding in natural populations appears greatest when breeding seasons are extremely short (Millar, 1984; Millar and Innes, 1985). This varying intensity of reproduction under different seasonal conditions is reflected in the proportion (average and maximum) of females pregnant being negatively related to duration of the breeding season (May, 1979). The low intensity of reproduction in long-season environments may be the result of either a low level of reproduction during much of the breeding season (unimodal breeding pattern) or bimodal peaks in breeding (Batzli, 1977; Brown, 1964a; Burt, 1940; Cornish and Bradshaw, 1978; Drickamer, 1978; Rintamaa et al., 1976; Svendsen, 1964).

TABLE 6.—Average number of litters per breeding season and average time between litters within populations of Peromyscus.

Species	Location	Breeding season (weeks)	Average no. litters per year	Days between litters	Source
P. maniculatus	Great Slave Lake, North West Territories	8	1.8	29.3	May, 1979; Gyug, 1979
P. maniculatus	Pinawa, Manitoba	16	—	25.8	Gyug, 1979
P. maniculatus	Kluane, Yukon Territory	9	2.5	—	Krebs and Wingate, 1985
P. maniculatus	Kananaskis, Alberta	10	1.9	29.6	Millar and Innes, 1983
P. maniculatus	Giles Co., Virginia	29	1.8	—	Wolff, 1985
P. maniculatus	San Francisco, California	29	4.0	67.3	McCabe and Blanchard, 1950
P. leucopus	London, Ontario	24	2.0	24.8	Gyug, 1979
P. leucopus	Giles Co., Virginia	28	2.0	—	Wolff, 1985
P. truei	San Francisco, California	27	3.4	65.0	McCabe and Blanchard, 1950
P. californicus	San Francisco, California	29	3.2	63.7	McCabe and Blanchard, 1950
P. mexicanus	Monteverde, Costa Rica	41	2.5	—	Anderson, 1982

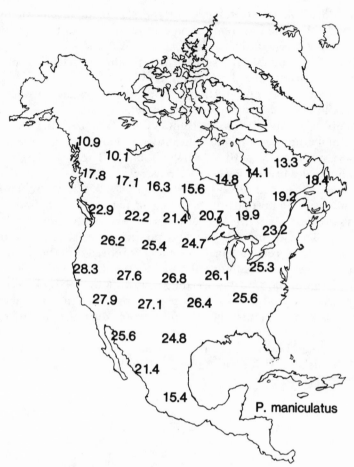

FIG. 3.—Geographic trends in duration (weeks) of the breeding season in *P. maniculatus*. Duration based on Equation 3.

The period of the year over which breeding takes place in *P. maniculatus* decreases with latitude and longitude with the best descriptor of breeding season length being

$$Y = -33.0 + 2.79\ X + 0.0748\ Z - 0.0370\ X^2, \qquad (3)$$

where Y = duration of reproduction in weeks, X = latitude, and Z = longitude (r = 0.58, df = 71, P < 0.01; Fig. 3). The best geographic descriptor of length of the breeding season in *P. leucopus* is

$$Y = 30.5 - 0.853\ X + 0.423\ Z, \qquad (4)$$

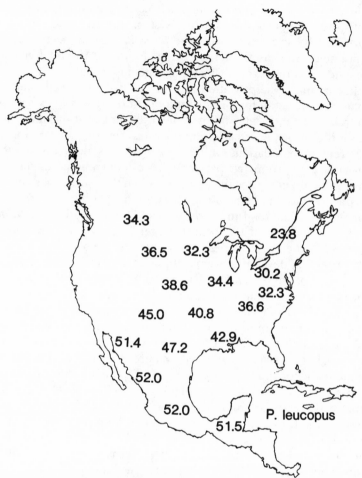

Fig. 4.—Geographic trends in duration (weeks) of the breeding season in *P. leucopus*. Duration based on Equation 4.

where Y = duration of reproduction in weeks, X = latitude, and Z = longitude (r = 0.82, df = 25, P < 0.02; Fig. 4).

Discussion

The ease with which *Peromyscus* are captured, handled, and maintained in captivity, and the extent to which they have been studied indicates that they should provide a model system for considering the relationship between reproductive and developmental patterns and environmental conditions. Such considerations require that both the relationships within the system and the environmental correlates be understood. In some ways, *Peromyscus* meet these requirements, but in others they either do not, or the relationships depend upon the level at which the system is considered.

Intrapopulational variation in reproductive and developmental characteristics is an important component of such studies because all traits are genetically and physiologically linked at this level, and this is the level at which selection operates. Several patterns appear inherent in populations of *Peromyscus*. One is that considerable variation in reproductive parameters is evident among females, although we do not know the extent to which this variation is genetic or phenotypic. There are strong indications that females breeding for the first time tend to produce small litters. In addition, large-bodied females tend to produce larger litters than small-bodied females of the same species. Large litters contain relatively small offspring that develop into small adults. Large litters place greater energic demands on females than do small litters. This sets the stage for potential explanations of variable litter sizes within populations (Leamy 1981) but clearly leaves numerous questions about the interdependence of developmental characteristics and the ecological consequences of these differences. Ontogeny begins with conception, but the relationships between prenatal and postnatal development are not known. The effects of different gestation periods or uterine crowding on neonate weights are not known. The relationships between growth and morphological development are not well known, and the consequences of early development on subsequent reproductive performance as an adult are poorly understood. This lack of detailed information about patterns of variation within populations must be remedied before the consequences of selection for specific traits can be considered. The consequences of variation can only be realistically judged from studies done under natural conditions.

Interpopulational variation in reproductive and developmental characteristics is better documented than intrapopulational patterns, although primarily among populations of the two most widespread species (*P. maniculatus* and *P. leucopus*). Interpopulational variation in reproductive characteristics is relatively low in both species (Table 7); this likely contributes to the lack of significant interrelationships of traits among populations. Neonatal weights vary with maternal size among *P. leucopus* and *P. maniculatus* populations, and with litter size among populations of *P. maniculatus*.

TABLE 7.—*Coefficients of variation (%) in reproductive characteristics and weight among populations and species of* Peromyscus, *based on Appendix I–IV. Missing data represents N < 7.*

Characteristic	P. maniculatus	P. leucopus	Species
Gestation	4	—	11
Neonate weight	16	7	39
Nestling growth	12	—	28
Age eyes open	10	7	21
Litter size	16	14	31
Adult weight	17	13	50

Relative energy demands for reproduction vary with average litter size among populations.

Litter size is not related to adult size among populations of *P. maniculatus* or *P. leucopus*, and growth rates do not vary with litter size or adult size among populations of *P. maniculatus*.

Relatively low variability in developmental traits in particular has several consequences for future studies at the interpopulational level. One is that developmental patterns cannot be considered to be finely tuned to local environmental conditions. Rather, developmental patterns appear to be "good enough" to permit the persistence of populations under a wide range of environmental conditions. Second, comparisons of traits between different environments must be interpreted with caution. Major differences in traits among local environments or habitats should not be expected when only minor differences occur among major ecological zones. Local comparisons should involve replicate samples to avoid minor differences due to sampling biases or founder effects. It should be remembered that organisms do not respond to latitude, longitude, or elevation per se. Comparisons should involve meaningful measures of the environment.

The reasons for low variability of reproductive and developmental traits among diverse ecological conditions are far from clear. Speculations based on resistance to change, such as design constraints or the slowness with which selection operates, can be made. Alternatively, *Peromyscus* breed only when conditions are suitable for the successful growth and development of young. Thus it can be argued that all developmental patterns are adapted to relatively benign environmental conditions.

Interpopulational differences are evident in the duration and frequency of breeding, patterns of maturation, and litter size. Seasonal breeding appears to be finely tuned to environmental conditions and some effects of local conditions have been documented (Drickamer, 1978; Gashwiler, 1979). Geographic trends in seasonal breeding generally correlate with latitude, but not strictly so for *P. maniculatus* or *P. leucopus* (Figs. 3 and 4). Conditions suitable for reproduction are complex, as are the cues that trigger breeding. For these reasons, only studies that deal with the specific timing of breeding in relation to environmental and demographic conditions (Fairbairn, 1977) will contribute significantly to our understanding of seasonal breeding in *Peromyscus*. Our current understanding of intrapopulational patterns provides background information for such studies.

Patterns of seasonal reproduction set the stage for patterns of maturation, whereby young of the year do not mature in the summer of their birth when breeding seasons are extremely short. To date, delayed maturation has been documented only for northern populations of *P. maniculatus*. This delayed maturation does not appear to be associated with any specific trends in other reproductive and developmental characteristics but obviously has demographic consequences for these populations. More widespread are apparent seasonal differences in patterns of maturation, and differences in age at maturation between males and females in temperate populations. Laboratory data can never be used to model natural populations. Specific field experiments will be necessary before seasonal patterns of maturation and differences in the age at maturation between males and females can be understood.

Seasonal breeding also appears to have consequences in terms of the frequency of postpartum pregnancy (Millar, 1984 and Table 6), although samples to date are hardly sufficient to warrant firm conclusions.

Litter size, like seasonal breeding, shows significant geographic trends, although the patterns for both *P. maniculatus* and *P. leucopus* are not simple (Figs. 1 and 2). Despite litter size being the most frequently reported reproductive characteristic of *Peromyscus*, the most likely explanation for differences in litter size among populations is derived from negative, rather than positive relationships. Litter size does not vary with adult size among populations, does not vary predictably with length of the breeding season and opportunities for reproduction, and does not vary with demographic conditions. The only logical alternative is that litter size varies with resource levels. The fact that litter size does not respond to supplemental feeding or natural variation in food availability does not disprove this hypothesis. Rather, these findings indicate that litter size is not phenotypically plastic in *Peromyscus*. Detailed comparisons of food availability and foraging activity will be needed to test this hypothesis.

The fact that both coevolved complexes of traits and geographic patterns are not clear at the interpopulational level contributes to our understanding of previously reported comparisons that have given conflicting results. For example, Lackey (1978) found no latitudinal difference in litter size in *P. leucopus*, not because litter size did not vary among populations, but because litter size happened to be similar in the two areas studied. Similarly, differences in elevational trends reported by different authors may be the result of different local effects.

The coevolved complexes of reproductive and developmental characteristics are clearer at the interspecific level than among populations. Traits are more variable among species than among populations, (Table 7) and large-bodied species show a clear pattern of larger neonates, smaller litters, longer gestation, and slower growth and development than small-bodied species. This pattern is similar to that found within and among populations of the same species in that large females produce large neonates, but the relationship between litter size and adult size is opposite to that found within intraspecific populations. This has been a complicating factor in studies that have included multiple samples of the more common species in making geographic comparisons (Drickamer and Bernstein, 1972; Millar, 1984; Millar *et al.*, 1979; Modi, 1984). In addition, most species, and all of the large-bodied species, are restricted to southern North America and Central America. Only the small-bodied species occur in northern North America. This precludes any meaningful consideration of broad geographic trends in reproductive and developmental characteristics among species, but opens possibilities for relating clear differences in life cycle characteristics to specific environmental conditions. Some progress in this direction has been made by Glazier (1980), who noted that the largest-bodied species (with the lowest reproductive rates) had the most restricted ranges. He speculated that geographically restricted species are confined to stable species-rich environments where fecundity may be restrained by severe competition for limited resources. Modi (1984) has also speculated on the environmental basis for differences in life history patterns. More work that involves the quantification of environmental conditions within species habitats is needed to confirm or refute their suggestions. A major problem in interpreting differences among species, however, will be the fact that the environmental factors that shaped existing life cycle patterns may have done so during the early phylogeny of the species. Thus, current environmental conditions in any one locality may not accurately reflect the conditions under which these species evolved.

Taken together, current data on *Peromyscus* generally support Layne's (1968) conclusion that "the reproductive features of *Peromyscus* are readily modified in response to environmental conditions. . . ." but must be qualified to some extent. The degree of modification depends to a considerable extent on the level of organization being considered. Little modification in basic reproductive and developmental patterns is evident within species that occupy diverse environments; differences in the timing of reproductive events constitute the major adaptation to different environments at the intraspecific level. Major modifications are evident at the interspecific level, with body size and associated offspring size being the most variable traits, and prenatal and developmental traits being the most conservative.

ACKNOWLEDGMENTS

This review was supported by the Natural Sciences and Engineering Research Council of Canada. D. Burkholder, G. Hickling, and D. Innes provided valuable assistance with the compilation and analysis of data. J. Layne and S. Anderson kindly provided unpublished data. D. Innes, P. Myers, J. Layne, and G. Kirkland reviewed the manuscript.

LITERATURE CITED

ANDERSON, S. D. 1982. Comparative population ecology of *Peromyscus mexicanus* in a Costa Rican wet forest. Unpubl. Ph.D. dissert., Univ. of Southern California, Los Angeles, California, 324 pp.

ANDERSON, S. 1972. Mammals of Chihuahua. Bull. Amer. Mus. Nat. Hist., 148: 149–410.

BABB, T. E., AND C. R. TERMAN. 1982. The influence of social environment and urine exposure on sexual maturation of male prairie deer mice (*Peromyscus maniculatus bairdi*). Res. Population Ecol., 24: 318–328.

BAILEY, V. 1931. Mammals of New Mexico. North Amer. Fauna, 53: 1–412.

BAKER, R. H., AND J. K. GREER. 1962. Mammals of the Mexican state of Durango. Publ. Mus. Michigan State Univ. Biol. Series, 2: 125–154.

BATZLI, G. O. 1977. Population dynamics of the white-footed mouse in flood plain and upland forests. Amer. Midland Nat., 97: 18–31.

BEDIZ, G. M., AND J. M. WHITSETT. 1979. Social inhibition of sexual maturation in male prairie deer mice. J. Comp. Physiol. Psychol., 93: 493–500.

BEER, J. R., AND C. F. McLEOD. 1966. Seasonal changes in the prairie deer mouse. Amer. Midland Nat., 76: 277–289.

BEER, J. R., C. F. MacLEOD, AND L. D. FRENZEL. 1957. Prenatal survival and loss in some cricetid rodents. J. Mamm., 38: 392–402.

BEIDLEMAN, R. G.   1954.   October breeding in *Peromyscus* in north central Colorado.   J. Mamm., 35:118.

BENDELL, J. F.   1959.   Food as a control of a population of white-footed mice, *Peromyscus leucopus noveboracensis* (Fischer).   Canadian J. Zool., 37:173–209.

BIGLER, W. J., AND J. H. JENKINS.   1975.   Population characteristics of *Peromyscus gossypinus* and *Sigmodon hispidus* in tropical hammocks of South Florida.   J. Mamm., 56: 633–643.

BIRDSALL, D. A., AND D. NASH.   1973.   Occurrence of successful multiple insemination of females in natural populations of deer mice (*Peromyscus maniculatus*).   Evolution, 27:106–110.

BLAIR, W. F.   1940.   A study of prairie deer mouse populations in southern Michigan.   Amer. Midland Nat., 24:273–305.

———.   1958.   Effects of X-irradiation of a natural population of deer-mouse (*Peromyscus maniculatus*).   Ecology, 39:113–118.

BLANK, J. L., AND C. DESJARDINS.   1984.   Spermatogenesis is modified by food intake in mice.   Biol. Reprod., 30:410–415.

BORELL, A. E., AND R. ELLIS.   1934.   Mammals of the Ruby Mountains region of north eastern Nevada.   J. Mamm., 15:12–44.

BRADFORD, D. F.   1974.   Water stress of free living *Peromyscus truei*.   Ecology, 55:1407–1414.

———.   1975.   The effects of an artificial water supply on free-living *Peromyscus truei*.   J. Mamm., 56:705–707.

BRAND, L. R., AND R. E. RYCKMAN.   1968.   Laboratory life history of *Peromyscus eremicus* and *Peromyscus interparietalis*.   J. Mamm., 49:495–501.

BROWN, H. L.   1945.   Evidence of winter breeding of *Peromyscus*.   Ecology, 26: 308–309.

BROWN, L. N.   1964a.   Reproduction of the brush mouse and white-footed mouse in the central United States.   Amer. Midland Nat., 72:226–240.

———.   1964b.   Dynamics in an ecologically isolated population of the brush mouse.   J. Mamm., 45:436–442.

———.   1966.   Reproduction of *Peromyscus maniculatus* in the Laramie Basin of Wyoming.   Amer. Midland Nat., 76:183–189.

BURT, W. H.   1940.   Territorial behavior and populations of some small mammals in southern Michigan.   Misc. Publ. Mus. Zool., Univ. Michigan, 45:6–58.

CALDWELL, L. D.   1964.   An investigation of competition in natural populations of mice.   J. Mamm., 45:12–30.

CALDWELL, L. D., AND J. B. GENTRY.   1965.   Natality in *Peromyscus polionotus* populations.   Amer. Midland Nat., 74:168–175.

CANHAM, R. P.   1969.   Early cessation of reproduction in an unusually abundant population of *Peromyscus maniculatus borealis*.   Canadian Field-Nat., 83:279.

CARMON, J. L., F. B. GOLLEY, AND R. G. WILLIAMS.   1963.   An analysis of the growth and variability in *Peromyscus polionotus*.   Growth, 27:247–254.

CASE, T. J.   1978.   On the evolution and adaptive significance of post-natal growth rates in the terrestrial vertebrates.   Quart. Rev. Biol., 53:243–282.

CATLETT, R. H., AND R. Z. BROWN.   1961.   Unusual abundance of *Peromyscus* at Gothic Colorado.   J. Mamm., 42:415.

CLARK, F. H.   1938.   Age of sexual maturity in mice of the genus *Peromyscus*.   J. Mamm., 19:230–234.

CLUTTON-BROCK, T. H., AND P. HARVEY.   1984.   Comparative approaches to investigating adaptation.   Pp. 7–29, *in* Behavioural Ecology (J. R. Krebs and

N. B. Davies, eds.). Second edition, Sinauer Associates, Massachusetts, 493 pp.

CORNISH, L. M., AND W. N. BRADSHAW. 1978. Patterns in twelve reproductive parameters for the white-footed mouse (*Peromyscus leucopus*). J. Mamm., 59:731–739.

COVENTRY, A. F. 1937. Notes on the breeding of some Cricetidae in Ontario. J. Mamm., 18:489–496.

DAPSON, R. W. 1979. Phenologic influences on cohort-specific reproductive strategies in mice (*Peromyscus polionotus*). Ecology, 60:1125–1131.

DAVENPORT, L. B., JR. 1964. Structure of two *Peromyscus polionotus* populations in old-field ecosystems at the AEC Savannah River Plant. J. Mamm., 45: 95–113.

DAVIS, D. E. 1956. A comparison of natality rates in white footed mice for four years. J. Mamm., 37:513–516.

DAVIS, D. E., AND D. J. DAVIS. 1947. Notes on the reproduction of *Peromyscus eremicus* in a laboratory colony. J. Mamm., 25:181–183.

DEACON, J. E., W. G. BRADLEY, AND M. K. LARSEN. 1964. Ecological distribution of the mammals of Clark Canyon, Charleston Mountains, Nevada. J. Mamm., 45:397–409.

DESJARDINS, C., AND LOPEZ, M. J. 1983. Environmental cues evoke differential responses in pituitary testicular function in deer mice. Endocrinology, 112:1398–1406.

DEWSBURY, D. A., D. J. BAUMGARDNER, R. L. EVANS, AND D. G. WEBSTER. 1980. Sexual dimorphism for body mass in 13 taxa of muroid rodents under laboratory conditions. J. Mamm., 61:146–149.

DICE, L. R. 1954. Breeding of *Peromyscus floridanus* in captivity. J. Mamm., 35:260.

DICE, L. R., AND R. M. BRADLEY. 1942. Growth in the deer mouse, *Peromyscus maniculatus*. J. Mamm., 23:416–427.

DOUGLAS, C. L. 1969. Comparative ecology of pinyon mice and deer mice in Mesa Verde National Park, Colorado. Univ. Kansas Publ., Mus. Nat. Hist., 18:421–504.

DRICKAMER, L. C. 1978. Annual reproduction patterns in populations of two sympatric species of *Peromyscus*. Behavioral Biol., 23:405–408.

DRICKAMER, L. C., AND J. BERNSTEIN. 1972. Growth in two subspecies of *Peromyscus maniculatus*. J. Mamm., 53:228–231.

DRICKAMER, L. C., AND B. M. VESTAL. 1973. Patterns of reproduction in a laboratory colony of *Peromyscus*. J. Mamm., 54:523–528.

DUDLEY, D. 1974a. Paternal behavior in the California mouse, *Peromyscus californicus*. Behav. Biol., 11:247–252.

———. 1974b. Contributions of paternal care to the growth and development of the young in *Peromyscus californicus*. Behav. Biol., 11:155–166.

DUNMIRE, W. W. 1960. An altitudinal survey of reproduction in *Peromyscus maniculatus*. Ecology, 41:174–182.

EGOSCUE, H. J. 1964. Ecological notes and laboratory life history of the canyon mouse. J. Mamm., 45:387–396.

ELEFTHERIOU, B. E., AND M. X. ZARROW. 1961. A comparison of body weight and thyroid gland activity in two subspecies of *Peromyscus maniculatus* from birth to 70 days of age. Gen. Comp. Endocrinology, 1:534–540.

FAIRBAIRN, D. J. 1977. Why breed early? A study of reproductive tactics in *Peromyscus*. Canadian J. Zool., 55:862–871.

FLAKE, L. D. 1974. Reproduction of four rodent species in a shortgrass prairie of Colorado. J. Mamm., 55:213–216.

FLEMING, T. H., AND R. J. RAUSCHER. 1978. On the evolution of litter size in *Peromyscus leucopus*. Evolution, 32:45–55.

FOLTZ, D. W. 1981. Genetic evidence for long term monogamy in a small rodent, *Peromyscus polionotus*. Amer. Nat., 117:665–675.

FORRESTER, D. J. 1975. Reproductive performance of a laboratory colony of the deer mouse, *Peromyscus maniculatus gambelii*. Lab. Anim. Sci., 25:448–449.

FULLER, W. A. 1969. Changes in numbers of three species of small rodent near Great Slave Lake, N.W.T., Canada, 1964–1967, and their significance for general population theory. Ann. Zool. Fennici, 6:113–144.

GASHWILER, J. S. 1979. Deer mouse reproduction and its relationship to a tree seed crop. Amer. Midland Nat., 102:95–104.

GILBERT, B. S., AND C. J. KREBS. 1981. Effects of extra food on *Peromyscus* and *Clethrionomys* populations in the southern Yukon. Oecologia, 51:326–331.

GLAZIER, D. S. 1979. An energetic and ecological basis for different reproductive rates in five species of *Peromyscus* (mice). Unpubl. Ph.D. dissert., Cornell Univ., Ithaca, New York, 162 pp.

———. 1980. Ecological shifts and the evolution of geographically restricted species of North American *Peromyscus* (mice). J. Biogeogr., 7:63–83.

GUETZOW, D. D., AND F. W. JUDD. 1981. Postnatal growth and development in a subtropical population of *Peromyscus leucopus texanus*. Southwestern Nat., 26:183–191.

GYUG, L. W. 1979. Reproductive and developmental adjustments to breeding season length in *Peromyscus*. Unpubl. M.Sc. thesis, Univ. Western Ontario, London, Ontario, 84 pp.

GYUG, L. W., AND J. S. MILLAR. 1980. Fat levels in a subarctic population of *Peromyscus maniculatus*. Canadian J. Zool., 58:1341–1346.

———. 1981. Growth of seasonal generations in three natural populations of *Peromyscus*. Canadian J. Zool., 59:510–514.

HAIGH, G. R. 1983a. Reproductive inhibition and recovery in young female *Peromyscus leucopus*. J. Mamm., 64:706.

———. 1983b. Effects of inbreeding and social factors on the reproduction of young female *Peromyscus maniculatus bairdii*. J. Mamm. 64:48–54.

HALFPENNY, J. C. 1980. Reproductive strategies: intra and interspecific comparison within the genus *Peromyscus*. Unpubl. Ph.D. dissert., Univ. Colorado, Fort Collins, Colorado, 160 pp.

HALL, E. R., AND W. W. DALQUEST. 1963. The mammals of Veracruz. Univ. Kansas Publ. Mus. Nat. Hist., 14:165–362.

HANSEN, L., AND G. O. BATZLI. 1978. The influence of food availability on the white-footed mouse: populations in isolated woodlots. Canadian J. Zool., 56:2530–2541.

HARLAND, R. M., AND J. S. MILLAR. 1980. Activity of breeding *Peromyscus leucopus*. Canadian J. Zool., 55:313–316.

HARLAND, R. M., P. J. BLANCHER, AND J. S. MILLAR. 1979. Demography of a population of *Peromyscus leucopus*. Canadian J. Zool., 57:323–328.

HELM, J. D., C. S. HERNANDEZ, AND R. H. BAKER. 1974. Observaciones sobre los ratones de las marismas, *Peromyscus perfulvus* Osgood (Rodentia, Criceti-dae). An. Inst. Biol. Univ. Nac. Auton. Mex. Ser. Zool., 45:141–146. [cited by Modi (1984)]

HILL, J. L. 1974. *Peromyscus*: effect of early pairing on reproduction. Science, 186:1042–1044.

HILL, R. W. 1972. The amount of maternal care in *Peromyscus leucopus* and its thermal significance for the young. J. Mamm., 53:774–790.

HOFFMANN, R. S. 1955. A population high for *Peromyscus maniculatus*. J. Mamm., 36:571–572.

HONACKI, J. H., K. E. KINMAN, AND J. W. KOEPPL, EDS. 1982. Mammal species of the world. Allen Press, Inc., Lawrence, Kansas, 694 pp.

HORNER, B. E. 1947. Parental care of young mice of the genus *Peromyscus*. J. Mamm. 28:31–36.

HOWARD, W. E. 1949. Dispersal, amount of inbreeding, and longevity of a local population of prairie deer mice on the George Reserve, southern Michigan. Contr. Lab. Vert. Biol., Univ. Michigan, 43:1–52.

HUGGETT, A. ST. G., AND W. F. WIDDAS. 1951. The relationship between mam-malian foetal weight and conception age. J. Physiol., London, 114: 306–317.

JACKSON, W. B. 1952. Population of the wood mouse *Peromyscus leucopus* subjected to the application of D.D.T. and Parathiol. Ecol. Monogr., 22:259–281.

JAMESON, E. W., JR. 1953. Reproduction of deer mice (*Peromyscus maniculatus* and *P. boylei*) in the Sierra Nevada, California. J. Mamm., 34:44–58.

JONES, J. K., JR. 1964. Distribution and taxonomy of mammals of Nebraska. Univ. Kansas Publ., Mus. Nat. Hist., 16:1–356.

JUDD, F. W., J. HERRERA, AND M. WAGNER. 1978. The relationship between lipid and reproductive cycles of a subtropical population of *Peromyscus leucopus*. J. Mamm., 59:669–676.

JUDD, F. W., G. CARPENTER, AND M. WAGNER. 1984. Variation in reproduction of a subtropical population of *Peromyscus leucopus*. Pp.125–135, *in* Contribu-tions in mammalogy in honor of Robert L. Packard (R. E. Martin and B. R. Chapman, eds.). Spec. Publ. Mus., Texas Tech Univ., 22:1–234.

KAUFMAN, G. A., AND D. W. KAUFMAN. 1977. Body composition of the old-field mouse *Peromyscus polionotus*. J. Mamm., 58:429–434.

KING, J. A. 1958. Maternal behavior and behavioral development in two sub-species of *Peromyscus maniculatus*. J. Mamm., 39:177–190.

———. 1970. Ecological Psychology: An approach to motivation. Nebraska Symposium on Motivation, 1–33.

KING, J. A., J. C. DESHAIES, AND R. WEBSTER. 1963. Age of weaning of two sub-species of deer mice. Science, 139:483–484.

KING, J. A., E. O. PRICE, AND P. L. WEBER. 1968. Behavioral comparisions within the genus *Peromyscus*. Michigan Acad. Sci., Arts and Letters, Vol. *LIII*: 113–136.

KIRKLAND, G. L., JR., AND A. V. LINZEY. 1973. Observations on the breeding suc-cess of the deer mouse *Peromyscus maniculatus nubiterrae*. J. Mamm., 54: 254–255.

KREBS, C. J., AND I. WINGATE. 1985. Population fluctuations in the small mam-mals of the Kluane Region, Yukon Territory. Canadian Field-Nat., 99: 51–61.

LACKEY, J. A.   1973.   Reproduction, growth and development in high-latitude and low-latitude populations of *Peromyscus leucopus* (Rodentia).   Unpubl. Ph.D. dissert., Univ. Michigan, Ann Arbor, Michigan, 128 pp.

———.   1976.   Reproduction, growth and development in the Yucatan deer mice, *Peromyscus yucatanicus*.   J. Mamm., 57:638–655.

———.   1978.   Reproduction, growth and development in high latitude and low-latitude populations of *Peromyscus leucopus* (Rodentia).   J. Mamm., 59: 69–83.

LAFFODAY, S. K.   1957.   A study of prenatal and postnatal development in the old field mouse, *Peromyscus polionotus*.   Unpubl. Ph.D. dissert., Univ. Florida, Gainsville, Florida, 124 pp. [cited by Layne, 1968].

LAMPE, R. P., J. K. JONES, JR., R. S. HOFFMANN, AND E. C. BIRNEY.   1974.   The mammals of Carter Co., southeastern Montana.   Occas. Papers Mus. Nat. Hist., Univ. Kansas, 25: 1–30.

LAWLOR, T. E.   1965.   The Yucatan deer mouse, *Peromyscus yucantanicus*.   Univ. Kansas Publ., Mus. Nat. Hist., 16:421–438.

LAWTON, A. D., AND J. M. WHITSETT.   1979.   Inhibition of sexual maturation by a urinary pheromone in male prairie deer mice.   Horm. Behav., 13: 128–138.

LAYNE, J. N.   1966.   Postnatal development and growth of *Peromyscus floridanus*.   Growth, 30:25–45.

———.   1968.   Ontogeny.   Pp. 148–253, *in* Biology of *Peromyscus* (Rodentia) (J. A. King, ed.).   Spec. Publ., Amer. Soc. Mamm., 2: 1–593.

LEAMY, L.   1981.   The effect of litter size on fertility in *Peromyscus leucopus*.   J. Mamm., 62:692–697.

LEWIS, A. W.   1972.   Seasonal population changes in the cactus mouse, *Peromyscus eremicus*.   Southwestern Nat., 17:85–93.

LINDUSKA, J. P.   1942.   Winter rodent populations in field-shocked corn.   J. Wildl. Manag., 6:353–363.

LINZEY, A. V.   1970.   Postnatal growth and development of *Peromyscus maniculatus nubiterrae*.   J. Mamm., 51:152–155.

LINZEY, D. W., AND A. V. LINZEY.   1968.   Mammals of the Great Smoky Mountains National Park.   J. Elisha Mitchell Sci. Soc., 84:384–414.

LLEWELLYN, J. B.   1978.   Reproductive patterns in *Peromyscus truei truei* in a pinyon-juniper woodland of western Nevada.   J. Mamm., 59; 449–451.

LOMBARDI, J. R., AND J. M. WHITSETT.   1980.   Effects of urine from conspecifics on sexual maturation in female prairie deer mice, *Peromyscus maniculatus bairdii*.   J. Mamm., 61:766–768.

LONG, C. A.   1965.   The mammals of Wyoming.   Univ. Kansas Publ., Mus. Nat. Hist., 14:493–758.

———.   1968.   Populations of small mammals on railroad right-of-way in prairie of central Illinois.   Trans. Illinois State Acad. Sci., 61:139–145.

———.   1973.   Reproduction in the white-footed mouse at the northern limits of its geographic range.   Southwestern Nat., 18:11–20.

LONG, W. S.   1940.   Notes on the life histories of some Utah mammals.   J. Mamm. 21:170–180.

LYNCH, G. R., HEATH, H. W., AND JOHNSTON, C. M.   1981.   Effect of geographical origin on the photoperiodic control of reproduction in the white-footed mouse *Peromyscus leucopus*.   Biol. Reprod., 25:475–480.

MACMILLEN, R. E.   1964.   Population ecology, water relations and social behavior of a southern California semidesert rodent fauna.   Univ. California Publ. Zool., 71 : 1–66.

MANVILLE, R. H.   1952.   A late breeding cycle in *Peromyscus*.   J. Mamm., 33 : 389.

MARTELL, A. M.   1983.   Demography of southern red-backed voles (*Clethrionomys gapperi*) and deer mice (*Peromyscus maniculatus*) after logging in northcentral Ontario.   Canadian J. Zool., 61 : 958–969.

MAY, J. D.   1979.   Demographic adjustments to breeding season length in *Peromyscus*.   Unpubl. M.Sc. thesis, University of Western Ontario, London, Ontario, 63 pp.

MCCABE, T. T., AND B. D. BLANCHARD.   1950.   Three species of *Peromyscus*.   Rood Associates, Santa Barbara, California, 136 pp.

MCKEEVER, S.   1964.   Variation in the weight of the adrenal, pituitary and thyroid gland of the white-footed mouse, *Peromyscus maniculatus*.   Amer. J. Anat., 114 : 1–15.

MCLAREN, S. B., AND G. L. KIRKLAND, JR.   1979.   Geographic variation in litter size of small mammals in the central Appalachian region.   Proc. Pennsylvania Acad. Sci., 53 : 123–126.

MERRITT, J. F., AND J. M. MERRITT.   1980.   Population ecology of the deer mouse in the front range of Colorado.   Ann. Carnegie Mus., 49 : 113–130.

MERSON, M. H., AND R. L. KIRKPATRICK.   1981.   Relative sensitivity of reproductive activity and body-fat levels to food restriction in white-footed mice.   Amer. Midland Nat., 106 : 305–312.

MERSON, M. H., R. L. KIRKPATRICK, P. F. SCANLON, AND F. C. GWAZDAUSKAS.   1983.   Influence of restricted food intake on reproductive characteristics and carcass fat of male white-footed mice.   J. Mamm., 64 : 353–355.

METZGAR, L. H.   1979.   Dispersion patterns in a *Peromyscus* population.   J. Mamm., 60 : 129–145.

MIHOK, S.   1979.   Behavioral structure and demography of subarctic *Clethrionomys gapperi* and *Peromyscus maniculatus*.   Canadian J. Zool., 57 : 1520–1535.

MILLAR, J. S.   1975.   Tactics of energy partitioning in breeding *Peromyscus*.   Canadian J. Zool., 53 : 967–976.

———.   1978.   Energetics of reproduction in *Peromyscus leucopus*: the cost of lactation.   Ecology, 59 : 1055–1061.

———.   1979.   Energetics of lactation in *Peromyscus maniculatus*.   Canadian J. Zool., 57 : 1015–1019.

———.   1981.   Body composition and nutrient reserves of northern *Peromyscus leucopus*.   J. Mamm., 62 : 786–794.

———.   1982.   Life cycle characteristics of northern *Peromyscus maniculatus borealis*.   Canadian J. Zool., 60 : 510–515.

———.   1983.   Negative maternal effects in *Peromyscus maniculatus*.   J. Mamm., 64 : 540–543.

———.   1984a.   Reproduction and survival of *Peromyscus* in seasonal environments.   Pp. 253–266, *in* Winter ecology of small mammals. (J. F. Merritt, ed). Carnegie Mus. Nat. Hist., Spec. Publ., 10 : 1–380.

———.   1984b.   The role of design constraints in the evolution of mammalian reproductive rates.   Acta. Zool. Fennica, 171 : 133–136.

———.   1985.   Life cycle characteristics of *Peromyscus maniculatus nebrascensis*.   Canadian J. Zool., 63 : 1280–1284.

MILLAR, J. S., AND L. W. GYUG.   1981.   Initiation of breeding by northern *Peromyscus* in relation to temperature.   Canadian J. Zool., 59 : 1094–1098.

MILLAR, J. S., AND D. G. L. INNES. 1983. Demographic and life cycle character-istics of montane *Peromyscus maniculatus borealis*. Canadian J. Zool., 61: 574–585.

———. 1985. Breeding by *Peromyscus maniculatus* over an elevational gradient. Canadian J. Zool., 63:124–129.

MILLAR, J. S., F. B. WILLE, AND S. L. IVERSON. 1979. Breeding by *Peromyscus* in seasonal environments. Canadian J. Zool., 57:719–727.

MILLER, D. H., AND L. L. GETZ. 1977. Comparisons of population dynamics of *Peromyscus* and *Clethrionomys* in New England. J. Mamm., 58:1–16.

MODI, W. S. 1984. Reproductive tactics among deer mice of the genus *Pero-myscus*. Canadian J. Zool., 62:2576–2581.

MORRISON, P., R. DIETERICH, AND D. PRESTON. 1977. Body growth in sixteen rodent species and subspecies maintained in laboratory colonies. Physiol. Zool., 50:294–310.

MYERS, P., AND L. L. MASTER. 1983. Reproduction by *Peromyscus maniculatus*: size and compromise. J. Mamm., 64:1–18.

MYERS, P., L. L. MASTER, AND R. A. GARRETT. 1985. Ambient temperature and rainfall: an effect on sex ratio and litter size in deer mice. J. Mamm., 66: 289–298.

NADEAU, J. H., R. T. LOMBARDI, AND R. H. TAMARIN. 1981. Population structure and dispersal of *Peromyscus leucopus* on Muskeget Island. Canadian J. Zool., 59:793–799.

NICHOLSON, A. J. 1941. The homes and social habits of the wood mouse (*Pero-myscus leucopus noveboracensis*) in southern Michigan. Amer. Midland Nat., 25:196–223.

POURNELLE, G. H. 1952. Reproduction and early post-natal development of the cotton mouse, *Peromyscus gossypinus gossypinus*. J. Mamm., 33:1–20.

PRICE, E. 1967. The effect of reproductive perfomance on the domestication of the prairie deermice, *Peromyscus maniculatus bairdii*. Evolution, 21: 762–770.

RAND, A. L., AND P. HOST. 1942. Results of the Archbold expeditions. No. 45. Mammal notes from Highlands Co., Florida. Bull. Amer. Mus. Nat. Hist., 80:1–22.

REDFIELD, J. A., C. J. KREBS, AND M. J. TAITT. 1977. Competition between *Pero-myscus maniculatus* and *Microtus townsendii* in grasslands of coastal British Co-lumbia. J. Anim. Ecol., 46:607–616.

REED, E. B. 1955. January breeding of *Peromyscus* in North Central Colorado. J. Mamm., 36:462–463.

REICHMAN, O. J., AND K. M. VAN DE GRAAF. 1973. Seasonal activity and repro-duction patterns of five species of Sonoran Desert rodents. Amer. Mid-land Nat., 90:118–126.

RICKART, E. A. 1977. Reproduction, growth and development in two species of cloud forest *Peromyscus* from southern Mexico. Occas. Papers Mus. Nat. Hist., Univ. Kansas, 67:1–22.

RINTAMAA, D. L., P. A. MAZUR, AND S. H. VESSEY. 1976. Reproduction during two annual cycles in a population of *Peromyscus leucopus noveboracensis*. J. Mamm., 57:593–595.

ROBERTSON, P. B. 1975. Reproduction and community structure of rodents over a transect in southern Mexico. Unpubl. Ph.D. dissert., Univ. Kansas, Law-rence, Kansas, 113 pp.

ROGERS, J. G., JR., AND G. K. BEAUCHAMP. 1974. Relationship among three criteria of puberty in *Peromyscus leucopus noveboracensis.* J. Mamm., 55: 461–462.

———. 1976. Influence of stimuli from populations of *Peromyscus leucopus* on maturation of young. J. Mamm., 57:320–329.

ROOD, J. D. 1966. Observations on the reproduction of *Peromyscus* in captivity. Amer. Midland Nat., 76:496–503.

SADLEIR, R. M. F. S. 1974. The ecology of the deer mouse *Peromyscus maniculatus* in a coastal coniferous forest. II. Reproduction. Canadian J. Zool., 52: 119–131.

SADLEIR, R. M. F. S., K. D. CASPERSON, AND J. HARLING. 1973. Intake and requirements of energy and protein for the breeding of wild deermice, *Peromyscus maniculatus.* J. Reprod. Fert. Suppl., 19:237–252.

SAVIDGE, I. R. 1974. Maternal aggressiveness and litter survival in deer mice (*Peromyscus maniculatus bairdii*). Amer. Midland Nat., 91:449–451.

SCHEFFER, T. H. 1924. Notes on the breeding of *Peromyscus.* J. Mamm., 5: 258–260.

SCHMIDLY, D. J., M. R. LEE, W. S. MODI, AND E. G. ZIMMERMAN. 1985. Systematics and notes on the biology of Peromyscus hooperi. Occas. Papers Mus., Texas Tech Univ., 97:1–40.

SHEPPE, W. 1963. Population structure of the deer-mouse, *Peromyscus,* in the Pacific Northwest. J. Mamm., 44:180–182.

SKRYJA, D. D. 1978. Reproductive inhibition in female cactus mice (*Peromyscus eremicus*). J. Mamm., 59:543–550.

SLEEPER, R. A. 1980. Litter survival of deer mice (*Peromyscus maniculatus*) in nest boxes. Southwestern Nat., 25:259–260.

SMITH, A. T. 1982. Population and reproductive trends of *Peromyscus gossypinus* in the everglades of south Florida. Mammalia, 46:467–475.

SMITH, A. T., AND J. M. VRIEZE. 1979. Population structure of everglade rodents: responses to a patchy environment. J. Mamm., 69:778–794.

SMITH, M. H., AND J. T. McGINNIS. 1968. Relationships of latitude, altitude and body size to litter size and mean annual production of offspring in *Peromyscus.* Res. Population Biol., 10:115–126.

SMITH, W. K. 1939. An investigation into the early embryology and associated phenomena of *Peromyscus polionotus.* Unpubl. M.Sc. thesis, Univ. Florida, Gainsville, Florida, 35 pp. [cited by Layne (1968)].

SPENCER, A. W., AND H. W. STEINHOFF. 1968. An explanation of geographic variation in litter size. J. Mamm., 49:281–186.

STEARNS, S. C. 1983. The influence of size and phylogeny on patterns of covariation among life-history traits in the mammals. Oikos, 41:173–187.

STEBBINS, L. L. 1977. Energy requirements during reproduction of *Peromyscus maniculatus.* Canadian J. Zool., 55:1701–1704.

STORER, T. I., F. C. EVANS, AND F. G. PALMER. 1944. Some rodent populations in the Sierra Nevada of California. Ecol. Monogr., 14:165–192.

SULLIVAN, T. P. 1977. Demography and dispersal in island and mainland populations of the deer mouse, *Peromyscus maniculatus.* Ecology, 58:964–978.

———. 1979. Demography of populations of deer mice in coastal forest and clear-cut (logged) habitats. Canadian J. Zool., 57:1636–1648.

SULLIVAN, T. P., AND C. J. KREBS. 1981. An irruption of deer mice after logging of coastal coniferous forest. Canadian J. Forest Res., 11:586–592.

SULLIVAN, T. P., AND D. S. SULLIVAN. 1981. Responses of a deer mouse population to a forest herbicide application: reproduction, growth and survival. Canadian J. Zool., 59:1148–1154.

SVENDSEN, G. 1964. Comparative reproduction and development in two species of mice in the genus *Peromyscus*. Trans. Kansas Acad. Sci., 67:527–538.

SVIHLA, A. 1932. A comparative life history study of the mice of the genus *Peromyscus*. Misc. Publ. Mus. Zool., Univ. Michigan, 24:1–39.

———. 1934. Development and growth of deermice (*Peromyscus maniculatus artemisiae*). J. Mamm., 15:99–104.

———. 1935. Development and growth of the prairie deer mouse, *Peromyscus maniculatus bairdii*. J. Mamm., 16:109–115.

———. 1936. Development and growth of *Peromyscus maniculatus oreas*. J. Mamm., 17:132–137.

TAITT, M. J. 1981. The effect of extra food on small rodent populations: l. Deermice (*Peromyscus maniculatus*). J. Anim. Ecol., 50:111–124.

TEAGUE, R. A., AND E. L. BRADLEY. 1978. The existence of a puberty accelerating pheromone in the urine of the male prairie deermice (*Peromyscus maniculatus bairdii*). Biol. Reprod., 19:314–317.

TERMAN, C. R. 1965. A study of population growth and control exhibited in the laboratory by prairie deermice. Ecology, 46:890–895.

———. 1968. Inhibition of reproductive maturation and function in laboratory populations of prairie deermice: a test of pheromone influence. Ecology, 49:1169–1172.

———. 1969. Weights of selected organs of deermice (*Peromyscus maniculatus bairdii*) from asymptotic laboratory populations. J. Mamm., 50:311–320.

———. 1973. Recovery of reproductive function by prairie deermice (*Peromyscus maniculatus bairdii*) from asymptotic populations. Anim. Behav., 21:443–448.

———. 1979. Inhibition of reproductive development in laboratory populations of prairie deermice (*Peromyscus maniculatus bairdii*): influence of tactile cues. Ecology, 60:455–458.

———. 1980. Social factors influencing delayed reproductive maturation in prairie deermice (*Peromyscus maniculatus bairdii*) in laboratory populations. J. Mamm., 61:219–223.

TERMAN, C. R., AND E. L. BRADLEY. 1981. The influence of urine from asymptotic laboratory populations on sexual maturation in prairie deermice. Res. Population Ecol., 23:168–176.

TEVIS, L., JR. 1956. Responses of small mammal populations to logging of Douglas-fir. J. Mamm., 37:189–196.

THOMAS, B. 1971. Evolutionary relationships among *Peromyscus* from the Georgia strait, Gordon, Goletas and Scott islands of British Columbia, Canada. Unpubl. Ph.D. dissert., Univ. British Columbia, Vancouver, British Columbia, 159 pp.

VAN HORNE, B. 1981. Demography of *Peromyscus maniculatus* populations in seral stages of coastal coniferous forest in southeast Alaska. Canadian J. Zool., 59:1045–1061.

VAUGHAN, T. A. 1969. Reproduction and population densities in a montane small mammal fauna. Misc. Publ. Mus. Nat. Hist., Univ. Kansas, 51:51–74.

VESTAL, B. M., W. C. COLEMAN, AND P. R. CHU. 1980. Age of first leaving the nest in two species of deer mice (*Peromyscus*). J. Mamm., 61:143–146.

WILLIAMS, R. G., J. L. CARMON, AND F. B. GOLLEY. 1965a. Effect of sequence of pregnancy on litter size and growth in *Peromyscus polionotus*. J. Reprod. Fert., 9:257–260.

WILLIAMS, R. G., F. B. GOLLEY, AND J. L. CARMON. 1965b. Reproductive performance of a laboratory colony of *Peromyscus polionotus*. Amer. Midland Nat., 73:101–110.

WOLFF, J. O. 1985. Comparative population ecology of *Peromyscus leucopus* and *Peromyscus maniculatus*. Canadian J. Zool., 63:1548–1555.

APPENDIX I

Gestation periods and fetal growth rates within and among species of Peromyscus. Means are presented with ±SE; ranges are in parentheses. Fetal growth rates are based on the formula of Huggett and Widdas (1951)

Species Location	Adult weight (g)	Neonate weight (g)	Litter size	Gestation (days) Non-lactating	Gestation (days) Lactating	Fetal growth rate Non-lactating	Fetal growth rate Lactating	Source
P. maniculatus								
Douglas Co., Kansas	20	1.8	3.8	22.4±0.1 (22–23)	24.1±0.3 (22–27)	0.090	0.084	Svendsen, 1964
Ann Arbor, Michigan	21	1.7	6.0	23	27	0.086	0.073	Myers and Master, 1983
Columbia Co., Washington	—	1.7	4.5	23.4±0.1 (22–26)	30.6±0.8 (27–34)	0.085	0.064	Svihla, 1932
Alamogordo, New Mexico	—	1.8	3.8	23.6±0.2 (22–25)	25.3±0.4 (23–29)	0.086	0.080	Svihla, 1932
NE Utah	—	1.7	4.6	23.6±0.2 (22–27)	27.0±1.1 (22–35)	0.084	0.074	Svihla, 1932
California & New Mexico	—	1.7	4.3	23.5±0.1 (23–24)	26.6±0.7 (23–32)	0.084	0.075	Svihla, 1932
Livingston Co., Michigan	—	—	—	(24–?)[1]	—	—	—	Howard, 1949
Smoky Mts., Tennessee	18	1.8	4.1	—	28.5±0.8 (23–?)	—	0.071	Linzey, 1970
Great Slave Lake, North West Territories	19	1.9	5.0	(23–31)	26.3±0.8	—	0.078	Millar, 1982
Ann Arbor, Michigan	19	—	4.7	(21–?)	—	—	—	Rood, 1966; King, 1970

APPENDIX I
(Continued)

Species Location	Adult weight (g)	Neonate weight (g)	Litter size	Gestation (days)		Fetal growth rate		Source
				Non-lactating	Lactating	Non-lactating	Lactating	
N Michigan	24	—	4.8	— (23–?)	—	—	—	Rood, 1966; King, 1970
SE Alberta	20	1.9	5.1	25.5±0.3 (23–26)	29.5±1.4 (24–35)	0.079	0.069	Millar, 1985
Average	20	1.8	4.6	23.6	26.9	0.085	0.074	
P. leucopus								
Asheville, North Carolina	—	1.9	3.4	23.2±0.4 (22–24)	28.6±0.8 (23–34)	0.088	0.072	Svihla, 1932
Michigan, New York, Missouri	—	1.9	4.4	23.0±0.1 (22–25)	30.0±0.7 (23–37)	0.089	0.069	Svihla, 1932
Douglas Co., Kansas	27	2.2	4.3	22.5±0.1 (22–23)	23.9±0.2 (22–28)	0.096	0.090	Svendsen, 1964
Lansing, Michigan	21	—	4.7	—	— (20–?)	—	—	Rood, 1966; King, 1970
Campeche, Mexico	22	2.0	4.8	—	— (22–?)	—	—	Lackey, 1978
Ann Arbor, Michigan	20	1.9	4.3	—	24 (22–24)	—	0.086	Lackey, 1978
Average	22	2.0	4.3	22.9	26.6	0.091	0.079	
P. attwateri								
S.W. Missouri	29	—	3.4	23	29.3 (26–32)	—	—	Brown, 1964a, 1964b
P. crinitus								
Bernardino Co., California	—	—	3.0	—	— (25–?)	—	—	Rood, 1966

Toole Co., Utah	15	2.2	3.1	24.5	30 (27–31)	0.088	0.072	Egoscue, 1964
Average	15	2.2	3.0	24.5	30	0.088	0.072	
P. interparietalis								
Baja, California	21	2.7	2.4	—	31.9	—	0.072	Brand and Ryckman, 1968
P. eremicus								
Baja, California	19	2.2	2.2	—	35.0	—	0.062	Brand and Ryckman, 1968
San Diego, California	20	—	2.6	—	(27–?)	—	—	Rood, 1966; King, 1970
Carrizoza, New Mexico	—	2.5	2.6	21.0	—	0.107	—	Svihla, 1932
Average	19	2.3	2.5	21.0	35.0	0.107	0.062	
P. californicus								
Berkeley, California	53	—	2.0	—	(24–?)	—	—	Rood, 1966; King, 1970
Corralites, California	—	4.9	1.9	23.6±0.4 (21–25)	—	0.119	—	Svihla, 1932
Average	53	4.9	1.9	23.6	—	0.119	—	
P. poliomotus								
Aiken, South Carolina	—	1.7	3.6	23.5	29.4±0.3 (23–41)	0.084	0.067	Williams et al., 1965b
Florida	15	—	3.3	23.8 (23–24)	28.0 (25–31)	—	—	Laffoday, 1957; Smith, 1939 (from Layne, 1968)
Ocala, Florida	13	—	3.9	—	(21–)	—	—	Rood, 1966; King, 1970
Average	14	1.7	3.6	23.6	28.7	0.084	0.067	

APPENDIX I

(*Continued*)

Species Location	Adult weight (g)	Neonate weight (g)	Litter size	Gestation (days)		Fetal growth rate		Source
				Non-lactating	Lactating	Non-lactating	Lactating	
P. truei New Mexico, Colorado	—	2.3	2.8	26.2±0.3 (25−27)	40	0.084	0.055	Svihla, 1932
P. oreas Mount Baker, Washington	—	1.6	—	23	25	0.085	0.078	Svihla, 1936
P. yucatanicus Campeche, Mexico	29	2.5	2.8	27.8±0.2 (27−28)	31.6±0.4 (31−33)	0.081	0.071	Lackey, 1976
P. gossypinus N Florida	29	2.2	3.7	23	30	0.094	0.072	Pournelle, 1952
P. melanocarpus Oaxaca, Mexico	59	4.5	2.3	30	37.1 (31−34)	0.091	0.074	Rickart, 1977
P. mexicanus Oaxaca, Mexico	60	4.4	2.6	30	35 (31−39)	0.091	0.078	Rickart, 1977
P. floridanus Delray Beach, Florida	38	—	1.7	—	(33−?)	—	—	Rood, 1966; King, 1970

[1] Lactation not recorded

Appendix II

Neonate Weights of Peromyscus.

Species Location	Neonate weight (g)	Adult weight (g)	Litter size	Source
P. maniculatus				
Columbia Co., Washington	1.71	21	4.4	Svihla 1932, 1934
Columbia Co., Washington	1.81	21	4.4	Svihla 1932, 1935
Michigan, North Dakota, Iowa	1.67	15	3.0	Svihla 1932, 1934
Michigan, North Dakota, Iowa	1.62	15	3.0	Svihla 1932, 1935
Alamogordo, New Mexico	1.80	—	3.8	Svihla 1932
Dagett Co., Utah	1.72	—	4.6	Svihla 1932
Colorado, New Mexico	1.67	—	4.3	Svihla 1932
San Francisco, California	1.40	19	5.0	McCabe and Blanchard, 1950
unknown origin	2.00	16	—	Eleftheriou and Zarrow, 1961
unknown origin	2.00	18	—	Eleftheriou and Zarrow, 1961
Brighton, Colorado	1.83	21	4.1	Halfpenny, 1980
Mt. Evans, Colorado	1.82	21	6.4	Halfpenny, 1980
Ward, Colorado	1.75	23	5.1	Halfpenny, 1980
Smoky Mts., Tennessee	1.80	18	4.1	Linzey, 1970
Pinawa, Manitoba	1.65	17	—	Millar, 1979
Pinawa, Manitoba	1.57	20	5.3	Millar *et al.*, 1979
Great Slave Lake, North West Territories	1.87	19	5.0	Millar, 1982
Kananaskis, Alberta	1.70	21	5.2	Millar and Innes, 1983
Southeastern Alberta	1.95	20	5.1	Millar, 1985
Ann Arbor, Michigan	1.68	21	6.0	Myers and Master, 1983
Douglas Co., Kansas	1.81	20	3.8	Svendsen, 1964
Balaclava Island, British Columbia	2.24	25	4.0	Thomas, 1971
Vansittart Island, British Columbia	2.30	25	4.0	Thomas, 1971
Cox Island, British Columbia	2.40	28	4.0	Thomas, 1971
Hope Island, British Columbia	2.07	24	4.5	Thomas, 1971
Nigei Island, British Columbia	2.83	30	2.0	Thomas, 1971
Average	1.87	21	4.4	

Appendix II

(Continued)

Species Location	Neonate weight (g)	Adult weight (g)	Litter size	Source
P. leucopus				
Asheville, North Carolina	1.85	22	3.4	Svihla, 1932
Michigan, New York, Missouri	1.87	21	4.4	Svihla, 1932
East Lansing, Michigan	1.73	23	4.8	Fleming and Rauscher, 1978
Campeche, Mexico	1.99	22	4.8	Lackey, 1978
Ann Arbor, Michigan	1.89	20	4.3	Lackey, 1978
London, Ontario	1.80	22	5.0	Millar, 1975
London, Ontario	1.79	22	4.6	Millar, 1978
London, Ontario	1.92	22	4.9	Millar *et al.*, 1979
Douglas Co., Kansas	2.18	27	4.3	Svendsen, 1964
Rio Grande, Texas	1.80	20	3.6	Guetzow and Judd, 1981
Average	1.88	22	4.4	
P. polionotus				
Highlands Co., Florida	1.3	13	4.0	Rand and Host, 1942
Florida	1.61	15	3.3	Laffoday, 1957
Aiken, South Carolina	2.15	15	—	Carmon *et al.*, 1963
Aiken, South Carolina	1.59	—	—	Kaufman and Kaufman, 1977
Average	1.66	14	3.6	
P. gossypinus				
North Florida	2.19	29	3.7	Pournelle, 1952
Highlands Co., Florida	1.60	29	3.5	Rand and Host, 1942
Average	1.90	29	3.6	
P. truei				
New Mexico and Colorado	2.31	—	2.8	Svihla, 1932
San Francisco, California	2.34	27	3.4	McCabe and Blanchard, 1950
Average	2.32	27	3.1	
P. californicus				
Corralitos, California	4.92	38	1.9	Svihla, 1932
San Francisco, California	4.31	39	1.9	McCabe and Blanchard, 1950
Average	4.61	38	1.9	
P. oreas				
Pullman, Washington	1.63	—	—	Svihla, 1936
P. crinitus				
Tooele Co., Utah	2.20	15	3.1	Egoscue, 1964
P. megalops				
Mexico	3.90	71	1.6	Layne, 1968

Appendix II

(Continued)

Species Location	Neonate weight (g)	Adult weight (g)	Litter size	Source
P. eremicus				
Carrizoza, New Mexico	2.54	20	2.6	Svihla, 1932
Riverside Co., California	2.23	19	2.2	Brand and Ryckman, 1968
Average	2.38	19	2.4	
P. floridanus				
Alachua Co., Florida	2.40	27	3.1	Layne, 1966
P. thomasi				
Mexico	4.50	77	3.5	Layne, 1968
P. interparietalis				
Baja California, Mexico	2.74	21	2.4	Brand and Ryckman, 1968
P. yucatanicus				
Campeche, Mexico	2.50	29	2.8	Lackey, 1976
P. melanocarpus				
Oaxaca, Mexico	4.50	59	2.3	Rickart, 1977
P. mexicanus				
Oaxaca, Mexico	4.40	61	2.6	Rickart, 1977
Monteverde, Costa Rica	3.60	45	2.8	Anderson, 1982
Average	4.00	53	2.7	

APPENDIX III

Growth of Peromyscus. *Growth represented by von Bertalanffy growth constants for nestlings and independent young.*

Species Location	Adult weight (g)	Litter size	Growth constant Nestings	Growth constant Post-weaning	Source
P. maniculatus					
Pinawa, Manitoba	17	—	0.045	—	Millar, 1979
Pinawa, Manitoba	20	5.3	0.037	—	Millar et al., 1979
Great Slave Lake, North West Territories	19	5.0	0.042	—	Millar, 1982
Southeastern Alberta	20	5.1	0.048	—	Millar, 1985
Smoky Mts., Tennessee	18	4.1	0.045	0.056	Linzey, 1970
Atotonilco, Mexico	25	4.5	♂0.041 ♀0.044	♂0.063 ♀0.040	Drickamer and Bernstein, 1972
North Platte Valley, Nebraska	19	3.7	♂0.044 ♀0.041	♂0.086 ♀0.056	Drickamer and Bernstein, 1972
San Francisco, California	19	5.0	0.042	0.060	McCabe and Blanchard, 1950
Brighton, Colorado	21	4.1	0.048	—	Halfpenny, 1980
Ward, Colorado	23	5.1	0.047	—	Halfpenny, 1980
Mt. Evans, Colorado	22	6.4	0.029	—	Halfpenny, 1980
Midwest United States	21[1]	4.4	♂0.035 ♀0.039	— —	Morrison et al., 1977
Edmonton, Alberta	20[1]	3.9	♂0.040 ♀0.042	— —	Morrison et al., 1977
Columbia Co., Washington	21	4.4	0.046	0.030	Layne, 1968; Svihla, 1934
Michigan, North Dakota, Iowa	15	3.0	0.043	0.091	Layne, 1968; Svihla, 1935
Average	20	4.6	0.042	0.060	
P. leucopus					
Campeche, Mexico	22	4.8	0.040	0.052	Lackey, 1973, 1978
Ann Arbor, Michigan	20	4.3	0.038	0.054	Lackey, 1973, 1978
Lower Rio Grande, Texas	20	3.5	0.075	0.038	Guetzow and Judd, 1981
London, Ontario	22	4.9	0.039	—	Millar et al., 1979
London, Ontario	22	4.6	0.037	—	Millar, 1978
Average	21	4.4	0.046	0.048	
P. polionotus					
Aiken, South Carolina	15	— —	♂0.038 ♀0.041	♂0.049 ♀0.068	Carmon et al., 1963
Highlands Co., Florida	13	4.0	0.055	0.032	Layne, 1968; Rand and Host, 1942
Florida	15	3.3	0.045	0.070	Layne, 1968; Laffoday, 1957
Average	15	3.6	0.050	0.051	
P. yucatanicus					
Yucatan, Mexico	29	2.8	0.030	0.041	Lackey, 1976
P. mexicanus					
Oaxaca, Mexico	61	2.6	0.030	0.036	Rickart, 1977
P. melanocarpus					
Oaxaca, Mexico	59	2.3	0.029	0.029	Rickard, 1977
P. gossypinus					
N Florida	29	3.7	0.041	0.047	Pournelle, 1952
P. truei					
San Francisco, California	27	3.4	0.031	0.036	McCabe and Blanchard, 1950
P. californicus					
San Francisco, California	39	1.9	0.036	0.042	McCabe and Blanchard, 1950
P. megalops					
Mexico	71	1.6	0.020	0.022	Layne, 1968
P. floridanus					
Alachua Co., Florida	27	3.2	0.030	0.042	Layne, 1966; pers. comm.
P. thomasi					
Mexico	77	3.5	0.021	0.024	Layne, 1968

[1] At 20 weeks of age

APPENDIX IV

Age (days) at developmental events in nestling Peromyscus.

Species Location	Adult weight (g)	Litter size	Neonate weight (g)	Ear pinnae	Lower incisors	Upper incisors	Auditory meatus	Eyes open	Source
P. maniculatus									
North Platte Valley, Nebraska	19	3.7	—	—	—	—	—	13.2	Drickamer and Bernstein, 1972
Atotonilco, Mexico	25	4.5	—	—	—	—	—	14.0	Drickamer and Bernstein, 1972
N Michigan	—	5	—	—	5.7	—	—	16.8	King, 1958
Ann Arbor, Michigan	—	5	—	—	5.2	—	—	12.1	King, 1958
Ann Arbor, Michigan	19	—	—	—	—	—	—	13.0	King, 1970; King et al., 1968
N Michigan	24	—	—	—	—	—	—	16.1	King, 1970; King et al., 1968
North Platte Valley, Nebraska	19	—	—	—	—	—	—	13.1	King, 1970; Drickamer and Bernstein, 1972
Atotonilco, Mexico	25	—	—	—	—	—	—	13.9	King, 1970; Drickamer and Bernstein, 1972
Columbia Co., Washington	—	4.4	1.7	—	—	—	—	14.5	Svihla, 1932
Michigan, North Dakota, Iowa	—	3.0	—	—	—	—	—	13.7	Svihla, 1932
Alamogordo, N. Mexico	—	3.8	1.8	—	—	—	—	14.4	Svihla, 1932
La Jolla, California	—	3.2	1.7	—	—	—	—	13.3	Svihla, 1932
NE Utah	—	4.6	1.8	—	—	—	—	13.7	Svihla, 1932
Carlotta, California	—	—	1.7	—	—	—	—	15.0	Svihla, 1932
California, New Mexico	—	4.3	1.7	—	—	—	—	15.1	Svihla, 1932

APPENDIX IV
(Continued)

Species Location	Adult weight (g)	Litter size	Neonate weight (g)	Ear pinnae	Lower incisors	Upper incisors	Auditory meatus	Eyes open	Source
Sandpoint, Michigan	—	4.0	1.6	3	—	—	—	14.0	Svihla, 1935
Adams Co., Washington	21	—	1.8	3	—	—	—	15.0	Svihla, 1934, 1935
Cocke Co., Tennessee	18	4.1	1.8	3.7	6.6	7.9	10.9	15.4	Linzey, 1970
Pinawa, Manitoba	17	5.3	1.7	—	—	—	—	13.5	Millar, 1979
Great Slave Lake, North West Territories	19	5.0	1.9	—	—	—	—	13.5	Millar, 1982
Hope Island, British Columbia	24	4.5	2.1	2	—	—	—	14.0	Thomas, 1971
Balaclava Island, British Columbia	25	4.0	2.2	2	—	—	—	16.0	Thomas, 1971
Cox Island, British Columbia	28	4.0	2.4	2	—	—	—	17.5	Thomas, 1971
Texada Island, British Columbia	27	3.3	—	3	—	—	—	16.5	Thomas, 1971
Vansittart Island, British Columbia	25	4.0	2.3	3.5	—	—	—	16.0	Thomas, 1971
Nigei Island, British Columbia	31	2.0	2.8	3	—	—	—	—	Thomas, 1971
East Lansing, Michigan	—	—	—	—	—	—	—	11.9	Thomas, 1971
East Lansing, Michigan	—	—	—	—	—	—	—	13.0	Thomas, 1971
Average	23	4.1	1.9	2.8	5.8	7.9	10.0	14.4	

									Reference
P. leucopus									
Lower Rio Grande, Texas	20	3.5	1.8	—	—	—	10.3	12.2	Guetzow and Judd, 1981
Lansing, Michigan	21	—	—	—	—	—	—	12.5	King, 1970; Drickamer and Vestel, 1973
Asheville, North Carolina	—	3.4	1.9	—	—	—	—	13.1	Svihla, 1932
Michigan, New York, Missouri	—	4.4	1.9	—	—	—	—	13.4	Svihla, 1932
Campeche, Mexico	22	4.8	2.0	2.3	(4.0)[1]	—	9.6	10.9	Lackey, 1978
Ann Arbor, Michigan	20	4.3	1.9	2.5	(4.6)[1]	—	10.4	12.2	Lackey, 1978
London, Ontario	23	5.0	1.8	—	—	—	—	13.5	Millar, 1975
East Lansing, Michigan	—	—	—	—	—	—	—	11.4	Vestal et al., 1980
Average	21.2	4.2	1.9	2.4	(4.3)[1]	—	10.1	12.4	
P. eremicus									
Baja California, Mexico	19	2.2	2.2	1.0	2.5	2.5	10.5	14.0	Brand and Ryckman, 1968
San Diego Co., California	20	—	—	—	—	—	—	11.7	King, 1970; Drickamer and Vestel, 1973
Carrizoza, New Mexico	—	2.6	2.5	—	—	—	—	15.5	Svihla, 1932
Average	19	2.4	2.3	1.0	2.5	2.5	10.5	13.7	
P. crinitus									
Toole Co., Utah	15	3.1	2.2	—	—	—	—	16.0	Egoscue, 1964
San Bernardino Co., California	—	—	—	—	—	—	—	12.3	King, 1970; Drickamer and Vestel, 1973
Average	15	3.1	2.2	—	—	—	—	14.1	

APPENDIX IV
(Continued)

Species Location	Adult weight (g)	Litter size	Neonate weight (g)	Ear pinnae	Lower incisors	Upper incisors	Auditory meatus	Eyes open	Source
P. poliomotus									
Ocala, Florida	13	—	—	—	—	—	—	13.5	King, 1970; Drickamer and Vestel, 1973
Highlands Co., Florida	13	4.0	1.3	4.5	6	8	—	14.0	Rand and Host, 1942
Florida	15	3.3	1.6	3.7	6	—	—	13.6	Layne, 1968, from Laffoday, 1957
Average	13.7	3.6	1.4	4.1	6	8	—	13.7	
P. californicus									
Berkeley, California	53.1	—	—	—	—	—	—	14.7	King, 1970; Drickamer and Vestel, 1973
Corralites, California	—	1.9	4.9	—	—	—	—	13.7	Svihla, 1932
Average	53.1	1.9	4.9	—	—	—	—	14.2	
P. mexicanus									
Oaxaca, Mexico	60	2.6	4.4	4.6	10.5	12.0	15.5	19.3	Rickart, 1977
Monteverde, Costa Rica	45	2.8	3.6	5.6	11.2	13.9	16.5	20.9	Anderson, 1982
Average	52.5	2.7	4.0	5.1	11.3	12.9	16.0	20.1	
P. floridanus									
Delray Beach, Florida	38	—	—	—	—	—	—	14.2	King, 1970; Drickamer and Vestel, 1973
Alachua Co., Florida	27	3.1	2.4	3.9	6.8	9.0	12.2	16.5	Layne, 1966
Average	32	3.1	2.4	3.9	6.8	9.0	12.2	15.3	

P. interparietalis								
Baja California, Mexico	21	2.4	2.7	1	4.5	6	10	11.5	Brand and Ryckman, 1968
P. truei								
New Mexico, Colorado	—	2.8	2.3	—	—	—	—	17.5	Svihla, 1932
P. oreas								
Mt. Baker, Washington	—	—	1.6	4	—	—	—	16.0	Svihla, 1936
P. yucatanicus								
Campeche, Mexico	29	2.8	2.5	3.3		(8)[1]	14.2	17.5	Lackey, 1976
P. gossypinus								
N Florida	29	3.7	2.2	4		(6)[1]	—	13.0	Pournelle, 1952
P. melanocarpus								
Oaxaca, Mexico	59	2.3	4.5	4.7	11.6	14.8	17.6	21.9	Rickart, 1977
P. thomasi								
Mexico	77	3.5	4.5	3.5	—	—	14	19.5	Layne, 1968
P. megalops								
Mexico	71	1.6	3.9	5.8	9.3	—	17.7	22.8	Layne, 1968

[1] Upper and lower incisors not given separately

Appendix V

Litter size among populations of Peromyscus.

Species Location	Adult wt. (g)	Mean litter size	N	Source
P. maniculatus				
Chihuahua, Mexico	25	3.6	21	Anderson, 1972
Lake Co., Minnesota	—	5.3	71	Beer *et al.*, 1957
Dakota Co., Minnesota	—	5.1	217	Beer *et al.*, 1957
Livingston Co., Michigan	—	4.0	43	Blair, 1940
Williamson Co., Texas	—	5.0	31	Blair, 1958
NE Nevada	—	5.3	11	Borell and Ellis, 1934
Hays, Kansas	—	3.6	17	Brown, 1945
Laramie Basin, Wyoming	—	5.3	25	Brown, 1966
Temagami, Ontario	—	5.4	50	Coventry, 1937
Clarke Canyon, Nevada	—	4.8	5	Deacon *et al.*, 1964
SW Colorado	—	4.6	20	Douglas, 1969
North Platte Valley, Nebraska	19	3.7	25	Drickamer and Bernstein, 1972
Atotonilco, Mexico	25	4.5	21	Drickamer and Bernstein, 1972
Ann Arbor, Michigan	—	4.7	2285	Drickamer and Vestal, 1973
N Michigan	—	4.6	711	Drickamer and Vestal, 1973
Mono Co., California	—	4.0	40	Dunmire, 1960
NE Colorado	—	4.7	71	Flake, 1974
Siskiyou Co., California	—	4.3	568	Forrester, 1975
Great Slave Lake, North West Territories	17	5.7	109	Fuller, 1969
Clackamas Co., Oregon	—	4.4	427	Gashwiler, 1979
Bar Harbor, Maine	14	4.3	10	Glazier, 1979
Great Slave Lake, North West Territories	20	5.4	29	Gyug, 1979; May, 1979
Kluane, Yukon Territory	26	5.7	61	Krebs and Wingate, 1985
Brighton, Colorado	21	4.3	4	Halfpenny, 1980
Mt. Evans, Colorado	22	6.4	7	Halfpenny, 1980

APPENDIX V

(Continued)

Species Location	Adult wt. (g)	Mean litter size	N	Source
Ward, Colorado	23	5.7	6	Halfpenny, 1980
Livingston Co., Michigan	—	4.3	25	Howard, 1949
Plumas Co., California	19	4.6	96	Jameson, 1953
Nebraska	22	4.5	62	Jones, 1964
Cocke Co., Tennessee	—	4.2	33	Kirkland and Linzey, 1973
Somerset Co., Pennsylvania	—	5.5	29	Kirkland and Linzey, 1973
SE Montana	—	5.1	11	Lampe *et al.*, 1974
Smoky Mts., Tennessee	18	4.1	31	Linzey, 1970
Cocke Co., Tennessee	17	3.5	14	Linzey and Linzey, 1968
NW Wyoming	—	5.6	9	Long, 1965
Carbon Co., Wyoming	—	5.6	29	Long, 1965
SW Utah	—	4.9	9	Long, 1940
Los Angeles, California	20	4.3	11	MacMillen, 1964
Manitouwadge, Ontario	—	5.4	27	Martell, 1983
San Francisco, California	19	5.0	89	McCabe and Blanchard, 1950
Lassen Co., California	—	4.4	71	McKeever, 1965
New York	—	5.4	43	McLaren and Kirkland, 1979
Pennsylvania	—	4.3	195	McLaren and Kirkland, 1979
West Virginia	—	3.9	29	McLaren and Kirkland, 1979
Great Slave Lake, North West Territories	—	5.9	74	Mihok, 1979
Pinawa, Manitoba	17	5.3	80	Millar *et al.*, 1979
Great Slave Lake, North West Territories	19	5.0	98	Millar, 1982
SE Alberta	20	5.1	104	Millar, 1985
Kananaskis, Alberta	20	5.3	102	Millar and Innes, 1983
Edmonton, Alberta	20	3.9	206	Morrison *et al.*, 1977
Midwest United States	21	4.4	194	Morrison *et al.*, 1977

Appendix V

(Continued)

Species Location	Adult wt. (g)	Mean litter size	N	Source
Ann Arbor, Michigan	21	6.0	307	Myers and Master, 1983
East Lansing, Michigan	—	4.5	92	Price, 1967
Ann Arbor, Michigan	—	4.7	610	Rood, 1966
N Michigan	—	4.8	70	Rood, 1966
Vancouver, British Columbia	—	4.5	19	Sadleir, 1974
E Washington	—	5.1	50	Sheffer, 1924
S British Columbia	—	5.5	53	Sheppe, 1963
San Juan Mts., Colorado	—	5.5	15	Sleeper, 1980
Weld and Larimer Cos., Colorado	—	4.6	140	Spencer and Steinhoff 1968
S Alberta	—	5.4	8	Stebbins, 1977
Douglas Co., Kansas	20	3.8	84	Svendsen, 1964
Columbia Co., Washington	—	4.4	42	Svihla, 1932
Alamogordo, New Mexico	—	3.8	40	Svihla, 1932
NE Utah	—	4.6	44	Svihla, 1932
California and New Mexico	—	4.3	27	Svihla, 1932
Michigan, Iowa, North Dakota	—	3.0	21	Svihla, 1932
La Jolla, California	—	3.2	5	Svihla, 1932
Michigan	—	4.0	10	Svihla, 1935
Humboldt Co., California	—	3.4	65	Tevis, 1956
Hope Island, British Columbia	24	4.5	5	Thomas, 1971
Triangle Island, British Columbia	30	4.7	3	Thomas, 1971
Cox Island, British Columbia	28	4.0	5	Thomas, 1971
Texada Island, British Columbia	27	3.3	3	Thomas, 1971
Prince of Wales Island, Alaska	—	4.9	9	Van Horne, 1981
Grande Co., Colorado	—	5.6	111	Vaughan, 1969
Giles Co., Virginia	17	3.4	52	Wolff, 1985
Average	21.0	4.7		

Appendix V

(Continued)

Species Location	Adult wt. (g)	Mean litter size	N	Source
P. leucopus				
Chihuahua, Mexico	25	3.0	6	Anderson, 1972
Lake Opinicon, Ontario	20	5.2	10	Bendell, 1959
		5.5	23	Bendell, 1959
SW Missouri	20	3.7	42	Brown, 1964a, 1964b
Livingston Co., Michigan	—	4.3	39	Burt, 1940
Algonquin, Ontario	—	5.0	50	Coventry, 1937
Seabrook, New Jersey	—	4.5	116	Davis, 1956
Lansing, Michigan	—	4.7	871	Drickamer and Vestal, 1973
East Lansing, Michigan	23	4.8	59	Fleming and Rauscher, 1978
Ithaca, New York	20	3.9	11	Glazier, 1979
Lower Rio Grande, Texas	20	3.5	60	Guetzow and Judd, 1981
Vera Cruz, Mexico	—	4.8	5	Hall and Dalquest, 1963
Ann Arbor, Michigan	—	5.3	9	Hill, 1972
Lower Rio Grande, Texas	20	3.5	60	Guetzow and Judd, 1981
Lower Rio Grande, Texas	—	4.0	91	Judd *et al.*, 1984
Seabrook, New Jersey	21	4.2	135	Jackson, 1952
E. Nebraska	29	4.4	43	Jones, 1964
Campeche, Mexico	22	4.8	110	Lackey, 1973, 1978
Ann Arbor, Michigan	20	4.3	128	Lackey, 1973, 1978
Stevens Point, Wisconsin	—	4.8	31	Long, 1973
Central Illinois	—	4.3	31	Long, 1968
New York	—	4.7	31	McLaren and Kirkland, 1979
Pennsylvania	—	4.9	9	McLaren and Kirkland, 1979
West Virginia	—	3.6	6	McLaren and Kirkland, 1979
London, Ontario	23	5.0	65	Millar, 1975
London, Ontario	22	4.9	90	Millar, 1978
Lansing, Michigan	—	4.7	239	Rood, 1966
Douglas Co., Kansas	27	4.3	71	Svendsen, 1964

Appendix V

(Continued)

Species Location	Adult wt. (g)	Mean litter size	N	Source
Asheville, North Carolina	—	3.4	21	Svihla, 1932
Michigan, New York, Missouri	—	4.4	53	Svihla, 1932
Giles Co., Virginia	19	3.5	35	Wolff, 1985
Average	22.1	4.3		
P. truei				
Chihuahua, Mexico	27	3.3	11	Anderson, 1972
Clarke Canyon, Nevada	—	4.0	8	Deacon *et al.*, 1964
SW Colorado	—	4.0	13	Douglas, 1969
SW Utah	—	3.0	3	Long, 1940
San Francisco, California	27	3.4	49	McCabe and Blanchard, 1950
New Mexico & Colorado	—	2.8	19	Svihla, 1932
Average	27	3.4		
P. eremicus				
New Mexico	—	3.0	7	Bailey, 1931
Riverside Co., California	19	2.2	14	Brand & Ryckman, 1968
Tucson, Arizona	—	2.4	404	Davis and Davis, 1947
San Diego Co., California	—	2.8	372	Drickamer and Vestal, 1973
Portal, Arizona	21	2.4	12	Glazier, 1979
Maricopa Co., Arizona	—	2.6	13	Lewis, 1972
Los Angeles, California	19	2.9	21	MacMillen, 1964
San Diego Co., California	—	2.6	153	Rood, 1966
Carrizoza, New Mexico	—	2.6	5	Svihla, 1932
Average	19.7	2.6		
P. attwateri				
SW Missouri	29	3.4	23	Brown, 1964*a*, 1964*b*
P. boylei				
Chihuahua, Mexico	25	4.1	18	Anderson, 1972
Durango, Mexico		3.0	8	Baker and Greer, 1962
Plumas Co., California	23	3.1	42	Jameson, 1953
Oaxaca, Mexico	27	2.3	7	Robertson, 1975
Average	25	3.1		

Appendix V

(Continued)

Species Location	Adult wt. (g)	Mean litter size	N	Source
P. gossypinus				
S Florida	35	3.1	14	Bigler and Jenkins, 1975
N Florida	29	3.7	72	Pournelle, 1952
Highlands Co., Florida	29	3.5	4	Rand and Host, 1942
S Florida	35	4.7	16	Smith, 1982; Smith and Vrieze, 1979
Average	32	3.8		
P. californicus				
Berkeley, California	—	1.9	493	Drickamer and Vestal, 1973
Los Angeles, California	34	2.5	6	MacMillen, 1964
San Francisco, California	39	1.9	53	McCabe and Blanchard, 1950
Berkeley, California	—	2.0	190	Rood, 1966
Corralites, California	—	1.9	15	Svihla, 1932
Average	36.5	2.0		
P. yucatanicus				
Campeche, Mexico	29	2.8	5	Lackey, 1976
Yucatan, Mexico	—	3.0	13	Lawlor, 1965
Average	29	2.9		
P. polionotus				
Aiken, South Carolina	—	3.1	64	Caldwell, 1964
Aiken, South Carolina	—	3.1	172	Caldwell and Gentry, 1965
Ocala National Forest, Florida	—	2.9	23	Dapson, 1979
Ocala National Forest, Florida	—	3.8	636	Drickamer and Vestal, 1973
Lake Placid, Florida	13	3.6	11	Glazier, 1979
Highlands Co., Florida	13	4.0	13	Rand and Host, 1942
Ocala National Forest, Florida	—	3.9	217	Rood, 1966
Aiken, South Carolina	—	3.9	143	Williams *et al.*, 1965
Aiken, South Carolina	—	3.6	196	Williams *et al.*, 1965
Average	13	3.5		

Appendix V

(Continued)

Species Location	Adult wt. (g)	Mean litter size	N	Source
P. floridanus				
Gainesville, Florida	—	2.1	10	Dice, 1954
Delray Beach, Florida	—	2.4	239	Drickamer and Vestal, 1973
Lake Placid, Florida	42	2.2	4	Glazier, 1979
Alachua Co., Florida	27	3.2	57	Layne, 1966
Delray Beach, Florida	—	1.7	25	Rood, 1966
Average	34.5	2.3		
P. oreas				
S British Columbia	—	6.1	21	Sheppe, 1963
P. melanotis				
Chihuahua, Mexico	19	3.5	6	Anderson, 1972
P. interparietalis				
Baja California, Mexico	21	2.4	40	Brand and Ryckman, 1968
P. difficilis				
Mexico City, Mexico	31	2.8	88	Anderson 1972; Dric- kamer & Vestal, 1973
Durango, Mexico	30.5	3.0	2	Baker and Greer, 1962
Average	31	2.9		
P. melanophrys				
Zacatecas, Mexico	—	3.3	33	Drickamer and Vestal, 1973
P. thomasi				
Guerrero, Mexico	—	2.6	21	Drickamer and Vestal, 1973
Mexico	77	3.5	—	Layne, 1968
Oaxaca, Mexico	79	2.0	4	Robertson, 1975
Average	78	2.7		
P. crinitus				
Bernardino Co., California	—	2.9	206	Drickamer and Vestal, 1973
Tooele Co., Utah	15	3.1	130	Egoscue, 1964
Bernardino Co., California	—	3.0	122	Rood, 1966
Average	15	3.0		
P. mexicanus				
Monteverde, Costa Rica	45	2.8	12	Anderson, 1982
Veracruz, Mexico	—	2.4	10	Hall and Dalquest, 1963

Appendix V

Litter size among populations of Peromyscus.

Species Location	Adult wt. (g)	Mean litter size	N	Source
Mexico	—	2.5	37	Lackey, 1976
Oaxaca, Mexico	60	2.6	20	Rickart, 1977
Oaxaca, Mexico	53	2.6	17	Robertson, 1975
Average	53	2.6		
P. zarhynchus				
Mexico	—	2.0	14	Lackey, 1976
P. gymnotis				
Mexico	—	2.6	5	Lackey, 1976
P. perfulvus				
Michoacan, Mexico	44	2.6		Helm *et al.*, 1974
P. hooperi				
Coahuila, Mexico	21	2.9		Schmidly *et al.*, 1985
P. megalops				
Mexico	71	1.6	3	Layne, 1968
Oaxaca, Mexico	65	2.0	5	Robertson, 1975
Average	68	1.8		
P. melanocarpus				
Oaxaca, Mexico	59	2.3	30	Rickart, 1977
Oaxaca, Mexico	59	2.3	25	Robertson, 1975
Average	59	2.3		
P. lepturus				
Oaxaca, Mexico	48	1.9	8	Robertson, 1975
P. evides				
Oaxaca, Mexico	39	3.5	8	Robertson, 1975

APPENDIX VI

Length of the breeding season in Peromyscus.
Multiple estimates represent different years or trapping grids.

Species	Approximate duration		Source
Location	Months	Weeks	
P. maniculatus			
Kluane, Yukon Territory	June and July	5,5,5,6	Gilbert and Krebs, 1981
Kluane, Yukon Territory	June and July	8,7,11,9	Krebs and Wingate, 1985
Great Slave Lake, Northwest Territories	May– July	10,7	Fuller, 1969
Great Slave Lake, Northwest Territories	May– July	9,10	Mihok, 1979
Great Slave Lake, Northwest Territories	May– July	9	Gyug, 1979
Great Slave Lake, Northwest Territories	May– July	9	May, 1979
Prince of Wales Island, Alaska	May– August	14	Van Horne, 1981
Kananaskis, Alberta	May– July	9,14,8,9,8	Millar and Innes, 1983
Pinawa, Manitoba	April– August	21,14,18,12,18,15	Millar *et al.*, 1979
Vancouver, British Columbia	June– August	12	Sullivan, 1977
Vancouver, British Columbia	February– November	23,32	Fairbairn, 1977
Vancouver, British Columbia	April– September	23,13,16,16,22,20,15	Sullivan, 1979
Vancouver, British Columbia	March– September	22,30	Sullivan and Krebs, 1981
Vancouver, British Columbia	May– August	15,15,12	Sullivan and Sullivan, 1981
Vancouver, British Columbia	March– August	14,23	Sadleir, 1974
Vancouver, British Columbia	March– September	27,36,40	Redfield *et al.*, 1977
Samual Island, British Columbia	April– December	34	Sullivan, 1977
Saturna Island, British Columbia	May– September	18	Sullivan, 1977

Appendix VI

(Continued)

Species Location	Approximate duration		Source
	Months	Weeks	
Livingston Co., Michigan	March–November	33	Howard, 1949
Livingston Co., Michigan	March–October	31	Blair, 1940
Dakota Co., Minnesota	March–November	33	Beer and McLeod, 1966
Missoula, Montana	April–October	30	Metzgar, 1979
Douglas Co., Kansas	all year	52	Svendsen, 1964
NW Massachusetts	April–August	19	Drickamer, 1978
Giles Co., Virginia	March–October	29	Wolff, 1985
Laramie Basin, Wyoming	April–August	16	Brown, 1966
Brighton, Colorado	April–November	31	Halfpenny, 1980
Mt. Evans, Colorado	May–August	17	Halfpenny, 1980
Ward, Colorado	April–August	20	Halfpenny, 1980
Ward, Colorado	April–August	19	Merritt and Merritt, 1980
Grand Co., Colorado	May–August	14	Vaughan, 1969
NE Colorado	February–November	40	Flake, 1974
Clackamas Co., Oregon	all year	30,52	Gashwiler, 1979
Mono Co., California	May–August	11,16	Dunmire, 1960
Lassen Co., California	March–December	41	McKeever, 1964
Humboldt Co., California	May–December	35	Tevis, 1956
Plumas Co., California	February–December	27,36	Jameson, 1953
Bass Lake, California	April–November	30	Storer *et al.*, 1944
San Francisco, California	April–October	29	McCabe and Blanchard, 1950
E Washington	all year	52	Scheffer, 1924

Appendix VI

(Continued)

Species	Approximate duration		Source
Location	Months	Weeks	
SW Colorado	April–September	23	Douglas, 1969
Williamson Co., Texas	November–April	23	Blair, 1958
P. leucopus			
London, Ontario	April–October	27,27,24,24,20	Millar *et al.*, 1979
London, Ontario	April–September	24	Harland *et al.*, 1979
Ann Arbor, Michigan	April–October	23	Lackey, 1973
Piatt Co., Illinois	February–November	40	Batzli, 1977
Muskeget Island, Massachusetts	May–November	30	Nadeau *et al.*, 1981
Giles Co., Virginia	April–November	28	Wolff, 1985
Monongalia Co., West Virginia	February–October	36	Cornish and Bradshaw, 1978
Livingston Co., Michigan	March–October	32	Burt, 1940
NW Massachusetts	March–October	32	Drickamer, 1978
Champaign Co., Illinois	April–October	32,28	Hansen and Batzli, 1978
Stevens Point, Wisconsin	March–August	23	Long, 1973
Central Illinois	March–October	32	Long, 1968
Tolland Co., Connecticut	March–November	24,29	Miller and Getz, 1977
Wood Co., Ohio	March–October	27,32	Rintamaa *et al.*, 1976
SW Missouri	September–May	33	Brown, 1964a
Douglas Co., Kansas	all year	52	Svendsen, 1964
S Texas	all year	52	Guetzow & Judd, 1981
S Texas	all year	52	Judd *et al.*, 1978
S Texas	March–January	41	Judd *et al.*, 1984
Campeche, Mexico	all year	52	Lackey, 1973

Appendix VI

(Continued)

Species	Approximate duration		Source
Location	Months	Weeks	
Seabrook, New Jersey	June–September	18	Jackson, 1952
P. truei			
San Francisco	April–October	27	McCabe and Blanchard, 1950
Virginia City, Nevada	all year	26,52	Llewellyn, 1978
SW Colorado	May–September	20	Douglas, 1969
Monterey Co. California	all year	38,52	Bradford, 1974
P. eremicus			
Maricopa Co., Arizona	all year	52	Lewis, 1972
Pima Co., Arizona	all year	52	Reichman and Van de Graff, 1973
P. attwateri			
SW Missouri	September–June	37	Brown, 1964*a*
P. boylei			
Plumas Co., California	April–October	14,27	Jameson, 1953
Oaxaca, Mexico	all year	52	Robertson, 1975
Bass Lake, California	March–September	27	Storer *et al.,* 1944
P. gossypinus			
S Florida	May–January	33	Smith and Vrieze, 1979
S Florida	all year	33,52	Smith, 1982
S Florida	all year	52	Bigler and Jenkins, 1975
P. californicus			
San Francisco, California	April–October	29	McCabe and Blanchard, 1950
P. yucatanicus			
Campeche, Mexico	all year	52	Lackey, 1976
P. polionotus			
South Carolina	all year	52	Caldwell and Gentry, 1965
South Carolina	all year	52	Davenport, 1964
P. floridanus			
Alachua Co., Florida	June–March	40	Layne, 1966

APPENDIX VI

(Continued)

| Species | Approximate duration | | |
Location	Months	Weeks	Source
P. oreas			
S British	March–	20	Sheppe, 1963
Columbia	July		
P. melanocarpus			
Oaxaca, Mexico	all year	52	Rickart, 1977
Oaxaca, Mexico	all year	52	Robertson, 1975
P. mexicanus			
Oaxaca, Mexico	all year	52	Rickart, 1977
Oaxaca, Mexico	all year	52	Robertson, 1975
Monteverde,	April–	41	Anderson, 1982
Costa Rica	January		
P. lepturus			
Oaxaca, Mexico	all year	52	Robertson, 1975
P. melanurus			
Oaxaca, Mexico	all year	52	Robertson, 1975
P. evides			
Oaxaca, Mexico	all year	52	Robertson, 1975

POPULATION BIOLOGY

DONALD W. KAUFMAN AND GLENNIS A. KAUFMAN

Abstract.—Recent literature on *Peromyscus* encompasses both descriptive and experimental studies that have led to new insights in population biology, including issues in population genetics, population ecology, behavioral ecology, and community ecology. Genetic variation has been assessed in over 20 species of *Peromyscus*, with several studies focused on understanding patterns of macrogeographic differentiation, including island effects. Detailed analyses of the dynamics of microgeographic genetic structure are needed to better understand the organization and dynamics of populations. Additionally, little is known about the adaptive significance of biochemical variation. Social interactions in populations of *Peromyscus* are being examined in greater detail than before and are potentially important to the study of population regulation. Advances in understanding microhabitat selection and distribution, mostly involving *P. leucopus* and *P. maniculatus*, have been accomplished through use of multivariate statistical techniques and to lesser degree experimental manipulations. A major part of the increased effort in microhabitat analysis of *Peromyscus* species has been the result of studies of interspecific microhabitat partitioning. However, integrated studies of habitat selection, foraging, interspecific competition, and differential risk of predation are needed to better assess and describe interspecific resource partitioning within assemblages and the influence of age and sex on intraspecific resource partitioning. Ecological limits to population size, organization of populations, sex ratios, foraging behavior, and predation risk are less studied in *Peromyscus*, although some interesting observations have been made.

In the *Biology of* Peromyscus *(Rodentia)* (King, 1968), population biology was covered primarily in four chapters: "Genetics" by Rasmussen, "Population Dynamics" by Terman, "Home Range and Travels" by Stickel, and "Habitats and Distribution" by Baker. Comparison of these chapters to recent literature demonstrates the significant advances that have been made. The new body of literature is characterized by more exhaustive descriptions of the structure and organization of populations, as well as tests of conceptual or theoretical predictions. In recent years, major contributions have been made in population and evolutionary genetics and in community ecology, subjects that were only beginning to interest biologists studying *Peromyscus* in the mid-1960s.

An important factor in advancing our knowledge of population biology in *Peromyscus* has been the adoption of more rigorous research procedures designed to test specific hypotheses. In this paper, we have highlighted some of the advances in the population biology of *Peromyscus* during the last 20 years that were made by use of this methodology. Our coverage of population biology in *Pero-*

myscus does not include dispersal, territoriality, or reproduction, which are covered in this volume by Wolff and Millar. We were selective in studies covered and did not attempt to include all papers published on the population biology of *Peromyscus*. Also, we have tended to use recent publications whose cited literature should aid in developing reference lists for specific topics.

GENETIC VARIATION

Knowledge of genetic variation in *Peromyscus* in the mid-1960s was limited to mostly studies of morphological, physiological, and pigmentation traits (Rasmussen, 1968). Work on protein variation in *Peromyscus* had begun, but these studies dealt with individual polymorphic proteins (for example, Randerson, 1965) and gave no hint of the degree of genetic variation within or among individuals, populations, or species. Analyses of genetic variation in *Peromyscus* began with a study of *P. polionotus* in which over 50 percent of the assessed loci (17 of 32) were polymorphic in one or more of the 30 populations examined (Selander *et al.*, 1971). The genus soon became a focal point for studies of biochemical variation in mammals. As a result, analysis of genetic variation in *Peromyscus* has played a major role in the study of mammalian population genetics.

The genus *Peromyscus*, with an average heterozygosity per individual (H) of 0.04 for 23 species and average proportion of polymorphic loci per population (P) of 0.15 for 21 species, falls in the middle of the range of values for 12 genera of rodents (Table 1). In comparison, average values for all mammals were 0.04 for H and 0.19 for P. The degree of genetic variation among species within *Peromyscus* can be quite high with H ranging from 0.00 to 0.12 and P ranging from 0.00 to 0.44 (Table 2). This broad variation was partly affected by the inclusion of eight island species. These island species had lower levels of genetic variation (average H = 0.01, average P = 0.03) than continental ones (average H = 0.05, average P = 0.21). Loudenslager (1978) reported a positive relationship between heterozygosity and log of the relative area of the geographic range; however, five of 14 species he examined were island forms. No relationship is evident when only continental species are considered (Table 2).

It should be noted that many estimates of genetic heterozygosity, especially those for highly variable species, are likely to be low as some portion of the variability in proteins is hidden to standard electrophoretic procedures (for *Peromyscus* see Aquadro and Avise,

TABLE 1.—*Heterozygosity per individual (H) and proportion of polymorphic loci per population (P) for 12 genera of rodents. Values calculated from genetic data for individual species as summarized by Nevo* et al. *(1984).*

Genus	Heterozygosity (H)			Polymorphism (P)		
	Number of species	Range	Average	Number of species	Range	Average
Dipodomys	11	0.00–0.06	0.02	11	0.00–0.22	0.09
Geomys	6	0.00–0.05	0.04	5	0.00–0.23	0.17
Microtus	5	0.04–0.13	0.08	9	0.10–0.58	0.34
Mus	3	0.02–0.07	0.04	4	0.11–0.27	0.18
Neotoma	3	0.07–0.08	0.08	4	0.24–0.38	0.31
Perognathus	11	0.02–0.10	0.05	11	0.10–0.32	0.19
Peromyscus	23	0.00–0.12	0.04	21	0.00–0.44	0.15
Rattus	7	0.01–0.09	0.04	7	0.05–0.32	0.17
Sigmodon	2	0.03–0.04	0.03	2	0.13–0.18	0.16
Spermophilus	20	0.00–0.10	0.04	4	0.17–0.31	0.25
Thomomys	3	0.05–0.09	0.07	3	0.23–0.40	0.32
Zapus	2	0.01–0.02	0.01	2	0.05–0.09	0.07
Mammalia	184	0.00–0.18	0.04	181	0.00–0.80	0.19

1982). Previous underestimation of variation is more likely to have occurred for highly variable loci than monomorphic ones (Aquadro and Avise, 1982, and references therein). Although hidden variability may influence estimates of heterozygosity, it should not appreciably alter the proportion of polymorphic loci. It is uncertain whether the detection of additional genetic variation would reveal new patterns of geographic differentiation not evident in earlier studies.

ADAPTIVE SIGNIFICANCE OF GENETIC VARIATION

The adaptive nature of biochemical variation is an interesting and imposing problem because it is usually difficult to determine potential selective pressures (Zera *et al.*, 1985). However, Snyder (1981 and references therein) began a major study of hemoglobin in *P. maniculatus* in the mid-1970s because he expected a physiological cause for genetic variation in hemoglobin. In this case, deer mice that occur over a wide range of altitudes and at high altitudes experience an obvious environmental stress, hypoxia. This environmental stress was expected to select for genetic changes in hemoglobin compared to the hemoglobin of deer mice at lower altitudes. Although geographic patterns in hemoglobin variation

TABLE 2.—*Estimates of heterozygosity per individual (H) and proportion of polymorphic loci per population (P) for 23 species of* Peromyscus *(adapted from the summarization of electrophoretic data by Nevo* et al., *1984; original references used by Nevo* et al. *are listed). Data are listed by categories of range characteristics as follows: island ranges, small continental ranges (approximately 100,000 or fewer square miles), intermediate continental ranges (approximately 100,000 square miles to less than 1,000,000 square miles), and large continental ranges (approximately 1,000,000 or more square miles).*

Species	H	P	Original references[1]
Island			
P. caniceps	0.01	0.07	4
P. dickeyi	0.00	0.00	4
P. eva[2]	0.00	0.00	4
P. guardia	0.01	0.07	4
P. hooperi	0.03		14
P. interparietalis	0.00	0.00	4
P. sejugis	0.02	0.07	4
P. stephani	0.00	0.00	4
Average	0.01	0.03	
Small Continental			
P. californicus	0.07	0.33	4, 19, 20
P. commanche	0.07	0.40	9
P. floridanus	0.05	0.21	22
P. merriami	0.02	0.07	4
Average	0.05	0.25	
Intermediate Continental			
P. attwateri	0.01	0.06	2
P. difficilis	0.04	0.23	2, 9
P. eremicus	0.04	0.16	4
P. gossypinus	0.05		17
P. melanotis	0.04	0.14	2
P. pectoralis	0.04	0.18	3, 10, 11
P. polionotus	0.06	0.26	7, 15, 16, 18
Average	0.04	0.17	
Large Continental			
P. boylii	0.03	0.13	3, 10, 20
P. leucopus	0.07	0.20	6, 17
P. maniculatus	0.12	0.44	1, 5, 8, 12, 13, 20, 21
P. truei	0.03	0.12	2, 9, 20
Average	0.06	0.22	
Continental			
Average for 15 species	0.05	0.21	

[1] References.—1. Aquadro and Kilpatrick, 1981, 2. Avise *et al.*, 1979, 3. Avise *et al.*, 1974*a*, 4. Avise *et al.*, 1974*b*, 5. Baccus *et al.*, 1980, 6. Browne, 1977, 7. Garten, 1976, 8. Gill, 1976, 9. Johnson and Packard, 1974, 10. Kilpatrick and Zimmerman, 1975, 11. Kilpatrick and Zimmerman, 1976, 12. Loudenslager, 1978, 13. Massey and Joule, 1981, 14. Nevo *et al.*, 1984, 15. Peck and Biggers, 1975, 16. Selander, 1970, 17. Selander *et al.*, 1974, 18. Selander *et al.*, 1971, 19. Smith, 1979, 20. Smith, 1981, 21. Smith *et. al.*, 1978, 22. Smith *et al.*, 1973.

[2] *P. eva* is represented by one island sample and none from mainland so we have included this species only with the island forms.

were complex, findings generally supported a hypothesis of selection for hemoglobin types based on differences in oxygen availability at high and low altitudes (Snyder, 1981). Such analyses are difficult, and Snyder's hemoglobin work is the only documented example of a physiological basis of adaptive significance of protein variation in *Peromyscus*.

Understanding protein variation also has involved searches for correlations between life history traits and genetic data such as single locus genotypes, multilocus genotypes, and levels of heterozygosity. Correlations have been reported (Table 3); however, caution should be used when interpreting such patterns. For example, the relationship between heterozygosity and both male body weight and aggressive behavior in *P. polionotus* occurs across macrogeographic space (Garten, 1976). It is possible that these are not cause-effect relationships, but rather that behavior, size, and heterozygosity are independently adapted to environmental conditions that change over the geographic range of the species. Additionally, within samples collected from microgeographic areas, successes in finding significant relationships between life history traits and genotypes at individual loci or heterozygosity have been outweighed by the failures to demonstrate correlations (consult references listed in Table 3). Standardization of proteins examined and ecological procedures used is needed, although the adaptive advantage of different levels of heterozygosity can only be suggested by this approach. Studies are required that can actually demonstrate relationships between levels of heterozygosity and individual fitness.

Genetic variation may be related to niche breadth also (Van Valen, 1965); however, Smith (1981) found no relationship between habitat niche width (diversity of habitats occupied) and heterozygosity (based on $27-29$ loci and two or three populations per species) using *P. boylii* (H = $0.02-0.04$), *P. californicus* (H = $0.03-0.07$), *P. maniculatus* (H = $0.11-0.15$), and *P. truei* (H = $0.02-0.04$). It is possible that an interspecific analysis such as this may mask any intraspecific patterns because levels of heterozygosity varied among species.

Selection component analysis was developed to test for the presence of gametic selection, sexual selection, fecundity selection, zygotic selection, random mating, or a combination of factors (Christiansen and Frydenberg, 1973, 1976; Nadeau and Baccus, 1981). Use of selection component analysis suggested that selection acts on most of the polymorphic loci studied in *P. maniculatus* at differ-

TABLE 3.—*Life history correlates of protein variation in* Peromyscus.

Species	Characteristic	Genetic trait	Reference
P. gossypinus	Availability of food	Esterase-1	Smith *et al.*, 1984
P. maniculatus	Density	Transferrin	Canham, 1969
	Male body weight	Heterozygosity	Massey and Joule, 1981
	Aggressive behavior	3-locus genotype	Fairbairn, 1978
P. poliomotus	Density	Heterozygosity	Smith *et al.*, 1975
	Reproductive rate	Heterozygosity	Smith *et al.*, 1975
	Male body weight	Heterozygosity	Smith *et al.*, 1975; Garten, 1976
	Aggressive behavior	Heterozygosity	Smith *et al.*, 1975; Garten, 1976

ent altitudes in Colorado (Baccus *et al.*, 1980; Nadeau and Baccus, 1981). However, selection pressures maintaining a polymorphism are not identified by this method, and no attempt was made to look for possible selective pressures related to genetic variation in *P. maniculatus*.

GENETIC STRUCTURE OF POPULATIONS

Macrogeographic variation in allele frequencies has been examined for several species of *Peromyscus*, including, *P. polionotus* (Selander *et al.*, 1971), *P. floridanus* (Smith *et al.*, 1973), *P. maniculatus* (Avise *et al.*, 1979), *P. californicus* (Smith, 1979), and *P. leucopus* (Robbins *et al.*, 1985). In these studies, each species exhibited geographic variation in allele frequencies. Additionally, heterogeneity in allele frequencies differed among alleles and loci, and the spatial distribution of certain alleles indicated regional genetic subdivision within species.

In *P. maniculatus*, for example, a high level of geographic variation was evident for some loci and alleles, including a range in frequency of 0.01 to 0.82 for allele 100 at the 6-phosphogluconate dehydrogenase-1 locus (6-Pgd-1^{100}), 0.39 to 1.00 for allele 137 at the alcohol dehydrogenase-1 locus (Adh-1^{137}), and 0.08 to 0.58 for allele 130 at the glutamate oxalate transaminase-1 locus (Got-1^{130}; Avise *et al.*, 1979). Rather than using the range of allele frequencies, heterogeneity in allele frequency among samples is more effectively compared by Wright's (1965) F_{ST} value, which is a standardized variance of allele frequency ($F_{ST} = \sigma_p^2/(\bar{p}\bar{q})$, where σ_p^2 is the variance in p (allele frequency) among samples weighted by sample size and $\bar{p}$ is the average allele frequency and $\bar{q} = 1 - \bar{p}$). For *P. maniculatus*, geographic subdivision is indicated by an average F_{ST} of 0.16, with F_{ST} for individual alleles ranging from 0.04 to 0.38 (Avise *et al.*, 1979). Regardless of the level of differentiation, allele frequencies were significantly heterogeneous in *P. maniculatus*. Macrogeographic differentiation was generally similar for other *Peromyscus* species, for example, average F_{ST} was 0.22, 0.18, and 0.06 for *P. leucopus* from Texas to Tennessee (calculated from Price and Kennedy, 1980), western New York (calculated from Browne, 1977), and Pennsylvania to Vermont (Van Deusen and Kaufman, 1985), respectively, and 0.21 for *P. californicus* across its species range (calculated from Smith, 1979).

Although allele frequencies often vary significantly across species ranges, it is important to note that these populations were genetically quite similar. Thus in *P. maniculatus*, one allele predominated in loci that were monomorphic or only slightly polymorphic within populations, and the same alleles tended to fall into high, intermediate, or low frequency classes for most samples of highly polymorphic loci (Avise *et al.*, 1979). This high degree of similarity among populations was also evident in *P. californicus* (Smith, 1979) and *P. leucopus* (Robbins *et al.*, 1985).

Finally, regional geographic structuring, that is, relatively high frequency of an allele in one portion of the species range and low frequency or absence of the allele in another portion, was evident for some alleles in different species of *Peromyscus*. For example, 6-Pgd-1 [122] was present in samples of *P. maniculatus* from the Northwest Territories to Montana, but absent from other localities in the species range (Avise *et al.*, 1979). Similar structuring occurred in *P. polionotus* (Selander *et al.*, 1971) and *P. californicus* (Smith, 1979).

As noted above, island species were less variable than mainland species (Table 2). However, this trend was less apparent in island and mainland populations within a species (Table 4). No reduction in variation was evident for *P. eremicus* on islands in the Gulf of California, *P. leucopus* on islands in Lake Erie, *P. maniculatus* on California's Channel Islands, or *P. polionotus* on east coast islands of Florida. In contrast, a reduction in H and P was found for *P. polionotus* on islands and peninsulas on the west coast of Florida and in H (but not P) for *P. maniculatus* on Maine islands. However, *P. maniculatus* on Matinicus Island, the most isolated of the Maine islands, exhibited less genetic variation than *P. maniculatus* from other coastal islands along the coast of Maine, the adjacent mainland, and North America. Decreased genetic variability on islands, especially when correlated with distance from the mainland, was probably caused by some combination of founder effect, reduced gene flow from the mainland, and genetic drift (Aquadro and Kilpatrick, 1981).

At a microgeographic scale, *P. leucopus* exhibited significant genetic differentiation (Krohne and Baccus, 1985), although generally less than that observed on a macrogeographic scale (Table 5). Additional studies of *Peromyscus* species are needed to elucidate the relationship between degree of differentiation and size of geographic area. For example, it is difficult to know whether an F_{ST} value of 0.07 (Table 5) for *P. maniculatus* from an altitudinal gradient in Colorado (Massey and Joule, 1981) is typical for population differentiation over a 25 kilometer geographic scale in *Peromyscus* species. Finally, Krohne and Baccus (1985) documented

TABLE 4.—*Comparison of levels of genetic variation in continental and island populations of Peromyscus using average heterozygosity per individual (H) and average proportion of polymorphic loci per population (P; 99% criterion unless otherwise noted).*

Locality	Number of samples	Number of Loci	H	P	References
P. eremicus[1]					
Two islands, Gulf of California	2	25	0.01	0.09	Avise et al., 1974b
Baja California	1	25	0.01	0.04	Avise et al., 1974b
California	13	25	0.01	0.02	Avise et al., 1974b
Nevada, Arizona, Sonora	27	25	0.06	0.19	Avise et al., 1974b
Texas	3	25	0.01	0.00	Avise et al., 1974b
P. leucopus					
Three islands, Lake Erie	3	28	0.07	0.17	Browne, 1977
Ontario	1	28	0.08	0.32	Browne, 1977
Ohio	2	28	0.08	0.18	Browne, 1977
North America	21	25	0.07	0.20	Robbins et al., 1985
P. maniculatus					
Four channel islands, California	4	28	0.08	0.31	Gill, 1976
Six coastal islands, Maine	6	29	0.04	0.21	Aquadro and Kilpatrick, 1981
Matinicus Island, Maine	1	29	0.00	0.04	Aquadro and Kilpatrick, 1981
Maine	1	29	0.08	0.28	Aquadro and Kilpatrick, 1981
North America	18	21	0.09	0.21	Avise et al., 1979
P. polionotus					
Santa Rosa Island, Florida	2	32	0.02	0.06	Selander et al., 1971
Western beach peninsulas, Florida	4	32	0.03	0.10	Selander et al., 1971
Florida panhandle	6	32	0.05	0.22	Selander et al., 1971
Two eastern islands, Florida	2	32	0.09	0.22	Selander et al., 1971
Peninsular Florida	6	32	0.09	0.29	Selander et al., 1971

[1] Polymorphic loci based on 95% criterion.

TABLE 5.—F_{ST} values for samples taken over a range of macrogeographic to microgeographic scales for Peromyscus leucopus and P. maniculatus.

Geographic area	Number of loci	Number of samples	Average F_{ST}	Reference
P. leucopus				
Texas to Tennessee	4	14	0.22	Price and Kennedy, 1980
West Virginia to Vermont	4	10	0.06	Van Deusen and Kaufman, 1985
Western New York	6	6	0.18	Browne, 1977
Average			0.15	
Indiana, <50 ha., year 1	4	5	0.17	Krohne and Baccus, 1985
Same area, year 2	4	5	0.02	Krohne and Baccus, 1985
Indiana, <15 ha., year 1	4	3	0.01	Krohne and Baccus, 1985
Same area, year 2	4	3	0.03	Krohne and Baccus, 1985
Average			0.06	
P. maniculatus				
North America	6	18	0.16	Avise et al., 1979
Pennsylvania to Vermont	3	8	0.21	Van Deusen and Kaufman, 1985
Average			0.18	
Maine, coastal islands	9	5	0.25	Aquadro and Kilpatrick, 1981
Colorado, <25 km. gradient	7	5	0.07[1]	Massey and Joule, 1981

[1] Average of seasonal samples

temporal fluctuation in genetic differentiation for *P. leucopus*, but additional work is required to determine the extent of temporal variation in microgeographic genetic structure of *Peromyscus* populations.

Another question concerning microgeographic genetic structure, as outlined by Krohne and Baccus (1985), is whether the genetic units revealed by analysis of differentiation correspond to demographic units. Krohne *et al.* (1984) and Krohne and Baccus (1985) compared demographic characteristics and allele frequencies of *P. leucopus* populations on five grids (2.4–4.9 ha.) scattered over approximately 50 hectares. In addition, movements of individual *P. leucopus* within and among the five grids were monitored. Although *P. leucopus* was structured genetically (Table 5), genetic and demographic units did not coincide. For example, four of the five grids encompassed a single demographic unit that contained more than a single genetic unit (F_{ST} analysis), whereas three grids without genetic differentiation constituted two different demographic units.

POPULATION DYNAMICS

Early live-trapping studies of *Peromyscus* were directed toward analyses of density and dynamics as well as age structure and sex ratio of populations (summarized in Terman, 1968). Recent studies have continued to use capture-mark-recapture procedures to analyze population dynamics in other species and habitats (for example, Adler and Tamarin, 1984; Briggs, 1986; Wolff, 1985). One general advance is the greater use of statistical procedures which better describe both annual and multi-year variability in density. Techniques for estimating population size from data on marked and unmarked individuals continue to be studied and improved (for example, Nichols, 1986). Although various techniques are used to calculate population size, absolute densities are often estimated by assuming that a population occupies an area only slightly larger than the trapping grid. Regardless of refinements in population estimators, it is apparent that the area trapped must be estimated to obtain absolute density values and techniques have and are being developed and tested (O'Farrell *et al.*, 1977; Wilson and Anderson, 1985).

We will not attempt to catalog absolute and relative density estimates from the literature in this review, rather we will highlight work on ecological factors that may limit population density and review research on social behavior that relates to population regulation.

Quantity and quality of foods, availability of water, number and distribution of nest sites, architecture of living and dead vegetation, and depth and density of litter are some of the ecological factors that could limit the density of small mammal populations. The only one examined in detail for *Peromyscus* is the availability of food; this was done using food supplementation experiments. In these studies, positive responses were found for *P. gossypinus* (Smith *et al.*, 1984), *P. leucopus* (Bendell, 1959; Briggs, 1986; Hansen and Batzli, 1978), *P. maniculatus* (Fordham, 1971; Gilbert and Krebs, 1981; Taitt, 1981), and *P. polionotus* (Smith, 1971). Hansen and Batzli (1979) found no increase in density of *P. leucopus* with supplemental food in contrast to their 1978 report, but suggested that a high mast crop during the study may have reduced any effect of their experimentally added food. The generality of a positive response in these studies suggests that *Peromyscus* species often must be limited by the availability of food, or at least high quality food. Changes in dispersal, reproduction, survival, or various combinations of these bring about the positive response to food, but results of the various experiments do not point to a single, straightforward cause.

In some situations, habitat features rather than food may limit population density, but no field experiments in which a single feature is altered have been attempted. The positive response of grassland *P. maniculatus* to prairie fires in areas left unburned for several years (for example, Kaufman *et al.*, 1983*a*) is, however, apparently related to removal of litter and standing dead vegetation as no other features change so drastically and quickly. In this case, the litter layer is probably a physical barrier that limits density by making seeds less available than they would be in the absence of litter. After a fire and, therefore, the removal of prairie litter, the pattern of density increase of *P. maniculatus* is consistent with patterns seen after most cases of food supplementation.

Our review of the population biology of *Peromyscus* suggests that some of the most important advances in understanding the dynamics of *Peromyscus* populations have come from numerous laboratory and field studies of social behavior, and not from studies of density alone. Wolff has reviewed recent advances in the study of social behavior in *Peromyscus* elsewhere in this volume. It is important to note that improved understanding of two components of social behavior, dispersal and inhibition of maturation, have direct relevance to the study of population regulation. It is now apparent that the likelihood of dispersal and the timing of maturation of a young mouse are partially dependent on the number and status of the other mice with which it interacts. As a result, a young female may be able to stay in close proximity to its mother, but not breed until its mother dies or leaves the area because the young female is inhibited from maturing by pheromonal cues (for example, Haigh, 1983). Additionally, the likelihood and distance of dispersal for young males and females are partially determined by social interactions. Thus social interactions can regulate numbers relative to carrying capacity set by other factors such as food. Under extremely favorable conditions, environmental limits may be so high, however, that social interactions could determine the carrying capacity. Continued study of social behavior will make a major contribution to understanding the dynamics and organization of *Peromyscus* populations.

ORGANIZATION OF POPULATIONS

Given that animals are nonrandomly distributed across habitats, genetic subdivision occurs, and heterosis may occur, Smith *et al.* (1975, 1978) proposed a conceptual model for population organization in small mammals. This model, partially based on information from *Peromyscus* species, postulated the existence of two types of populations: (1) primary populations where recruitment occurs principally from reproduction and (2) secondary populations where recruitment occurs primarily from immigration. The hypothesized driving force in this model is the creation of new and vigorous primary populations in empty suitable habitat by dispersers derived from two or more older primary populations. The proposed vigor of these new primary populations results from heterosis because offspring of founders should be "hybrids" of individuals from different primary populations. Microgeographic genetic differentiation in *P. leucopus* (Krohne and Baccus, 1985) supports the basic concept of spatially structured populations. Whether this genetic structure is due to an array of primary and secondary populations and whether heterosis is the driving force behind such structure is still open to question for *Peromyscus* species.

SEX RATIO

Field-trapped samples of *Peromyscus* species are usually male biased with the usual explanation being that males move farther than females and, therefore, males are trapped from a larger area than that from which females are trapped (Stickel, 1968; Terman, 1968). Unfortunately, this explanation has been invoked in the absence of documentation of the true sex ratios of populations. Recent studies

TABLE 6.—*Sex ratios by body size for* P. maniculatus *(Kaufman and Kaufman, 1982*a*)*
and P. leucopus *(Kaufman and Kaufman, 1982*b*).*

Body length (mm)	P. maniculatus		P. leucopus	
	Number	Percent males	Number	Percent males
61–70	8	50	6	50
71–80	34	32*	39	51
81–90	203	57*	145	60*
91–100	46	56	262	57*
101–110	0		50	24**
111–120	0		1	0
All mice	291	54	503	54

*Percent males significantly different from 50% at $P < 0.05$, using Chi-square test
**Percent males significantly different from 50% at $P < 0.01$, using Chi-square test

have demonstrated differences in sex ratios among body-size categories of *P. leucopus* and *P. maniculatus* (Table 6; Kaufman and
Kaufman, 1982*a*, 1982*b*). Additionally, proportions of males in
populations of *P. maniculatus* were negatively related to density for
12 sites (relative densities of three to eight individuals per trapline:
69–87 percent males; nine to 26 individuals: 50–67 percent males;
Kaufman and Kaufman, 1982*a*). These patterns for body size and
density suggest that the determinants of sex ratios for trapped
samples include other factors beside mobility and trappability.

In the laboratory, secondary sex ratios (ratios at birth) suggest an
initial bias in favor of males in at least two species of *Peromyscus*
(Table 7). The assumption that secondary sex ratio, whether one to
one or not, is constant among different females in a population has
recently been called into question (Burley, 1982; Silk, 1983; Trivers
and Willard, 1973, and references therein). Little information is
available for sex ratios at birth or at the end of parental investment
in *Peromyscus*, but recent observations suggest that secondary sex
ratio may vary with litter size or the physiological condition of females. In a laboratory study of *P. polionotus*, secondary sex ratio
varied with litter size (Kaufman and Kaufman, 1982*c*). In this case,
17 percent of six individuals in six litters of one neonate were
males, 49 percent of 670 individuals in 167 litters of two to six neonates were males, and 68 percent of 31 individuals in four litters of
seven to eight neonates were males. These categories also exhibited
decreased offspring size at birth, with the average weight of individuals ranging from 1.70 grams for litter size one, 1.51–1.58
grams for litter sizes two to six, and 1.41–1.44 for litter sizes seven
to eight (Kaufman and Kaufman, 1987).

TABLE 7.—*Secondary sex ratios recorded for laboratory maintained* Peromyscus.

Species	Number of litters	Number of young	Percent males	Reference
P. maniculatus				
Laboratory populations	255	1004	54**	Terman and Sassaman, 1967
Laboratory pairs	564	1775	54**	Terman and Sassaman, 1967
Wild- and laboratory-conceived		642	54*	Canham, 1970
P. polionotus				
Wild pairs		622	51	Williams *et al.*, 1965
Lab-raised pairs (first generation)		296	53	Williams *et al.*, 1965
Lab-raised pairs (multi-generation)	177	707	49	Kaufman and Kaufman, 1982*c*

*$P < 0.05$, using Chi-square test
**$P < 0.01$, using Chi-square test

For *P. maniculatus*, Myers *et al.* (1985) reported that short-term variation in weather conditions affected litter size and secondary sex ratio. Free-living females exposed to high temperature in the spring produced large litters with an increase in female offspring, whereas females exposed to high temperatures in autumn produced small litters with a decrease in male offspring. Observations of females exposed to high temperatures under laboratory conditions suggested differences between male and female embryos in the timing of their sensitivity to high temperature (Myers *et al.*, 1985). The adaptive reasons, if any, for the patterns of sex ratios in *P. maniculatus* and *P. polionotus* are not known; however, such patterns point to the need for further research to understand the significance of secondary sex ratios in *Peromyscus* species.

MICROHABITAT DISTRIBUTION AND SELECTION

Studies of microhabitat selection and distribution can be grouped into analyses of differential association or use of (1) qualitatively defined microhabitats, (2) quantitatively defined microhabitats, (3) specific microsites, features, or both, and (4) experimentally manipulated (or simulated) microsites or features. Qualitatively defined microhabitats are described from visual observations by the investigator without detailed measurements of environmental features. Using this method, Kirkland and Griffin (1974) demonstrated that the distribution of *P. maniculatus* was nonrandom within an ecotonal area between deciduous and coniferous forests in the Adirondack Mountains. Similarly, differential use of trap stations by *P. leucopus* in a riparian forest and grassland site in eastern Kansas was related to both the presence and complexity of woody vegetation (Kaufman *et al.*, 1983*b*). This approach is useful for demonstrating associations of species with relatively broad categories of microhabitats (Table 8), but is inappropriate when examining the response of species to specific conditions or features.

Analysis of quantitatively defined microhabitats uses detailed measurements of environmental conditions and features (Table 8). Subsequently, multivariate statistical techniques are employed to identify major patterns of associated habitat variables that are characteristic of broader environmental patterns. For example, *P. leucopus* in eastern Tennessee occurred at sites that were characterized by the presence of deciduous trees, low density of trees, high density of shrub understory, and low shrub evergreenness, or, in general terms, *P. leucopus* preferred shrubby, deciduous sites (Dueser and Shugart, 1978). An advantage of this technique over the qualitative one is that it is repeatable and unbiased and can reveal patterns of environmental characteristics that are not apparent to the investigator (Dueser and Shugart, 1978).

Studies of the association of small mammals with particular features or microsites (for example, preference of *P. leucopus* for sites situated near trees, logs, rocks, and stumps; Barry and Francq, 1980) are usually based on the distribution of trap captures, although other techniques can be used (Table 8). Microsite preferences have been examined frequently in *Peromyscus* in wooded habitats that have readily recognizable microsites or features (Barry and Francq, 1980; Kaufman *et al.*, 1983*b*, 1985). In contrast, this procedure seems less suitable for the study of grassland species because of the compressed vertical space and closed understory which makes identification of distinct microsites or features difficult.

The effects of experimental manipulations and simulations of habitat features on populations or individuals have been studied in both field and laboratory (Table 8). Experimental manipulations in the field have included alterations in density, size, dispersion of selected features, or a combination thereof. In some cases, treatments are juxtaposed so that the situation is a choice experiment for individual mice. For example, M'Closkey (1976*b*) used artificial shrubs so that influence of branch density and angle on shrub use by *P. leucopus* could be tested. Habitat manipulations have also

TABLE 8.—*Examples of studies of interspecific habitat partitioning in small mammals with one or more species of* Peromyscus *as a major component of the assemblage. Small mammals listed are only those included in the analysis.*

Description of study	Species	References

I. Qualitatively defined microhabitats

A. Upland, mixed, and cedar associations, southeastern Ontario (live-trapping on grid)

Peromyscus leucopus
Peromyscus maniculatus

Smith and Speller, 1970

Remarks.—*P. leucopus* exhibited significant microhabitat selection for upland forest sites, whereas *P. maniculatus* did not demonstrate significant selection among the three microhabitats. Patterns of selection were significantly different between the two species with greater use of cedar forest and mixed forest sites and lesser use of upland forest sites by *P. maniculatus* in comparison to *P. leucopus.*

B. Deciduous and coniferous forest ecotone, Adirondack Mountains, New York (live-trapping on grid)

Peromyscus maniculatus
Blarina brevicauda
Clethrionomys gapperi
Napeozapus insignis
Tamias striatus

Kirkland and Griffin, 1974

Remarks.—Significant differences in microhabitat use. *P. maniculatus* avoided coniferous zone, wet trap stations, and *C. gapperi* capture sites in 1972 but not 1971 (no wet stations on this grid).

C. Grass-sage, ecotone, and piñon-juniper, north-central New Mexico (live-trapping on grid)

Peromyscus maniculatus
Peromyscus truei

Holbrook, 1978

Remarks.—Significant differences in microhabitat use. *P. maniculatus* was associated with grass-sage and *P. truei* with piñon-juniper.

D. Grass-sage, piñon-juniper, oak, and mixed shrub, north-central New Mexico (live-trapping on grid)

Peromyscus boylii
Peromyscus difficilis
Peromyscus maniculatus
Peromyscus truei

Holbrook, 1978

Remarks.—Significant differences in microhabitat use. In 1972, associations were *P. boylii* with oak; *P. difficilis* with mixed shrub and to a lesser extent oak; *P. maniculatus* with grass-sage and to a lesser extent piñon-juniper; and *P. truei* with piñon-juniper, although it was more evenly distributed across the four microhabitats than other species. In 1973, *P. maniculatus,* the only resident on the grid, shifted its activity to piñon-juniper with relatively high use of other three categories, especially grass-sage and oak.

II. Quantitatively defined microhabitats

A. Wet prairie site, southern Ontario (live-trapping on grid)

Peromyscus leucopus
Microtus pennsylvanicus

M'Closkey and Fieldwick, 1975

Remarks.—Significant differences in microhabitat use based on foliage height diversity, density of trees, and depth of perennial grass.

TABLE 8.—*(Continued)*.

Description of study	Species	References
B. Coastal sage scrub, California (live-trapping on grid)	*Peromyscus californicus* *Peromyscus eremicus* *Peromyscus maniculatus* *Dipodomys agilis* *Neotoma fuscipes* *Neotoma lepida* *Reithrodontomys megalotis*	M'Closkey, 1976*a*

Remarks.—P. eremicus, D. agilis, N. fuscipes, and *N. lepida,* permanent residents, exhibited significant microhabitat separation. Other species were seasonal and showed greater microhabitat overlap. *P. eremicus* and *P. maniculatus* were the most similar-sized pair of species and demonstrated seasonal and microhabitat separation.

C. Grass, sagebrush, and piñon-juniper, north-central New Mexico (snap-trapping, 30 localities)	*Peromyscus boylii* *Peromyscus difficilis* *Peromyscus maniculatus* *Peromyscus truei*	Holbrook, 1978

Remarks.—Vegetation structure was not a major determinant of the distribution of individual species, but measures of vegetation complexity were correlated with numbers of coexisting species (microsympatry) of *Peromyscus.*

D. Mixed deciduous-evergreen forest, eastern Tennessee (live-trapping on three grids)	*Peromyscus leucopus* *Blarina brevicauda* *Ochrotomys nuttalli* *Tamias striatus*	Dueser and Shugart, 1979

Remarks.—Significant differences in microhabitat use. *P. leucopus* and *T. striatus* were associated with deciduous forest but separated by shrubbiness (*P. leucopus* > *T striatus*), *O. nuttalli* with evergreen forest and heavy undergrowth, and *B. brevicauda* with midsuccessional mixed forest and heavy undergrowth.

E. Mixed deciduous-evergreen forest, eastern Tennessee (live-trapping on two grids)	*Peromyscus leucopus* *Blarina brevicauda* *Ochrotomys nuttalli* *Tamias striatus*	Kitchings and Levy, 1981

Remarks.—Significant differences in microhabitat use. This study site was in early successional stages in contrast to Dueser and Shugart's (1979) study which was in later forest stages. Specific patterns of microhabitat association changed: *P. leucopus* in more open areas with thick herbaceous understory, *B. brevicauda* in deciduous woodland but not heavy undergrowth, and *O. nuttalli* less associated with evergreen overstory.

F. Farm shelterbelts, southwestern Minnesota (live-trapping, five shelterbelts)	*Peromyscus leucopus* *Blarina brevicauda* *Clethrionomys gapperi* *Microtus pennsylvanicus* *Sorex cinereus*	Yahner, 1982

Remarks.—Significant differences in microhabitat use. *P. leucopus* was associated with dense woody understory and large trees, *C. gapperi* with dense woody understory and small trees, *M. pennsylvanicus* with open areas away from woody vegetation, and *B. brevicauda* and *S. cinereus* with intermediate levels of woody understory and tree size.

TABLE 8.—*(Continued)*.

Description of study	Species	References
G. Cedar glade, eastern Tennessee (live-trapping on grid)	*Peromyscus leucopus* *Ochrotomys nuttalli*	Seagle, 1985

Remarks.—Significant differences in use of microhabitats. *P. leucopus* in non-shrubby, woody sites and *O. nuttalli* in edge sites. *P. leucopus* had narrower niche breadth than *O. nuttalli*.

| H. Deciduous forest, eastern Tennessee (live-trapping on grid) | *Peromyscus leucopus* *Blarina brevicauda* | Seagle, 1985 |

Remarks.—Significant differences in use of microhabitats. *P. leucopus* associated with denser understory sites and *B. brevicauda* with sparser understory sites. Niche breadth greater for *P. leucopus* than *B. brevicauda*.

| I. Deciduous forest, southern Vermont (live-trapping, scattered sites) | *Peromyscus leucopus* *Peromyscus maniculatus* | Parren and Capen, 1985 |

Remarks.—Significant differences in microhabitat use. *P. maniculatus* was positively associated with relatively open sites with large trees and less low vegetation in contrast to *P. leucopus*.

III. Specific microsites and/or features

| A. Deciduous forest, southwestern Virginia (radiotelemetry) | *Peromyscus leucopus* *Peromyscus maniculatus* | Wolff and Hurlbutt, 1982 |

Remarks.—Significant difference in use of tree and ground refuges. *P. leucopus* chose ground refuges (73%), whereas *P. maniculatus* chose tree refuges (92%).

| B. Deciduous forest, southwestern Virginia (live-trapping on lines) | *Peromyscus leucopus* *Peromyscus maniculatus* | Barry *et al.*, 1984 |

Remarks.—Significant differences in use of ground and elevated (trees and shrubs) microhabitats were found between *P. leucopus* and *P. maniculatus*. Pattern was one of greater use of elevated microsites by *P. maniculatus* than *P. leucopus*. Additionally, *P. maniculatus* were significantly more likely to climb trees than were *P. leucopus* as part of their escape routes following release by investigators.

| C. Sand dunes, Great Basin desert, southwestern Nevada (live-trapping and sand-tracking) | *Peromyscus maniculatus* *Dipodomys deserti* *Dipodomys merriami* *Microdipodops pallidus* | Kotler, 1985*a* |

Remarks.—Significant difference in use of open microsites and those under bushes between *P. maniculatus* and the heteromyids. Two methods gave different results for *P. maniculatus* (43% in open based on captures; 10% in open based on tracks); methodology should be of concern to those studying microhabitat use in small mammals.

IV. Experimental manipulations

| A. Fertilization of short-grass prairie, north-eastern Colorado (live-trapping on grids; water, nitrogen, water + nitrogen, and control) | *Peromyscus maniculatus* *Microtus ochrogaster* *Onychomys leucogaster* *Spermophilus tridecemlineatus* | Grant *et al.*, 1977 |

Remarks.—Differential responses to changes in vegetation with fertilization. *P. maniculatus* and *S. tridecemlineatus* present on all plots, *M. ochrogaster* chose wet treatments, and *O. leucogaster* chose dry treatments.

TABLE 8.—*(Continued).*

Description of study	Species	References
B. Simulated nest sites below, at, or above substrate in the laboratory, mouse colony originally derived from deciduous forest populations, northeastern USA.	*Peromyscus leucopus* *Peromyscus maniculatus*	Stah, 1980

Remarks.—Both species preferred to nest aboveground; however, when tested together *P. leucopus* shifted to floor nest and *P. maniculatus* continued to nest aboveground.

C. Addition of cardboard shelters, Mojave desert, southern California (live-trapping on four grids)	*Peromyscus crinitus* *Peromyscus maniculatus* *Dipodomys merriami* *Onychomys torridus* *Perognathus longimembris*	Thompson, 1982

Remark.—Addition of cardboard structures resulted in invasion of these grids by *P. crinitus* and *P. maniculatus.*

D. Deciduous forest, western North Carolina (live-trapping in clear-cut openings, first year succession)	*Peromyscus leucopus* *Peromyscus maniculatus*	Buckner and Shure, 1985

Remarks.—Density of *P. leucopus* was negatively related to size of opening, whereas density of *P. maniculatus* was positively related to size of opening. Differences in use of openings were related to differences in microhabitat conditions among openings; *P. maniculatus* associated with high foliage height diversity and slash diversity and *P. leucopus* with large volumes of logs and stumps and low-lying cover.

E. Burned and unburned tall-grass prairie, eastern Kansas (live-trapping on grid)	*Peromyscus maniculatus* *Reithrodontomys megalotis* *Sigmodon hispidus*	Peterson *et al.*, 1985

Remarks.—Analysis of patch use revealed *P. maniculatus* chose burned upland, *R. megalotis* unburned lowland, and *S. hispidus* lowland with no response to fire.

F. Burned and unburned tall-grass prairie, eastern Kansas (live-trapping on 20 traplines)	*Peromyscus maniculatus* *Reithrodontomys megalotis*	Kaufman *et al.*, 1988

Remarks.—Microhabitat use was significantly different between these two species. *P. maniculatus* selected breaks in the first year after spring fire and avoided lowland 5+ years after fire, whereas *R. megalotis* chose lowland 2–4 years after fire and avoided annually burned breaks. Use by *P. maniculatus* was negatively related to increasing litter and positively related to exposed soil; no significant relationships were found between habitat structure and use by *R. megalotis.*

included removal of woody vegetation in deciduous forest (Buckner and Shure, 1985), addition of fertilizer to short-grass prairie (Abramsky *et al.*, 1979; Grant *et al.*, 1977), and addition of cardboard shelters to desert sites (Thompson, 1982). Laboratory studies have examined the use of both natural features that can be brought into the laboratory and features that can be simulated under laboratory conditions. An example of this latter approach was a study of the effect of height on nest or refuge use by *P. leucopus* and *P. maniculatus* (Stah, 1980).

Most analyses of microhabitat selection by *Peromyscus* have employed quantitative microhabitat or experimental approaches to ascertain determinants of microhabitat use (Table 8). However, even for a given species in similar habitats, a lack of consistency in proposed determinants of microhabitat use usually occurred among studies. This variation reflects a lack of consistency in measurements of habitat characteristics, but even more so, the variability in microhabitat relationships. Lack of generality is likely related to (1) variability in habitat conditions, (2) low numbers of individuals observed in individual studies, (3) lack of replication across space, or (4) insufficient duration of studies to account for temporal variation. These problems suggest a need for studies with greater replication over both space and time. Additionally, differences among studies may be related to differences in sex-age structure of populations studied.

Most microhabitat studies of *Peromyscus* have not examined sex or age effects; however, this seems necessary to better understand population structure and organization. For example, Van Horne (1982) examined habitat preferences of juvenile and adult *P. maniculatus* within seral stages of coastal coniferous forest in Alaska using quantitatively defined habitat conditions. Differential habitat use occurred between adults and juveniles; over-winter survival was related to habitat quality for both age groups.

Sexual differences in microhabitat use by *Peromyscus* were reported in two studies, one involving *P. maniculatus* (Bowers and Smith, 1979) and the other *P. leucopus* (Seagle, 1985). Bowers and Smith (1979) examined *P. maniculatus* in three sites, a *Pinus ponderosa* community in northeastern Oregon, an *Artemisia-Sarcobatus* community in southwestern Idaho, and an *Atriplex-Eurotia* community in southwestern Utah. In the two heterogeneous communities, *P. ponderosa* and *Artemisia-Sarcobatus*, male and female deer mice were spatially segregated with males occupying more xeric sites as defined in terms of water potential for plants.

Seagle (1985) examined use of quantitatively defined microhabitats by *P. leucopus* in both cedar glade and deciduous forest sites in Tennessee. No differences in microhabitat use between males and females were found in the cedar glade; however, in the deciduous forest, males were associated with microsites having more herbaceous ground cover than microsites used by females. Seagle (1985) suggested that the failure to find sexual differences in microhabitat use by deer mice in the cedar glade was related to interspecific partitioning of microhabitat conditions between *P. leucopus* and *Ochrotomys nutalli*, but this remains open to question. Kaufman *et al.* (1985) found no differences between male and female *P. leucopus* in use of microsites associated with trees, trees of different sizes, and microsites associated with shrubs.

Patterns reported by Van Horne (1982), Bowers and Smith (1979), and Seagle (1985) raise questions about the need for additional tests for differences among intrapopulation components. Is the lack of other published studies of age and sexual differences in habitat use the result of investigators failing to look for such differences or failing to report the absence of differences among population components? If investigators have generally failed to find patterns, we may be in need of more refined studies or studies with larger sample sizes. For example, Bowers and Smith (1979) examined over a thousand mice, whereas many studies of microhabitat in *Peromyscus* have dealt with less than a hundred different individuals.

In laboratory trials, Clarke (1983) showed that *P. maniculatus* were at greater risk to owl predation on bright than on dark nights, and that use of open microsites decreased on bright relative to dark nights. Laboratory experiments also revealed greater relative use of open than protected patches for foraging on dark nights than on bright nights (Travers *et al.*, 1988). These studies suggest that improved understanding of the influence of predators is needed to better explain microhabitat use by *Peromyscus* species under field conditions. For example, different patterns of microhabitat use found in different studies may be related to predation pressure as well as portion of dark and bright nights included in each study.

Interspecific Habitat Partitioning

Research on interspecific resource partitioning in rodents has dealt with the presence and significance of patterns of differential use of foods and microhabitat structure. Most studies that have examined small mammal assemblages in which *Peromyscus* species are major components have dealt with microhabitat partitioning. These studies on *Peromyscus* have tended to focus on ecosystems with complex vertical structure such as those dominated by trees and shrubs and to a lesser extent on those dominated by grasses and forbs (Table 8). In some cases, assemblages consisted totally or mostly of *Peromyscus* species, but in most instances the assemblages included an array of rodent and insectivore genera.

Presently, the most commonly used approach to questions dealing with interspecific habitat partitioning by small mammals in systems dominated or codominated by *Peromyscus* is to test for differential use of quantitatively defined microhabitats (M'Closkey, 1976a; Dueser and Shugart, 1978, 1979). In fact, the increasing number of such studies (Table 8) is probably related to the recognition that mammals respond to finer details than evidenced by qualitatively defined microhabitat categories and to the suitability of multivariate statistics for reducing the complexity of the analysis of pattern in habitat characteristics (Seagle, 1985). Investigators interested in describing habitat niches of *Peromyscus* species should consider comments and observations by Dueser and Shugart (1979, 1982), Carnes and Slade (1982), Porter and Dueser (1982), Van Horne and Ford (1982), Parren and Capen (1985), and Seagle (1985).

Analyses based on quantitatively defined microhabitats and, to a lesser extent, qualitatively defined microhabitats have yielded insights into differences in microhabitat requirements among species (Table 8). However, because such analyses are descriptive, the results should be viewed with caution. First, statistical separation of species on a habitat axis does not prove that the animals recognize and respond to the conditions measured; rather an unmeasured but correlated feature may be the actual determinant. Second, even though average positions along habitat axes occupied by a series of species are different, interspecific overlap may be considerable and must be explained. Third, most studies have not been replicated, either in time or space, so that the generality of any specific study is unknown. Fourth, differences in resource use do not necessarily elucidate the mechanisms that are ultimately responsible for the interspecific patterns observed. Finally, understanding cause and effect of observed patterns of habitat use will require experimental manipulations and not just additional studies using refined descriptions of habitats and their use by mammals.

Responses to microsites and experimental manipulations (examples in Table 8) can be used to examine patterns suggested by either qualitatively or quantitatively defined microhabitats. However, manipulations involving fire, fertilization, and so forth usually cause a multitude of changes that make it difficult to assess the actual factor that determines patterns seen in descriptive studies. Rather, experimental studies altering fewer and more specific features are needed to further our understanding of the determinants of microhabitat use and the environmental resources that are subdivided by species within a given assemblage. Experimental analysis using actual or simulated habitat features in laboratory environments is yet another method for improving our understanding of microhabitat requirements, niche breadth, and niche overlap of the small mammal species within particular assemblages.

INTERSPECIFIC COMPETITION

In 1968, acceptance of interspecific competition among *Peromyscus* was based on microhabitat segregation under field conditions and aggressive interactions between species in the laboratory (Baker, 1968). Reciprocal changes in numbers over time and in use of space by *Peromyscus* (Brown, 1964; Kirkland and Griffin, 1974; Petticrew and Sadleir, 1974; Whitaker, 1967) suggest but do not demonstrate that interspecific competition is an active, ongoing process. Changes in environmental conditions could produce the same differential responses by the two species as those expected from interspecific competition. Consistency of reciprocal changes in numbers is highly suggestive, however, and can direct field experimentation for testing for the presence and magnitude of interspecific competition (Grant, 1972, 1978; Schoener, 1983).

Sheppe (1961) found that *P. maniculatus* invaded an area from which an isolated population of *P. oreas* had been removed. Subsequently, *P. oreas* reinvaded the removal area; these individuals were from populations of *P. oreas* that were more isolated from the removal area than were *P. maniculatus* (Sheppe, 1967). Similarly, Sheppe (1967) removed *Mus musculus* from one portion of a farm building and found that *P. maniculatus* from elsewhere in the building then invaded the vacant space. Using a different approach, Caldwell (1964) placed *P. polionotus* and *M. musculus* into a one-acre field enclosure. Both were spatially segregated after the introduction; however, *P. polionotus* eventually invaded the area occupied by *M. musculus*, and *M. musculus* went extinct. Caldwell and Gentry (1965) repeated the experiment and obtained similar results. Unfortunately, an absence of replicated plots, single-species control plots, or both in these studies does not allow the elimination of other causes for the observed patterns.

Lack of proper controls and replication in earlier studies led Grant (summarized in Grant, 1972) to undertake an experimental study of competition involving *Microtus*, *Clethrionomys*, and *Peromyscus* in enclosed natural woodland and grassland, in southern Quebec. In four experiments using *M. pennsylvanicus* and woodland *P. maniculatus*, Grant (1971, 1972) found that (1) *Peromyscus* used both grassland and woodland in the absence of *Microtus*; (2) *Peromyscus* used grassland less when the two species were introduced together than when *Peromyscus* was introduced by itself; (3) removal of *Microtus* when both species were originally present resulted in increased use of grassland by *Peromyscus*; and (4) addition of *Microtus* to enclosures with *Peromyscus* resulted in decreased use of grassland by *Peromyscus*. Overall, these four responses demonstrated the presence of competitive interactions between the two species.

Redfield *et al.* (1977) tested for competitive exclusion of *P. maniculatus* by *Microtus townsendii* in grassland habitats in British Columbia. This study examined *P. maniculatus* in an area with *M. townsendii* removed, an area with female voles removed, and control areas. High densities of *P. maniculatus* in experimental areas and low densities in control areas as well as experimental areas following recolonization by *M. townsendii* indicated the presence of competitive interactions. Abramsky *et al.* (1979) also examined competition between *Microtus* and *Peromyscus* using *M. ochrogaster* and *P. maniculatus* from the short-grass prairie in northeastern Colorado. In fertilized and watered plots, *P. maniculatus* increased in density following removal of *M. ochrogaster* from one plot, whereas no increase occurred on the control plot.

Holbrook (1979) reported competitive interactions among *P. maniculatus*, *P. boylii*, and *Neotoma stephensi* in east-central Arizona. In these experiments, *P. maniculatus* and *N. stephensi* expanded microhabitat utilization following removal of *P. boylii*. *Peromyscus maniculatus* used an even greater range of microhabitats following removal of both *P. boylii* and *N. stephensi*, whereas *P. boylii* showed few changes following removal of *N. stephensi*.

Data on density of mammal species and habitat variables have been employed in multiple regression analysis to estimate the presence and magnitude of interspecific competitive interactions

(Crowell and Pimm, 1976; Hallett and Pimm, 1979; Schoener, 1974). By removing the effects of habitat selection from density patterns, competitive interactions were demonstrated by negative partial regression coefficients between pairs of species. Using this approach, Crowell and Pimm (1976) found that *P. maniculatus* was displaced from wooded habitats by *Clethrionomys gapperi*, but *P. maniculatus* displaced *Microtus pennsylvanicus* from shrubby habitats on coastal islands of Maine. In eastern deciduous forest, Dueser and Hallett (1980) found that both competitive ability and the specificity of habitat selection increased in the following order, *P. leucopus*, *Tamias striatus*, and *Ochrotomys nuttalli*. *Peromyscus leucopus*, a habitat generalist, was outcompeted by the other two species, whereas *O. nuttalli*, a habitat specialist, outcompeted the other two. Finally, Porter and Dueser (1982) reported competitive interactions of *P. leucopus* with *M. pennsylvanicus*, *M. musculus*, and *Zapus hudsonius* on Assateague Island, a barrier island off Maryland and Virginia.

Experimental manipulation and regression analysis suggest that interspecific competition can be a significant factor in structuring assemblages of small mammals as well as influencing patterns of microhabitat selection. Further study combining these approaches seems necessary to document the consistency of both methods for detecting competition. Additionally, studies of competition must now begin to examine the resource(s) that pairs of species are competing for and the form that the competition takes. Many laboratory studies suggest that aggressive interactions may be a major component in many cases (reviews by Grant, 1972, 1978); however, competition for food or other resources is undoubtedly important in some instances.

Foraging Behavior

Peromyscus species feed on seeds, nuts, fruits, and invertebrates (Baker, 1968; Flake, 1973; Meserve, 1976; Whitaker, 1966; and references therein) using olfaction to detect both surface and buried foods (based on experiments with *P. maniculatus*; Howard and Cole, 1967; Johnson and Jorgensen, 1981). The ability of *P. maniculatus* to detect buried seeds increased as soil moisture increased (Johnson and Jorgensen, 1981). Seeds are eaten often near the point of detection in *P. maniculatus* (Mittelbach and Gross, 1984); however, caching of seeds does occur (for example, *P. eremicus*, *P. leucopus*, and *P. maniculatus*; Barry, 1976, and references therein). Caching apparently occurs more commonly in autumn than at other times; and both temperature and photoperiod can affect caching in some species (Barry, 1976).

Food preferences have been examined in several species of *Peromyscus* using both laboratory and field experiments. For example, in a laboratory experiment using seven seeds and grains, woodland *P. leucopus* and grassland *P. maniculatus* from Michigan chose seeds in relation to ease of eating, handling, energy and carbohydrate content, or some combination of these (Drickamer, 1970). Unfortunately, effects of size and quality of foods could not be separated from the ease of handling and eating foods. When six types of native seeds were presented to *P. maniculatus* in a Michigan old-field, Mittelbach and Gross (1984) found that *P. maniculatus* chose the largest seeds; however, effect of differential quality of seeds was not considered.

Gray (1979) attempted to define the criteria that two desert rodents, *P. maniculatus* and *Microdipodops pallidus*, used in selecting their diets. Use of psychophysical scaling techniques generally showed that the two species chose different sizes of food items (peanuts, sunflower seeds, millet seeds, and wheat germ). Rank orders of food items selected by the two species were not consistent with ordinal ranks of food item quality (for example, calories, protein, specific amino acids, and vitamin content) except for phenylalanine and niacin. Along the size dimension, food items most preferred by *M. pallidus* were least preferred by *P. maniculatus*. Using four sizes of only one food item, Brazil nuts, the two species also ranked physical size of foods differently.

Reichman and Fay (1983) compared feeding strategies in *P. maniculatus*, typically a noncaching species, and *Perognathus intermedius*, a caching species. Unreplenished amounts of five types of seeds of different quality (calories) were provided to supply the energy requirements of each mouse for 10 days. Of the two species, *P. maniculatus* ate a less diverse diet, with seeds of the highest caloric content preferred. This pattern supported the prediction that individuals of noncaching species should gather and eat the highest quality food items first, whereas caching strategists should manage their resources such that an optimal diet can be maintained over an extended period.

Foraging behavior has also been shown to vary among individuals within populations of *Peromyscus*. For example, Tardif and Gray (1978) found that resident *P. leucopus* chose a lower diversity diet than did immigrants. This difference was consistent with predictions from models of resource use that animals in unfamiliar and, therefore, relatively unpredictable environments should be food generalists, whereas individuals in familiar and, therefore, relatively predictable environments should be food specialists. Also, using *P. leucopus*, Ebersole and Wilson (1980) experimentally examined the relationships of dietary diversity to food density, degree of hunger, and difficulty of food processing. Dietary diversity was unrelated to seed density, positively related to hunger, and negatively related to processing time.

PREDATOR-PREY INTERACTIONS

Reduced activity of nocturnal small mammals on brightly illuminated nights is assumed to be an adaptive response related to increased risk of predation by visually hunting predators (see review in Clarke, 1983). Using *P. maniculatus* and short-eared owls in laboratory experiments, Clarke (1983) found that hunting effectiveness of owls increased with increased illumination. Vulnerability of *Peromyscus* to predation also depends on the structural complexity of microsites. For example, Kaufman (1974) found that barn owls and screech owls were much less effective in capturing *P. polionotus* in small field enclosures when the native vegetation was dense (only 35 percent of trials with captures) than when the native vegetation was sparse (87 percent of trials with captures). Shifts in the spatial distribution of activity by *P. maniculatus* with reduced use of open areas on bright versus dark nights (Clarke, 1983) further support the importance of habitat complexity in influencing predator efficiency.

In addition to differences in risk of predation associated with microsites, it seems likely that predation risk would vary among individuals within populations as well as between populations of *Peromyscus*. In laboratory experiments, Metzgar (1967) demonstrated that transient *P. leucopus* were at greater risk to predation by screech owls than were resident individuals. He attributed this to several possible factors: (1) residents may be able to perceive the presence of a predator sooner than a transient because residents are less involved in exploration, (2) residents may be able to escape more effectively than transients because residents know their home ranges, and (3) residents may be less active than transients because residents are in their own home ranges. Metzgar's study also suggested that young, dispersing *Peromyscus* would be at greater risk than established adults.

No sexual differences in vulnerability to predation were noted in *P. maniculatus* (Clarke, 1983) or in *P. polionotus* (Kaufman, 1974). Given various observations on differences in activity between males and females, lack of sexual differences in vulnerability was unexpected and may be due to restrictive conditions of small pens or cages used in choice experiments. Sexual differences in susceptibility to predation may also vary with the type of predator. For example, mammalian predators hunt by scent as well as sight, whereas avian predators use sight, sound, or both, but not scent. Cushing (1984) found that the least weasel (*Mustela nivalis*) selected estrous over diestrous scent of *P. maniculatus* and that estrous *P. maniculatus* were at greater risk to predation than were diestrous individuals (Cushing, 1985). He also concluded that differences in behavior between estrous and diestrous females probably contributed to differences in vulnerability to the least weasel.

Finally, pelage coloration can also affect risk of predation in *Peromyscus*. Experimental analysis of owl predation in *P. polionotus* revealed that vulnerability to predation varied between two naturally occurring pelage phenotypes, the typical dark brown form and a recessive light brown form (Kaufman, 1974). Differential predation varied with soil coloration such that light brown mice were at greater risk on dark soils and dark brown mice were at greater risk on light soils.

Interspecific patterns of vulnerability to predation were examined in an assemblage of desert rodents that included *P. maniculatus* (Kotler, 1985*b*). Assessment of predation rates (from skeletal material in owl pellets) together with estimated densities of bipedal forms (*Dipodomys* and *Microdipodops*) and quadrupedal forms (*Perognathus, Peromyscus,* and *Reithrodontomys*) revealed that long-eared owls were selectively preying on the quadrupedal forms. These species, numerically dominated by *P. maniculatus*, lack adaptations, such as inflated auditory bullae and bipedal locomotion, for perceiving and escaping predators, especially when in the open. The greater risk to predation of *P. maniculatus* and other quadrupedal forms is also consistent with their preference for foraging under bushes rather than in the open (Kotler, 1984).

FUTURE RESEARCH DIRECTIONS

Although important advances in the population biology of *Peromyscus* have been made, many individual studies have been short-term, from one to three years, and often unreplicated or only weakly replicated. Therefore, replicated long-term studies of free-living populations are needed to enable researchers to assess and establish the generality of patterns, as well as the levels of spatial and temporal variability. Continued study of *Peromyscus*, using field observation, experimental manipulation, or both, should emphasize the testing of hypotheses and predictions developed from theoretical arguments and previous research rather than only describing observations of natural populations or the results of manipulations. Comparative analyses of populations, both intraspecific and interspecific studies, are also important, because these should aid in understanding the proximate and ultimate causes of patterns exhibited by different populations.

With regard to techniques, effort must be directed toward developing new methods for "observing" *Peromyscus* in both natural and manipulated field situations. Such approaches could include new ways to use radiotelemetry, nightviewing devices, video cameras, belowground radar, trapping or recording devices linked to computers, fluorescent dyes, radioisotopes, and genetic markers. Use of such techniques should enable the movements of individuals, interactions among individuals, foraging behavior, microhabitat patch selection, and mating events among others to be monitored and studied in greater detail than previously done. However, creative uses of traditional snap and live traps should not be ignored.

New, as well as established, investigators interested in the population biology of small mammals, including *Peromyscus*, need to use interdisciplinary approaches that integrate concepts, theories, and ideas from a broad array of disciplines, including population and evolutionary genetics; population, community, and ecosystem ecology; behavior; and physiology. Also important is knowledge of mathematical modeling techniques and modern statistical tests and procedures so that research designs are known to be appropriate and hypotheses testable before initiation of the research.

Considerable work remains to be done on the biology of *Peromyscus* in most if not all of the research areas discussed in this review. For example, little is known about the spatial and temporal dynamics of the genetic structure of populations of *Peromyscus*, although many studies have focused on assessing genetic variation. General patterns of population density, litter size, timing of reproduction, sex ratios of trappable adults, and habitat relationships are known for the commonly studied species, but why do these patterns occur? For example, how do dispersal and mortality of adult and juvenile mice interact to produce the temporal dynamics of population density, the spatial distribution of individuals, and the genetic structure of the population? How do female body size and physical environmental conditions influence conception, pre-natal survival, and post-natal survival of male and female offspring and, therefore, the evolutionary success of an individual female or lineage? Although short-term studies have shown that small mammals exhibit statistically significant habitat relationships, are these consistent through time? How do behavior, physical features, interspecific competition, and predation risk interact to create interspecific habitat patterns? Within the general habitat conditions, how do physical features, competition, and predation interact to influence microhabitat patch choice by foraging rodents? These are only some examples of issues in need of further study in *Peromyscus*. Consideration of this review, the original publications, and the concepts and theory of general population biology reveal many other opportunities for population research on the different species, and more finely divided taxa or genetic units, of *Peromyscus*.

Acknowledgments

We thank E. J. Finck and B. K. Clark for reading this manuscript and J. M. Posey, L. K. Claassen, and D. M. Kaufman for help with manuscript preparation. We also greatly appreciate the efforts of G. L. Kirkland and J. N. Layne in the development of the chapter and the presentation of this final written version. Partial support for manuscript preparation was provided by NSF grants DEB 8012166, BSR 8307571, and BSR 8514327 and by Project No. R-231, Kansas Agricultural Experiment Station. This is contribution no. 86-293-B, Division of Biology, Kansas Agricultural Experiment Station, Kansas State University, Manhattan.

Literature Cited

ABRAMSKY, Z., M. I. DYER, AND D. HARRISON. 1979. Competition among small mammals in experimentally perturbed areas of the shortgrass prairie. Ecology, 60:530–536.

ADLER, G. H., AND R. H. TAMARIN. 1984. Demography and reproduction in island and mainland white-footed mice (*Peromyscus leucopus*) in southeastern Massachusetts. Canadian J. Zool., 62:58–64.

AQUADRO, C. F., AND J. C. AVISE. 1982. An assessment of "hidden" heterogeneity within electromorphs at three enzyme loci in deer mice. Genetics, 102:269–284.

AQUADRO, C. F., AND C. W. KILPATRICK. 1981. Morphological and biochemical variation and differentiation in insular and mainland deer mice (*Peromyscus maniculatus*). Pp. 214–230, *in* Mammalian population genetics (M. H. Smith and J. Joule, eds.). Univ. of Georgia Press, Athens, Georgia, 380 pp.

AVISE, J. C., M. H. SMITH, AND R. K. SELANDER. 1974a. Biochemical polymorphism and systematics in the genus *Peromyscus*. VI. The *boylii* species group. J. Mamm., 55:751–763.

———. 1979. Biochemical polymorphism and systematics in the genus *Pero-*

myscus. VII. Geographic differentiation in members of the *truei* and *maniculatus* species groups. J. Mamm., 60:177–192.

AVISE, J. C., M. H. SMITH, R. K. SELANDER, T. E. LAWLOR, AND P. R. RAMSEY. 1974*b*. Biochemical polymorphism and systematics in the genus *Peromyscus*. Insular and mainland species of the subgenus *Haplomylomys*. Syst. Zool., 23: 226–238.

BACCUS, R., J. JOULE, AND W. J. KIMBERLING. 1980. Linkage and selection analysis of biochemical variants in *Peromyscus maniculatus*. J. Mamm., 61: 423–435.

BAKER, R. H. 1968. Habitats and distribution. Pp. 98–126, *in* Biology of *Peromyscus* (Rodentia) (J. A. King, ed.). Spec. Publ., Amer. Soc. Mamm., 2: 1–593.

BARRY, R. E., JR., M. A. BOTJE, AND L. B. GRANTHAM. 1984. Vertical stratification of *Peromyscus leucopus* and *P. maniculatus* in southwestern Virginia. J. Mamm., 65:145–148.

BARRY, R. E., JR., AND E. N. FRANCQ. 1980. Orientation to landmarks within the preferred habitat by *Peromyscus leucopus*. J. Mamm., 61:292–303.

BARRY, W. J. 1976. Environmental effects on food hoarding in deermice (*Peromyscus*). J. Mamm., 57:731–746.

BENDELL, J. F. 1959. Food as a control of a population of white-footed mice, *Peromyscus leucopus noveboracensis* (Fischer). Canadian J. Zool., 37:173–209.

BOWERS, M. A., AND H. D. SMITH. 1979. Differential habitat utilization by sexes of the deermouse, *Peromyscus maniculatus*. Ecology, 60:869–875.

BRIGGS, J. M. 1986. Supplemental food and two island populations of *Peromyscus leucopus*. J. Mamm., 67:474–480.

BROWN, L. N. 1964. Ecology of three species of *Peromyscus* from southern Missouri. J. Mamm., 45:189–202.

BROWNE, R. A. 1977. Genetic variation in island and mainland populations of *Peromyscus leucopus*. Amer. Midland Nat., 97:1–9.

BUCKNER, C. A., AND D. J. SHURE. 1985. The response of *Peromyscus* to forest opening size in the southern Appalachian Mountains. J. Mamm., 66: 299–307.

BURLEY, N. 1982. Facultative sex-ratio manipulation. Amer. Nat., 120:81–107.

CALDWELL, L. D. 1964. An investigation of competition in natural populations of mice. J. Mamm., 45:12–30.

CALDWELL, L. D., AND J. B. GENTRY. 1965. Interactions of *Peromyscus* and *Mus* in a one-acre field enclosure. Ecology, 46:189–192.

CANHAM, R. P. 1969. Serum protein variation and selection in fluctuating populations of cricetid rodents. Unpubl. Ph.D. dissert., Univ. Alberta, Edmonton, 121 pp.

———. 1970. Sex ratios and survival in fluctuating populations of the deer mouse, *Peromyscus maniculatus borealis*. Canadian J. Zool., 48:809–811.

CARNES, B. A., AND N. A. SLADE. 1982. Some comments on niche analysis in canonical space. Ecology, 63:888–893.

CHRISTIANSEN, F. B., AND O. FRYDENBERG. 1973. Selection component analysis of natural polymorphisms using population samples including mother-offspring combinations. Theoret. Pop. Biol., 4:425–445.

———. 1976. Selection component analysis of natural polymorphisms using mother-offspring samples of successive cohorts. Pp. 277–301, *in* Population genetics and ecology (S. Karlin and E. Nevo, eds.). Academic Press, New York, 832 pp.

CLARKE, J. A. 1983. Moonlight's influence on predator/prey interactions between short-eared owls (*Asio flammeus*) and deermice (*Peromyscus maniculatus*). Behav. Ecol. Sociobiol., 13:205–209.

CROWELL, K. L., AND S. L. PIMM. 1976. Competition and niche shifts of mice introduced onto small islands. Oikos, 27:251–258.

CUSHING, B. S. 1984. A selective preference by least weasels for oestrous versus dioestrous urine of prairie deer mice. Anim. Behav., 32:1263–1265.

———. 1985. Estrous mice and vulnerability to weasel predation. Ecology, 66:1976–1978.

DRICKAMER, L. C. 1970. Seed preferences in wild caught *Peromyscus maniculatus bairdii* and *Peromyscus leucopus noveboracensis*. J. Mamm., 51:191–194.

DUESER, R. D., AND J. G. HALLETT. 1980. Competition and habitat selection in a forest-floor small mammal fauna. Oikos, 35:293–297.

DUESER, R. D., AND H. H. SHUGART, JR. 1978. Microhabitats in a forest-floor small mammal fauna. Ecology, 59:89–98.

———. 1979. Niche pattern in a forest-floor small-mammal fauna. Ecology, 60:108–118.

———. 1982. Reply to comments by Van Horne and Ford and by Carnes and Slade. Ecology, 63:1174–1175.

EBERSOLE, J. P., AND J. C. WILSON. 1980. Optimal foraging: the responses of *Peromyscus leucopus* to experimental changes in processing time and hunger. Oecologia, 46:80–85.

FAIRBAIRN, D. J. 1978. Behaviour of dispersing deer mice (*Peromyscus maniculatus*). Behav. Ecol. Sociobiol., 3:265–282.

FLAKE, L. D. 1973. Food habits of four species of rodents on a short-grass prairie in Colorado. J. Mamm., 54:636–647.

FORDHAM, R. A. 1971. Field populations of deermice with supplemental food. Ecology, 52:138–146.

GARTEN, C. T., JR. 1976. Relationships between aggressive behavior and genic heterozygosity in the oldfield mouse, *Peromyscus polionotus*. Evolution, 30:59–72.

GILBERT, B. S., AND C. J. KREBS. 1981. Effects of extra food on *Peromyscus* and *Clethrionomys* populations in the southern Yukon. Oecologia, 51:326–331.

GILL, A. E. 1976. Genetic divergence of insular populations of deer mice. Biochem. Genet., 14:835–848.

GRANT, P. R. 1971. Experimental studies of competitive interaction in a two-species system. III. *Microtus* and *Peromyscus* species in enclosures. J. Anim. Ecol., 40:323–350.

———. 1972. Interspecific competition among rodents. Ann. Rev. Ecol. Syst., 3:79–106.

———. 1978. Competition between species of small mammals. Pp. 38–51, *in* Populations of small mammals under natural conditions (D. P. Snyder, ed.). Spec. Publ. Ser., Pymatuning Lab. Ecol., Univ. Pittsburgh Press, Pittsburgh, 5:1–237.

GRANT, W. E., N. R. FRENCH, AND D. M. SWIFT. 1977. Response of a small mammal community to water and nitrogen treatments in a shortgrass prairie ecosystem. J. Mamm., 58:637–652.

GRAY, L. 1979. The use of psychophysical unfolding theory to determine principal resource axes. Amer. Nat., 114:695–706.

HAIGH, G. R. 1983. Reproductive inhibition and recovery in young female *Peromyscus leucopus*. J. Mamm., 64:706.

HALLETT, J. G., AND S. L. PIMM. 1979. Direct estimation of competition. Amer. Nat., 113:593–600.

HANSEN, L. P., AND G. O. BATZLI. 1978. The influence of food availability on the white-footed mouse: populations in isolated woodlots. Canadian J. Zool., 56:2530–2541.

―――. 1979. Influence of supplemental food on local populations of *Peromyscus leucopus*. J. Mamm., 60:335–342.

HOLBROOK, S. J. 1978. Habitat relationships and coexistence of four sympatric species of *Peromyscus* in northwestern New Mexico. J. Mamm., 59:18–26.

―――. 1979. Habitat utilization, competitive interactions, and coexistence of three species of cricetine rodents in east-central Arizona. Ecology, 60:758–769.

HOWARD, W. E., AND R. E. COLE. 1967. Olfaction in seed detection by deer mice. J. Mamm., 48:147–150.

JOHNSON, G. L., AND R. L. PACKARD. 1974. Electrophoretic analysis of Peromyscus comanche Blair, with comments on its systematic status. Occas. Papers Mus., Texas Tech Univ., 24:1–16.

JOHNSON, T. K., AND C. D. JORGENSEN. 1981. Ability of desert rodents to find buried seeds. J. Range Manag., 34:312–314.

KAUFMAN, D. W. 1974. Adaptive coloration in *Peromyscus polionotus*: experimental selection by owls. J. Mamm., 55:271–283.

KAUFMAN, D. W., AND G. A. KAUFMAN. 1982a. Sex ratio in natural populations of *Peromyscus maniculatus*. Amer. Midland Nat., 108:376–380.

―――. 1982b. Sex ratio in natural populations of *Peromyscus leucopus*. J. Mamm., 63:655–658.

―――. 1982c. Secondary sex ratio in *Peromyscus polionotus*. Trans. Kansas Acad. Sci., 85:216–217.

―――. 1987. Reproduction by *Peromyscus polionotus*: number, size, and survival of offspring. J. Mamm., 68:275–280.

KAUFMAN, D. W., G. A. KAUFMAN, AND E. J. FINCK. 1983a. Effects of fire on rodents in tallgrass prairie of the Flint Hills region of eastern Kansas. Prairie Nat., 15:49–56.

KAUFMAN, D. W., M. E. PEAK, AND G. A. KAUFMAN. 1985. *Peromyscus leucopus* in riparian woodlands: use of trees and shrubs. J. Mamm., 66:139–143.

KAUFMAN, D. W., S. K. PETERSON, R. FRISTIK, AND G. A. KAUFMAN. 1983b. Effect of microhabitat features on habitat use by *Peromyscus leucopus*. Amer. Midland Nat., 110:177–185.

KAUFMAN, G. A., D. W. KAUFMAN, AND E. J. FINCK. 1988. Influence of fire and topography on habitat selection by *Peromyscus maniculatus* and *Reithrodontomys megalotis* in ungrazed tall-grass prairie. J. Mamm., 69:342–352.

KILPATRICK, C. W., AND E. G. ZIMMERMAN. 1975. Genetic variation and systematics of four species of mice of the *Peromyscus boylii* species group. Syst. Zool., 24:143–162.

―――. 1976. Biochemical variation and systematics of *Peromyscus pectoralis*. J. Mamm., 57:506–522.

KING, J. A. (ED.). 1968. Biology of *Peromyscus* (Rodentia). Spec. Publ., Amer. Soc. Mamm., 2:1–593.

KIRKLAND, G. L., JR., AND R. J. GRIFFIN. 1974. Microdistribution of small mammals at the coniferous-deciduous forest ecotone in northern New York. J. Mamm., 55:417–427.

KITCHINGS, J. T., AND D. J. LEVY. 1981. Habitat patterns in a small mammal community. J. Mamm., 62:814–820.

KOTLER, B. P. 1984. Risk of predation and the structure of desert rodent communities. Ecology, 65:689–701.

———. 1985a. Microhabitat utilization in desert rodents: a comparison of two methods of measurement. J. Mamm., 66:374–378.

———. 1985b. Owl predation on desert rodents which differ in morphology and behavior. J. Mamm., 66:824–828.

KROHNE, D. T., AND R. BACCUS. 1985. Genetic and ecological structure of a population of *Peromyscus leucopus*. J. Mamm., 66:529–537.

KROHNE, D. T., B. A. DUBBS, AND R. BACCUS. 1984. An analysis of dispersal in an unmanipulated population of *Peromyscus leucopus*. Amer. Midland Nat., 112:146–156.

LOUDENSLAGER, E. J. 1978. Variation in the genetic structure of *Peromyscus* populations. I. Genetic heterozygosity—its relationship to adaptive divergence. Biochem. Genet., 16:1165–1179.

M'CLOSKEY, R. T. 1976a. Community structure in sympatric rodents. Ecology, 57:728–739.

———. 1976b. Use of artificial microhabitats by white-footed mice, *Peromyscus leucopus*. Amer. Midland Nat., 96:467–470.

M'CLOSKEY, R. T., AND B. FIELDWICK. 1975. Ecological separation of sympatric rodents (*Peromyscus* and *Microtus*). J. Mamm., 56:119–129.

MASSEY, D. R., AND J. JOULE. 1981. Spatial-temporal changes in genetic composition of deer mouse populations. Pp. 180–201, *in* Mammalian population genetics (M. H. Smith and J. Joule, eds.). Univ. of Georgia Press, Athens, Georgia, 380 pp.

MESERVE, P. L. 1976. Food relationships of a rodent fauna in a California coastal sage scrub community. J. Mamm., 57:300–319.

METZGAR, L. H. 1967. An experimental comparison of screech owl predation on resident and transient white-footed mice (*Peromyscus leucopus*). J. Mamm., 48:387–391.

MITTELBACH, G. G., AND K. L. GROSS. 1984. Experimental studies of seed predation in old-fields. Oecologia, 65:7–13.

MYERS, P., L. L. MASTER, AND R. A. GARRETT. 1985. Ambient temperature and rainfall: an effect on sex ratio and litter size in deer mice. J. Mamm., 66:289–298.

NADEAU, J. H., AND R. BACCUS. 1981. Selection components of four allozymes in natural populations of *Peromyscus maniculatus*. Evolution, 35:11–20.

NEVO, E., A. BEILES, AND R. BEN-SHLOMO. 1984. The evolutionary significance of genetic diversity: ecological, demographic and life history correlates. Pp. 13–213, *in* Evolutionary dynamics of genetic diversity (G. S. Mani, ed.). Lecture notes in biomathematics, Vol. 53, 312 pp.

NICHOLS, J. D. 1986. On the use of enumeration estimators for interspecific comparisons, with comments on a trappability estimator. J. Mamm., 67:590–593.

O'FARRELL, M. J., D. W. KAUFMAN, AND D. W. LUNDAHL. 1977. Use of livetrapping with assessment line method for density estimation. J. Mamm., 58:575–582.

PARREN, S. G., AND D. E. CAPEN. 1985. Local distribution and coexistence of two species of *Peromyscus* in Vermont. J. Mamm., 66:36–44.

PECK, C. T., AND C. J. BIGGERS. 1975. Electrophoretic analysis of plasma proteins of Mississippi *Peromyscus*. J. Hered., 66:237–241.

PETERSON, S. K., G. A. KAUFMAN, AND D. W. KAUFMAN. 1985. Habitat selection by small mammals of the tall-grass prairie: experimental patch choice. Prairie Nat., 17:65–70.

PETTICREW, B. G., AND R. M. F. S. SADLEIR. 1974. The ecology of the deer mouse *Peromyscus maniculatus* in a coastal coniferous forest. I. Population dynamics. Canadian J. Zool., 52:107–118.

PORTER, J. H., AND R. D. DUESER. 1982. Niche overlap and competition in an insular small mammal fauna: a test of the niche overlap hypothesis. Oikos, 39:228–236.

PRICE, P. K., AND M. L. KENNEDY. 1980. Genic relationships in the white-footed mouse, *Peromyscus leucopus*, and the cotton mouse, *Peromyscus gossypinus*. Amer. Midland Nat., 103:73–82.

RANDERSON, S. 1965. Erythrocyte esterase forms controlled by multiple alleles in the deer mouse. Genetics, 52:999-1005.

RASMUSSEN, D. I. 1968. Genetics. Pp. 340–372, *in* Biology of *Peromyscus* (Rodentia) (J. A. King, ed.). Spec. Publ., Amer. Soc. Mamm., 2:1–593.

REDFIELD, J. A., C. J. KREBS, AND M. J. TAITT. 1977. Competition between *Peromyscus maniculatus* and *Microtus townsendii* in grasslands of coastal British Columbia. J. Anim. Ecol., 46:607–616.

REICHMAN, O. J., AND P. FAY. 1983. Comparison of the diets of a caching and a noncaching rodent. Amer. Nat., 122:576–581.

ROBBINS, L. W., M. H. SMITH, M. C. WOOTEN, AND R. K. SELANDER. 1985. Biochemical polymorphism and its relationship to chromosomal and morphological variation in *Peromyscus leucopus* and *Peromyscus gossypinus*. J. Mamm., 66:498–510.

SCHOENER, T. W. 1974. Competition and the form of habitat shift. Theoret. Pop. Biol., 6:265–307.

———. 1983. Field experiments on interspecific competition. Amer. Nat., 122:240–285.

SEAGLE, S. W. 1985. Patterns of small mammal microhabitat utilization in cedar glade and deciduous forest habitats. J. Mamm., 66:22–35.

SELANDER, R. K. 1970. Biochemical polymorphism in populations of the house mouse and old-field mouse. Symp. Zool. Soc. Lond., 26:73–91.

SELANDER, R. K., D. W. KAUFMAN, R. J. BAKER, AND S. L. WILLIAMS. 1974. Genic and chromosomal differentiation in pocket gophers of the *Geomys bursarius* group. Evolution, 28:557–564.

SELANDER, R. K., M. H. SMITH, S. Y. YANG, W. E. JOHNSON, AND J. B. GENTRY. 1971. Biochemical polymorphism and systematics in the genus *Peromyscus*. I. Variation in the old-field mouse (*Peromyscus polionotus*). Stud. Genet., 6:49–90.

SHEPPE, W. 1961. Systematic and ecological relations of *Peromyscus oreas* and *P. maniculatus*. Proc. Amer. Philos. Soc., 105:421–446.

————. 1967. Habitat restriction by competitive exclusion in the mice *Peromyscus* and *Mus*. Canadian Field-Nat., 81:81–98.

SILK, J. B. 1983. Local resource competition and facultative adjustment of sex ratios in relation to competitive abilities. Amer. Nat., 121:56–66.

SMITH, D. A., AND S. W. SPELLER. 1970. The distribution behavior of *Peromyscus maniculatus gracilis* and *Peromyscus leucopus noveboracensis* (Rodentia: Cricetidae) in a southeastern Ontario woodlot. Canadian J. Zool., 48:1187–1199.

SMITH, M. F. 1979. Geographic variation in genic and morphological characters in *Peromyscus californicus*. J. Mamm., 60:705–722.

————. 1981. Relationships between genetic variability and niche dimensions among coexisting species of *Peromyscus*. J. Mamm., 62:273–285.

SMITH, M. H. 1971. Food as a limiting factor in the population ecology of *Peromyscus polionotus* (Wagner). Ann. Zool. Fennici, 8:109–112.

SMITH, M. H., C. T. GARTEN, JR., AND P. R. RAMSEY. 1975. Genic heterozygosity and population dynamics in small mammals. Pp. 85–102, *in* Isozymes IV, genetics and evolution (C. L. Markert, ed.). Academic Press, New York, 965 pp.

SMITH, M. H., M. N. MANLOVE, AND J. JOULE. 1978. Spatial and temporal dynamics of the genetic organization of small mammal populations. Pp. 99–113, *in* Populations of small mammals under natural conditions (D. P. Snyder, ed.). Spec. Publ. Ser., Pymatuning Lab. Ecol., Univ. Pittsburgh Press, Pittsburgh, 5:1–237.

SMITH, M. H., R. K. SELANDER, AND W. E. JOHNSON. 1973. Biochemical polymorphism and systematics in the genus *Peromyscus*. III. Variation in the Florida deer mouse (*Peromyscus floridanus*), a Pleistocene relict. J. Mamm., 54:1–13.

SMITH, M. W., W. R. TESKA, AND M. H. SMITH. 1984. Food as a limiting factor and selective agent for genic heterozygosity in the cotton mouse *Peromyscus gossypinus*. Amer. Midland Nat., 112:110–118.

SNYDER, L. R. G. 1981. Deer mouse hemoglobins: is there genetic adaptation to high altitude? BioScience, 31:299–304.

STAH, C. D. 1980. Vertical nesting distribution of two species of *Peromyscus* under experimental conditions. J. Mamm., 61:141–143.

STICKEL, L. F. 1968. Home range and travels. Pp. 373–411, *in* Biology of *Peromyscus* (Rodentia) (J. A. King, ed.). Spec. Publ., Amer. Soc. Mamm., 2:1–593.

TAITT, M. J. 1981. The effects of extra food on small rodent populations: I. Deermice (*Peromyscus maniculatus*). J. Anim. Ecol., 50:111–124.

TARDIF, R. R., AND L. GRAY. 1978. Feeding diversity of resident and immigrant *Peromyscus leucopus*. J. Mamm., 59:559–562.

TERMAN, C. R. 1968. Population dynamics. Pp. 412–450, *in* Biology of *Peromyscus* (Rodentia) (J. A. King, ed.). Spec. Publ., Amer. Soc. Mamm., 2:1–593.

TERMAN, C. R., AND J. F. SASSAMAN. 1967. Sex ratio in deer mouse populations. J. Mamm., 48:589–597.

THOMPSON, S. D. 1982. Structure and species composition of desert heteromyid rodent species assemblages: effects of a simple habitat manipulation. Ecology, 63:1313–1321.

TRAVERS, S. E., D. W. KAUFMAN, AND G. A. KAUFMAN. 1988. Differential use of experimental habitat patches by foraging *Peromyscus maniculatus* on dark and bright nights. J. Mamm., 69:869–872.

TRIVERS, R. L., AND D. E. WILLARD. 1973. Natural selection of parental ability to vary the sex ratio of offspring. Science, 179:90–92.

VAN DEUSEN, M., AND D. W. KAUFMAN. 1985. Observations on genetic differentiation in *Peromyscus*. Trans. Kansas Acad. Sci., 88:154–158.

VAN HORNE, B. 1982. Niches of adult and juvenile deer mice (*Peromyscus maniculatus*) in seral stages of coniferous forest. Ecology, 63:992-1003.

VAN HORNE, B., AND R. G. FORD. 1982. Niche breadth calculation based on discriminant analysis. Ecology, 63:1172–1174.

VAN VALEN, L. 1965. Morphological variation and width of ecological niche. Amer. Nat., 99:377–390.

WHITAKER, J. O., JR. 1966. Food of *Mus musculus, Peromyscus maniculatus bairdi* and *Peromyscus leucopus* in Vigo County, Indiana. J. Mamm., 47:473–486.

———. 1967. Habitat relationships of four species of mice in Vigo County, Indiana. Ecology, 48:867–872.

WILLIAMS, R. G., F. B. GOLLEY, AND J. L. CARMON. 1965. Reproductive performance of a laboratory colony of *Peromyscus polionotus*. Amer. Midland Nat., 73:101–110.

WILSON, K. R., AND D. R. ANDERSON. 1985. Evaluation of two density estimators of small mammal population size. J. Mamm., 66:13–21.

WOLFF, J. O. 1985. Comparative population ecology of *Peromyscus leucopus* and *P. maniculatus*. Canadian J. Zool., 63:1548–1555.

WOLFF, J. O., AND B. HURLBUTT. 1982. Day refuges of *Peromyscus leucopus* and *Peromyscus maniculatus*. J. Mamm., 63:666–668.

WRIGHT, S. 1965. The interpretation of population structure by F-statistics with special regard to systems of mating. Evolution, 19:395–420.

YAHNER, R. H. 1982. Microhabitat use by small mammals in farmstead shelterbelts. J. Mamm., 63:440–445.

ZERA, A. J., R. K. KOEHN, AND J. G. HALL. 1985. Allozymes and biochemical adaptation. Pp. 633–674, in Comprehensive insect physiology, biochemistry and pharmacology (G. A. Kerkut and L. I. Gilbert, eds.). Pergamon Press, New York, New York, 10:1–715.

SOCIAL BEHAVIOR

Jerry O. Wolff

Abstract.—Social behavior of *Peromyscus* is reviewed, emphasizing the social organization of *Peromyscus leucopus* and *P. maniculatus*. The major topics covered include 1) home range and territory, 2) mating systems, 3) adult-juvenile relationships as they relate to aggression, dispersal, and colonization, and 4) nesting behavior. Social behavior is labile and varies with density and habitat. The social organization of *Peromyscus leucopus* and *P. maniculatus* is similar and is based on home-range overlap between sexes and mutually exclusive home ranges within sexes. Territoriality is density-dependent, occurring only when population density exceeds a particular threshold, above which home range overlap becomes extensive. At low densities, home ranges are maintained by mutual avoidance. For males during the breeding season, the proximate function of territories and home ranges is the same. Males use home ranges or defend territories to provide access to reproductive females. Territorial defense may also help prevent cuckoldry. Females use home ranges for feeding and to provide living space for their young. They defend territories to prevent infanticide. Mating systems range from monogamy in *P. polionotus* to promiscuity in *P. leucopus* and *P. maniculatus*. Following the breeding season, juvenile male *P. leucopus* disperse, whereas daughters inherit their natal home range. Pregnant post-lactating females frequently will abandon their original home range and establish a new home range in a nearby habitat. During summer, animals nest singly or in male-female pairs. Communal nesting occurs during winter, with the most common group size being two and with few groups larger than four. Aggression appears to play a major role in the social organization of *Peromyscus* only at high densities.

Several studies conducted in the early 1940s and 1950s, notably those of Blair (1940, 1951), Dice and Howard (1951), and Nicholson (1941), set the stage for what is known about *Peromyscus* social behavior today. These papers and others were reviewed by Eisenberg (1968) and Stickel (1968). Although these earlier studies were primarily descriptive, most of what they concluded was correct. The major advances in our understanding of *Peromyscus* social behavior over the last two decades have resulted from experimental manipulations used to test hypotheses generated from descriptive and observational studies and from the application of evolutionary theory to explain the adaptive significance of behavioral systems. The objectives of this paper are to review the current literature from field and laboratory studies on *Peromyscus* social behavior and to interpret the results with respect to *Peromyscus* population ecology. The major topics covered include home range and territory; mating systems; adult-juvenile relationships relative to aggression, dispersal and colonization; and nesting behavior. Because most of the research on *Peromyscus* social behavior has been conducted on *P. leucopus* and *P. maniculatus*, emphasis is placed on these species; reference is made to other *Peromyscus* species where data are available.

HOME RANGE AND TERRITORY

Although size and shape of home range are difficult to determine and interpret (Meserve, 1977; Metzgar, 1972; 1973*b*), knowledge of home range is important to our understanding of *Peromyscus* social organization. Home range, defined as the total activity area of an animal, and territory, a defended portion of a home range, have been studied by live-trapping, tracking, and radiotelemetry. Although Madison (1977) suggested that radiotelemetry probably gives the best indication of space use, home range sizes of *P. leucopus* and *P. maniculatus* estimated by live-trapping and radiotelemetry (Wolff, 1985*a*) and by live-trapping and tracking (Metzgar, 1973*c*) were comparable. Home range sizes of *P. leucopus* and *P. maniculatus* range from 242 square meters (Cranford, 1984) to over 3000 square meters (Metzgar, 1973*a*, 1973*b*; see Stickel, 1968 for review of studies on these and other species of *Peromyscus*). The vertical dimension of the home range also must be considered for arboreal species (Barry and Francq, 1980; Barry *et al.*, 1984; Graves *et al.*, 1988; Meserve, 1977). Three unresolved questions concerning home ranges and territories are: 1) What factors determine home range size? 2) Under what conditions does a home range become a territory (that is, defended)? 3) What are the functions of a home range or territory? Home range size often is associated with resource availability, principally that of food. Smaller home range sizes on food supplemented grids compared to control grids have been reported for *P. leucopus* (Miller and Getz, 1977) and for *P. maniculatus* (Fordham, 1971; Taitt, 1981). Females of both species often show a greater response to a food supplement than do males (Fordham, 1971; Gerzoff, 1984; Hansen and Batzli, 1978; Wolff, 1985*a*, 1986*a*). Other studies, however, have shown that home range size is not correlated with food availability. Metzgar (1973*a*) and Sheppe (1967) found that feeding and exploratory home ranges were the same size and were not necessarily correlated with food availability. However, when food was added to a home range, activity areas shifted within the home range, although the size and position of the home range did not change (Sheppe, 1967). Resident animals still explored the same area even though they did not feed in it. Food availability may partially explain variation in home range size, but it does not appear to be the factor determining minimum home range size.

Home range size is apparently only partially correlated with density. Larger home range sizes at low densities than at high densities have been reported for *P. leucopus* in Ontario (Sheppe, 1967) and in Virginia (Wolff, 1985a) and for *P. maniculatus* in British Columbia (Taitt, 1981). Home range sizes, however, were not affected by densities in other studies on *P. leucopus* (Metzgar, 1971) or *P. maniculatus* (Wolff, 1985a). The relationship between home range area and density is not clear, but apparently a minimum area is occupied independent of density. Further understanding of the relationship between home range size, food, and density requires examination of the concept of territoriality.

Numerous studies have demonstrated the existence of territories or mutually exclusive home ranges in *Peromyscus* (Hansen and Batzli, 1978; Madison, 1977; Metzgar, 1971, 1973a, 1979, 1980; Myton, 1974; Wolff, 1985a; Wolff et al., 1983), but little evidence has been provided to show whether exclusive home ranges are maintained by mutual avoidance or by aggressive territorial defense. Wolff et al. (1983) and Wolff (1985a) experimentally tested for territoriality in syntopic *P. leucopus* and *P. maniculatus* by conducting behavioral trials between resident and intruder adults of the same sex. At densities of 25 to 60 mice per hectare, home range overlap with neighbors averaged 37 percent among males and 14 percent among females, and resident animals were aggressive toward intruders, winning 83 percent (131 of 158) of the trials. At population densities of less than 25 mice per hectare, home range overlap was reduced to seven percent for both males and females, and aggression occurred in only one of 40 trials. Thus, there was a minimum threshold density above which animals became defensive and maintained territories by aggressive behavior. In this case, *P. leucopus* and *P. maniculatus* were interspecifically as well as intraspecifically territorial, and thus behaved ecologically as a single species (Wolff et al., 1983). One species did not dominate the other, but territory holders were aggressive toward and were able to exclude strangers of either species. Interspecific territoriality may be one mechanism for maintaining coexistence between these ecologically similar species.

Intensity of territorial defense varies with the location within the territory. When mice were territorial at high densities, resident mice won 86 percent of the trials conducted at the center of their territory but only 55 percent of the trials conducted on the periphery (Wolff *et al.*,1983). Territorial bouts with strangers were more intense than those with neighbors. Neighbor recognition through prior association appears to diminish aggressive and territorial defense toward neighbors in *P. leucopus* (Vestal and Hellack, 1978). The social organization of *P. maniculatus* also involves mutual recognition of neighbors and exclusion of strangers (Healey, 1967). In *P. leucopus* and *P. maniculatus*, animals occupying adjacent home ranges sometimes are relatives (Wolff and Lundy, 1985), which may explain in part their low levels of aggression.

Territories appear to result from populations reaching a threshold density above which increased home range overlap results in defense of exclusive core areas. The only distinguishing feature between a home range and a territory is that the latter is defended. Much of the inconsistency in the literature with regard to home ranges and territories is because few studies have demonstrated the role of aggression in the maintenance of exclusive areas.

The proximate functions of home ranges or territories in *Peromyscus* have not been demonstrated clearly. As noted above, territory or home range is not necessarily a function of food availability (Metzgar, 1973*a*; Sheppe, 1966; Wolff 1985*a*, 1986*a*) except perhaps for females (Bowers and Smith, 1979; Fordham, 1971; Hansen and Batzli, 1978; Nadeau *et al.*, 1981; Wolff, 1985*a*) and is correlated only marginally with density (Metzgar, 1971; Sheppe, 1967; Wolff, 1985*a*). Several characteristics of home ranges and territories of males and females suggest that their functions may be different for each sex. Home ranges and territories for males are larger than those of females and they exhibit greater overlap (Cranford, 1984; Fairbairn, 1977; Metzgar, 1971, 1973*b*, 1979, 1980; Mihok, 1979; Mineau and Madison, 1977; Myton, 1974; Taitt, 1981; Wolff, 1985*a*, 1986*a*; Wolff *et al.*, 1983). Furthermore, males spend more time on the periphery of their home ranges than do females (Myton, 1974), and they shift or move their home ranges to overlap those of additional adult females (Fairbairn, 1977; Sheppe, 1966; Wolff, 1985*b*). These differences indicate that, in addition to using home ranges for nesting and feeding, males use home ranges or territories to provide access to one or more reproductive females. Reproductive competition and defense of females is more intense at high densities than at low densities, but does not always prevent cuckoldry and multiple inseminations (discussed below). At low densities, males may abandon home ranges and become vagrants searching for reproductive females (Fairbairn, 1978; Nadeau *et al.*, 1981).

Home ranges and territories of females are used for feeding, nesting, and rearing young. Territorial defense by females may also function in part to protect young against infanticide (Wolff, 1985*c*; Wolff and Cicirello, 1989). Laboratory studies on infanticide and maternal aggression in *Peromyscus* have shown that lactating females are more aggressive towards intruders of either sex than are nonbreeding females (Ayer and Whitsett, 1980; Gleason *et al.*, 1980; Rowley and Christian, 1976; Whitsett *et al.*, 1979; Wolff, 1985*c*). In a laboratory study, Wolff (1985*c*) found that nursing *P. leucopus* and *P. maniculatus* females exhibited postpartum aggression and excluded strange females from their nest site. When maternal females were away from their nests, however, intruding females killed or attacked neonates in 42 of 43 trials. Thus, given the opportunity, intruding females killed strange pups. One instance of infanticide has been observed in the field: an intruding female killed five, eight-day-old pups in a nest box while the mother was detained in a trap (Wolff, 1986*b*). A recent field study on infanticide in *P. leucopus* has shown that adult females killed strange pups in 74 percent of 26 trials (Wolff and Cicirello, 1989). Wolff and Cicirello concluded that at high densities females may commit infanticide to compete for territories and nesting sites. Infanticide among female *Peromyscus* may be more common than previously recognized and warrants further investigation.

The fact that reproductive females are more aggressive than males and that their territories show less overlap with neighbors suggests that females have a greater investment in territorial defense than do males. Investment in this case refers to time, energy, and risk, which are measures of parental investment that ultimately affect lifetime reproductive success. Although protection of the young seems to be a logical reason for female territoriality, complementary or alternative hypotheses exist (discussed under Juvenile Dispersal). The occurrence of territoriality only during the breeding season further suggests that reproductive competition can favor territorial behavior among males and females.

MATING SYSTEMS

The mating systems in *Peromyscus* spp. have been inferred mainly from field studies of the spatial distribution of males and females and from behavioral studies conducted in the laboratory. The spatial arrangement of male and female home ranges varies from study to study, but the general pattern is one of overlap between males and females and mutually exclusive home ranges within sexes (Fairbairn, 1977; Metzgar, 1979, 1980; Mihok, 1979; Myton, 1974; Novak, 1983; Sheppe, 1966; Wolff, 1985a; Wolff *et al.*, 1983; but see Bowers and Smith, 1979). Mating systems inferred from spatial overlap range from a presumed polygynous system of one male and several females in *P. maniculatus* (Mihok, 1979) to a presumed polyandrous system of one female and several males in *P. leucopus* (Myton, 1974; Wolff, 1985a). Home range overlap of heterosexual pairs and the occurrence of family groups, which are monogamous characteristics, have been reported for *P. californicus, P. eremicus, P. maniculatus*, and *P. polionotus* (Kleiman, 1977). Dewsbury (1981) presented a series of correlates associated with monogamy: sexual monomorphism, latency to initiate copulation, allogrooming, small numbers of ejaculations, lack of the Coolidge Effect, small litter size, paternal behavior, and slower developmental and sexual maturation rates. Based on these criteria, 11 *Peromyscus* species were ranked on a scale from most likely to least likely to be monogamous (Table 1). Monogamy is most likely to occur in those species having a score of +5 or greater, which include *P. californicus, P. eremicus, P. mexicanus*, and *P. polionotus*. Although these data are inferential and do not provide a conclusive description of *Peromyscus* mating systems, field studies largely support these conclusions.

The best evidence for determining whether matings in *Peromyscus* species are monogamous or promiscuous is provided by elec-

TABLE 1.—*Comparisons of scores on the monogamy scale for species of* Peromyscus *(after Dewsbury, 1981). No. = number of correlates measured; see text for description of correlations.*

Species	Score	No.
P. californicus	+10	6
P. eremicus	+6	8
P. mexicanus	+5	8
P. polionotus	+5	8
P. floridanus	+3	3
P. melanocarpus	+3	5
P. crinitus	+2	4
P. leucopus	+1	8
P. melanophrys	+1	2
P. yucatanicus	+1	3
P. gossypinus	0	7
P. maniculatus	−5	9

trophoretic analysis of blood groups of mothers, pups, and putative fathers. In an electrophoretic paternity exclusion analysis in *P. polionotus*, no evidence was found for multiple paternity (Foltz, 1981). All the pups were sired by a sole male, and in burrows where a male was present with a female, all subsequent litters were sired by that male. This is in contrast to *P. maniculatus* where an estimated 19 to 43 percent of litters resulted from multiple inseminations (Birdsall and Nash, 1973; Merritt and Wu, 1975).

Mating in *P. leucopus* appears to range from monogamy to promiscuity, depending upon conditions. In a radiotelemetry study of a low-density population, Mineau and Madison (1977) found that *P. leucopus* formed long-term monogamous pairs. Using electrophoresis to determine paternity in a moderately dense population of 15 to 20 mice per hectare, Baccus and Wolff (in preparation) estimated that up to 30 percent of the litters resulted from multiple inseminations. Multiple copulations in the field, however, might not always show up in an electrophoretic multiple-paternity test because the probability of detecting mixed litters from multiple-male copulations is less than 50 percent (Dewsbury and Baumgardner, 1981). Also, multiply mated females have a lower probability of pregnancy than singly mated females (Dewsbury, 1982*a*). Thus, electrophoretic data on multiple-male paternity studies give an underestimate of the frequency of multiple-male copulations. Promiscuous mating is probably more common than is detectable by this technique.

Female promiscuity may be an adaptive mechanism to prevent infanticide by males. Among rodents, males will occasionally kill unrelated pups (see Labov *et al.*, 1985 for review). Several studies on rodents, however, have shown that males that have cohabited or recently copulated with a female will recognize her later and not kill her offspring (Brooks, 1984; Cicirello and Wolff, submitted; Elwood and Ostermeyer, 1984; Huck *et al.*, 1982; Labov, 1980; Mallory and Brooks, 1978). Male *P. leucopus* that have copulated are inhibited from killing any pups within their home ranges, whereas dispersing males and resident males that have not copulated will kill pups (Wolff and Cicirello, 1989). Thus, promiscuous mating by female *Peromyscus* with resident and neighboring males, especially at high densities, may be a strategy by females to confuse paternity and prevent infanticide by males.

In summary, the mating systems in *Peromyscus* appear to be extremely labile, ranging from promiscuity to facultative monogamy in *P. leucopus* and *P. maniculatus*, to obligatory monogamy in *P. po-*

lionotus, and perhaps *P. eremicus, P. mexicanus,* and *P. californicus,* as well.

JUVENILE DISPERSAL

Dispersal often is viewed at the population level as a mechanism for regulating density and avoiding over-exploitation of resources. However, dispersal also plays an important role in the social organization of *Peromyscus.* Considerable dispersal activity and general movement occur throughout the year in *Peromyscus,* but are most pronounced during the breeding season (Fairbairn, 1977, 1978; Harland *et al.,* 1979; Krohne *et al.,* 1984; Sullivan, 1977; Van Horne, 1981; Wolff, 1985*b*). The general conclusions are that males move more than females, and that juveniles move more than adults. Dispersal of pregnant, post-lactating females has been reported for *P. maniculatus* (Van Horne, 1981) and *P. leucopus* (Harland *et al.,* 1979; Wolff and Lundy, 1985; Wolff *et al.,* 1988).

In a test for differential dispersal of juvenile male and female *P. leucopus,* Wolff and Lundy (1985) found the mean lengths of residence for males and females were 5.7 and 10.6 weeks, respectively. They also recorded a larger number of juvenile and subadult males than females moving onto grids. None of the 26 males, but seven of 23 (30 percent) females, attained sexual maturity and bred within their natal home ranges. Twenty-four percent (seven of 29) of adult females that bred in the spring bred again on the same grid in the autumn. Four of these seven females moved the center of their home ranges about 15 meters after weaning their first litters. Eight of 20 (40 percent) adult males from the spring breeding population were on their same home ranges in the autumn.

Differential dispersal of male and female juveniles results in separation of mothers from sons, and brothers from sisters prior to subsequent breeding seasons. The relationship between fathers and daughters is less clear, but Wolff *et al.* (1988) found only three possible cases of father-daughter inbreeding in *P. leucopus* out of 135 matings. Wolff *et al.* concluded that male-biased dispersal of juvenile *P. leucopus* was a mechanism to prevent inbreeding. Krohne *et al.* (1984) and Nadeau *et al.* (1981) also concluded that dispersal in *P. leucopus* is more intimately related to social interactions and the avoidance of inbreeding than it is to local density.

In *Peromyscus* the presence of a parent inhibits or supresses attainment of juvenile sexual maturation (Bediz and Whitsett, 1979; Haigh, 1983*a,* 1983*b*; Haigh *et al.,* 1985; Lombardo and Terman,

1980; Skryja, 1978; Terman, 1980, 1984; Thomas and Terman, 1975). Thus, some mechanism by which weaned young separate from their parents insures the earliest onset of sexual maturation of juveniles and at the same time prevents parent-offspring matings. Reproductive inhibition also occurs among littermates (Dewsbury, 1982b; Hill, 1974). Consequently, the differential dispersal of male and female siblings minimizes maturation time and reduces the chance of sibling matings. An alternative possibility is that juvenile males may disperse to reduce competition with adult males and to increase mating opportunities; however, this has been difficult to test experimentally (but see Wolff et al., 1988).

Several studies of Peromyscus have concluded that aggression by resident adult males causes dispersal of subordinates and juveniles and inhibits their immigration (Fairbairn, 1977, 1978; Healey, 1967; Metzgar, 1971, 1979; Pettigrew and Sadleir, 1974; Sadleir, 1965; Savidge, 1974; Van Horne, 1981). Other investigations have suggested that females are more instrumental than males in regulating recruitment (Halpin, 1981; Hansen and Batzli, 1978; Harland et al., 1979; Metzgar, 1971; Nadeau et al., 1981; Wolff et al., 1988). In a field study, Healey (1967) found that survival of juvenile P. maniculatus was considerably higher in trapped-out areas than in areas where aggressive mice had been introduced. At high densities, juvenile P. maniculatus in southeast Alaska were displaced to suboptimal habitat by adults (Van Horne, 1981). At low densities both adults and juveniles had higher survivorship in favorable habitat. To test for aggression between adults and juveniles in P. leucopus, Wolff (1986c) conducted within-sex behavioral trials between adult and unrelated juvenile males and females in a natural population within the home ranges of the adults (see also Wolff et al., 1988). Aggression occurred in only 13 percent of the trials, whereas nonaggressive contact occurred in 84 percent of the trials. Animals frequently avoided each other, but few signs of overt aggression were observed. The lack of aggression between adults and unrelated juveniles suggests that dispersal was self-motivated and not forced by adult aggression. On grids where adult males had been removed, all juvenile male P. leucopus still dispersed within five weeks after weaning (Wolff et al., 1988), suggesting that the length of residency of juvenile males was not affected by the presence or absence of adult males. In three years, only 14 of 184 (eight percent) males remained in their natal home ranges. In all 14 cases, the mothers and sisters of such males were no longer present on the natal home range. Thus, the presence of female relatives may have a greater affect on dispersal of juvenile males than does the presence of adult males.

Tolerance of immigration of nonreproductive animals by resident adult males, which resulted in juvenile recruitment, has been observed in *P. leucopus* (Metzgar, 1971; Nadeau *et al.*, 1981; Wolff *et al.*, 1988) and in *P. maniculatus* (Herman, 1984; Metzgar, 1980; Mihok, 1979). The difference between results obtained in these studies and those mentioned previously might be related to density. At high densities, when the population is at the carrying capacity, aggression and territorial defense may inhibit immigration and colonization; however, at low densities, when vacant suitable habitat is available for colonization, aggression appears to be reduced. In optimal woodland habitats in the southern Appalachians, *P. leucopus* and *P. maniculatus* were not aggressive when population densities were less than 25 mice per hectare but were aggressive at densities greater than 25 mice per hectare (Wolff, 1985a). Colonization rates did not vary between high or low densities, but at high densities only 25 percent of spring-born males and 40 percent of spring-born females attained sexual maturity by autumn, whereas at low densities, all spring-born males and 88 percent of spring-born females attained sexual maturity by autumn (Wolff, 1986c). Thus, territorial aggression by resident adults may not only deter intrusion of adults but may also inhibit sexual maturation of juveniles. Once again, it appears that aggression by resident adults is density-dependent and occurs only when populations are above a particular threshold density (Wolff, 1985a), which can be expected to vary among habitats.

A major difficulty in interpreting the role of aggression in dispersal and population regulation is that most investigators have not analyzed dispersal in the context of food and space availability. Dispersal in this context may regulate populations, though this may not be its primary function. When aggression does occur, it occurs within and not between sexes. Males are not aggressive toward females and thus would not inhibit their recruitment. The number of breeding females in the population, however, may be limited by female aggression. The fact that females are more aggressive than males and their territories show less overlap than males supports this suggestion. Several studies have concluded that female territoriality limits the density of females and regulates the density of the breeding population (Hansen and Batzli, 1978; Metzgar, 1971, 1980); however, this has not been tested experimentally. Densities

as high as 109 and 163 mice per hectare have been reported for
P. maniculatus (Merkt, 1981; Sexton *et al.*, 1982), whereas estimates
of 115 mice per hectare have been reported for *P. leucopus* (Goundie and Vessey, 1986). These figures suggest that either intrinsic
mechanisms for population regulation are not always effective or
the carrying capacities in these cases were extremely high.

Nesting Behavior

Most natural nest sites of *Peromyscus* have been located by using
radiotelemetry (Madison, 1977; Madison *et al.*, 1984; Wolff and
Hurlbutt, 1982). Wolff and Hurlbutt found that among *P. leucopus*,
eight nests were in trees and 22 were underground. Madison (1977)
described 16 underground nest sites for *P. leucopus*; 10 were in
woodchuck (*Marmota monax*) burrows and six were in rock piles.
During early autumn, 13 nest sites of *P. leucopus* were in trees, either in gray squirrel (*Sciurus carolinensis*) nests or in tree hollows
(Madison *et al.*, 1984). Nests of *P. leucopus* have also been found in
logs, stumps, and in ground burrows (Mumford and Whitaker,
1982), and vary with microhabitat (Layne, 1969; Wolfe, 1970).
Twenty-two of 24 nest sites of *P. maniculatus nubiterrae* were in large
hollow trees and two were underground (Wolff and Hurlbutt,
1982). Males and females used similar nest sites. Nine of 22 *P. leucopus* and four of 20 *P. maniculatus* occupied more than one nest
during an eight-week period (Wolff and Hurlbutt, 1982), and eight
of 15 *P. leucopus* used more than one nest site in a two-month period (Madison, 1977). *Peromyscus leucopus* and *P. maniculatus* frequently moved between natural nest sites and nest boxes (Wolff
and Durr, 1986). Trudeau *et al.* (1980) found that in *P. leucopus*,
nest boxes were used most frequently during summer by solitary
males, followed by solitary females, females with litters, male-female
pairs, litters with no adults present, nonreproductive mixed-sex
groups, males with litters, and male-male pairs. Occasionally females nested with a second litter when their first litter was still in
the same nest.

In laboratory studies, *P. l. noveboracensis* and *P. m. gracilis* preferred elevated compared to ground nest sites (Stah, 1980); however, Tadlock and Klein (1979) found that *P. m. gracilis* evinced no
preference between ground and arboreal nests. *Peromyscus leucopus*
frequently use nest boxes located one to two meters above ground
in trees, but do not use ground nest boxes (C. Terman, personal
communication; Wolff, unpublished). The use of ground compared
to arboreal nest sites seems to be correlated with habitat and forag-

ing behavior. *Peromyscus polionotus* and *P. floridanus*, both ground-dwelling species, prefer ground nests, whereas the more arboreal *P. gossypinus* prefers elevated nest sites (Klein and Layne, 1978). In areas of sympatry, *P. maniculatus nubiterrae* nest almost exclusively in large hollow trees, whereas *P. leucopus noveboracensis* are more variable in their use of nest sites, frequently nesting underground, or in smaller trees and closer to the ground (Wolff and Durr, 1986; Wolff and Hurlbutt, 1982). Although competition for nest sites has been suggested between *P. l. noveboracensis* and *P. m. nubiterrae* and between *P. l. noveboracensis* and *P. m. gracilis* (Smith and Speller, 1970), location of nest sites may reflect species preferences rather than competition.

During the breeding season, *Peromyscus* nest singly or occasionally in pairs or family groups, whereas during winter they form aggregations and nest communally. In New York, *P. leucopus* nested singly at about 10 to 15 meters in trees in early October, began communal nesting at heights below four meters in trees in mid-October, and shifted to communal nesting underground by late October and November (Madison *et al.*, 1984). In Virginia, 60 *P. leucopus* used arboreal nest boxes in November through January, none in mid-February when temperatures averaged −5°C, and 27 used nest boxes in April and May (Wolff and Durr, 1986). Thirty-four *P. leucopus* were trapped at underground burrow sites in midwinter, whereas none was caught at arboreal sites during this period. At this same study site, 28 of 33 *P. m. nubiterrae* nested communally in arboreal nest sites in January and February. On the basis of these observations, it appears that *P. leucopus* takes advantage of the thermoregulatory benefits of nesting underground and under the snow; *P. maniculatus* apparently does not. In western Pennsylvania, *P. m. nubiterrae* nested underground in winter; however, no large hollow trees are present in the study area (J. Merritt, personal communication). *Peromyscus leucopus* and *P. maniculatus* have been recorded nesting together during winter on three occasions (J. Merritt, personal communication; Wolff and Durr, 1986; Wolff and Hurlbutt, 1982).

The bioenergetic benefits of huddling in communal nests seem obvious (Glaser and Lustick, 1975). The combination of a nest, huddling, and the occurrence of torpor can provide a 74 percent energetic saving compared to individually housed nontorpid mice without a nest (Vogt and Lynch, 1982). Energy savings attributable to the nest alone were 18 to 28 percent; to nest and torpor, 33 to 43 percent; to huddling, 16 to 30 percent; and to huddling and torpor, 58 percent. Average daily metabolic rate of *P. maniculatus* is also lower in winter than in summer, resulting in less energy expenditure during the cold season (Merritt, 1984; Stebbins, 1978). Vickery and Millar (1984) presented a model to show the energetic advantages of huddling whenever ambient and nest temperatures were below the thermoneutral zone. Their model predicts that at moderate temperatures a group size of two or three may be optimal, whereas on very cold nights, optimal group size would be larger. In field studies, mean group size during winter was 2.8 for *P. maniculatus* (Wolff and Durr, 1986) and ranged from 1.5 to 3.2 for *P. leucopus* with no increase with colder temperatures (Madison et al., 1984; Wolff and Durr, 1986). Competition for food, which might increase the foraging radius, may set the upper limit to group size in communal nests.

Peromyscus hoard food during the winter, especially in the northern part of their ranges. In the laboratory, *P. leucopus noveboracensis* stored more food than did *P. maniculatus gracilis* (Tadlock and Klein, 1979). *Peromyscus leucopus noveboracensis* and *P. m. bairdii* stored more food at cold temperatures and short photoperiods than did *P. l. castaneus*, *P. m. blandus*, or *P. eremicus eremicus* (Barry, 1976). The latter three taxa have more southern distributions than the first two. It is noteworthy that *Peromyscus maniculatus blandus* stored more food at 27°C than at 7°C, which perhaps reflected an adaptation to summer torpor.

SOCIAL BIOLOGY OF *PEROMYSCUS LEUCOPUS*

Because so little is known about social organization in species of *Peromyscus* other than *P. leucopus* and *P. maniculatus*, it is difficult to summarize the social biology of the genus as a whole. Therefore, the following description and interpretation of the typical annual behavioral cycle of *P. leucopus* inhabiting mixed deciduous forests in the southern Appalachians of Virginia is presented to characterize the social biology of a representative species of *Peromyscus*. During the spring breeding season which starts in early April, females occupy non-overlapping home ranges maintained by mutual avoidance at low densities or territories maintained by overt aggression at high densities. Males have home ranges that overlap those of one or more females and mate promiscuously with the females encountered within their home ranges. A resident male may cohabit with a female until parturition. At the time of parturition, the male is excluded from the nest and may move to another nest site within his home range (Xia and Millar, 1988). Maternal females are more aggressive than nonbreeding females, probably as a defense against infanticide. Following weaning, juvenile males and some juvenile females disperse, whereas other juvenile females establish home ranges within or near their maternal home range. Dispersal of juvenile males appears to be voluntary, rather than being forced by adult aggression. A mid-summer breeding hiatus occurs from July to mid-August (Cornish and Bradshaw, 1978; Wolff, 1985b, 1986a). During the September through October breeding season, young-of-the-year may enter the breeding population. The probability of attaining sexual maturity by autumn is inversely related to density. Following the autumn breeding season, some juvenile males disperse, whereas others remain within the maternal home range and nest communally with kin and non-kin throughout the winter (Wolff and Durr, 1986). Individuals or groups shift nest sites several times during the winter, and the group composition changes as individuals move between communal nests. When the spring breeding season starts again, individuals leave the communal group and occupy home ranges that overlap one or more nonrelated members of the opposite sex. Although variation from this pattern may be expected to occur in different habitats or under different environmental conditions, the preceeding description appears to represent the typical social organization of *P. leucopus* and is probably very similar to that of *P. maniculatus*.

FUTURE DIRECTIONS

In this review, I have emphasized the social behavior of *P. leucopus* and *P. maniculatus*. This reflects not only my personal research interests, which have focused on these species, but the strong bias towards these species in the literature of social behavior in *Peromyscus*. The large and diverse number of species in the genus provides an excellent opportunity for comparative ethological studies within a taxon; however, comparable studies on other species are largely lacking. Moreover, I encountered several problems in interpreting and evaluating the results from previous studies on *Peromyscus* behavior. These difficulties result from several problem areas which should be considered in future studies.

1) Because many interpretations of *Peromyscus* social behavior have been based on data collected incidental to studies on population ecology, rigorous testing of observations has not been possible. Studies should be experimentally designed to examine social behavior directly, rather than as a byproduct of some other study.

2) Most studies have been of short duration and have not been

replicated. As a consequence, spatial and temporal variation frequently have not been taken into account. For instance, aggression and territorial defense occurred in only one out of seven years in Virginia (Wolff, 1986c); thus, a one-year study would not have revealed this variation.

3) Most studies have been descriptive with *a posteriori* interpretations and have not been designed to test *a priori* predictions to discriminate between alternative hypotheses. Background data on *Peromyscus* behavior are now sufficient to permit predictions and experimental manipulation of systems to test specific hypotheses.

4) Most studies have been incomplete; not enough parameters have been measured to determine their interactive effects. For instance, one should not consider dispersal without relating it to density, food availability (or some other measure of carrying capacity), time of year, home range size (or spatial distribution of sexes and ages), reproductive condition, age, and perhaps a host of other factors.

5) Methods of study should be standardized, and the most accurate techniques used for collecting data. Currently, the most accurate methods for determining home ranges and nest locations are radiotelemetry and tracking with fluorescent pigments. Radionucleotide-electrophoresis and DNA fingerprinting techniques provide the most unequivocal evidence for determining parentage. The role of aggression in dispersal and dispersion should be studied in the field, not in the laboratory. Laboratory studies should be treated with caution but can provide valuable information when designed to simulate natural conditions or to examine stereotyped behavior that is not a laboratory artifact.

6) Lastly, one must apply evolutionary theory to explain the adaptive significance of behavior at the level of the individual before it can be applied to demographic and population phenomena. One should not look for the average individual in a population, but rather for differences between sexes, ages, and individuals to see how these differences affect an individual's survival and reproductive success.

I encourage behavioral biologists to pursue these avenues of research, in order to provide a more complete picture of social behavior in *Peromyscus*.

Acknowledgments

I thank D. Dewsbury, G. Kaufmann, R. Barry, G. Kirkland, J. Layne, and D. Cicirello for helpful comments on earlier drafts of this manuscript, and G. Kirkland and

J. Layne for inviting me to participate in the "Biology of *Peromyscus*" symposium at the IV International Theriological Congress in Edmonton, Alberta. The conclusions drawn in this paper are my own and do not necessarily reflect those of the reviewers or editors. This work was supported by NSF Grant 83–06619.

LITERATURE CITED

AYER, M. L., and J. M. WHITSETT. 1980. Aggressive behaviour of female prairie deer mice in laboratory populations. Anim. Behav., 28:763–771.

BARRY, R. E., JR., and E. N. FRANQ. 1980. Orientation to landmarks within the preferred habitat by *Peromyscus leucopus*. J. Mamm., 61:292–303.

BARRY, R. E., JR., M. A. BOTJE, and L. B. GRANTHAM. 1984. Vertical stratification of *Peromyscus leucopus* and *P. maniculatus* in Southwestern Virginia. J. Mamm., 65:145–148.

BARRY, W. J. 1976. Environmental effects of food hoarding in deermice (*Peromyscus*). J. Mamm., 57:731–746.

BEDIZ, G. M., and J. M. WHITSETT. 1979. Social inhibition of sexual maturation in male prairie deermice. J. Comp. Physiol. Psychol., 93:493–500.

BIRDSALL, D. A., and D. NASH. 1973. Occurrence of successful multiple-insemination of females in natural populations of deer mice (*Peromyscus maniculatus*). Evolution, 27:106–110.

BLAIR, W. F. 1940. A study of prairie deer-mouse populations in southern Michigan. Amer. Midland Nat., 24:273–305.

———. 1951. Population structure, social behavior, and environmental relations in a natural population of the beach mouse (*Peromyscus polionotus leucocephalus*). Contrib. Lab. Vert. Biol., Univ. Michigan, 48:1–47.

BOWERS, M. A., and H. D. SMITH. 1979. Differential habitat utilization by sexes of the deermouse, *Peromyscus maniculatus*. Ecology, 60:869–875.

BROOKS, R. J. 1984. Causes and consequences of infanticide in populations of rodents. Pp. 331–348, *in* Infanticide: comparative and evolutionary perspectives (G. Hausfater and S. B. Hrdy, eds.). New York, Aldine, 598 pp.

CICIRELLO, D. M., AND J. O. WOLFF. 1989. The effects of cohabitation and copulation on pup recognition and infanticide in white-footed mice. Manuscript.

CORNISH, L. M., and W. N. BRADSHAW. 1978. Patterns in twelve reproductive parameters for the white-footed mouse (*Peromyscus leucopus*). J. Mamm., 59:731–739.

CRANFORD, J. A. 1984. Population ecology and home range utilizations of two subalpine meadow rodents (*Microtus longicaudus* and *Peromyscus maniculatus*). Pp. 285–292, *in* Winter ecology of small mammals (J. F. Merritt, ed.). Spec. Publ. Carnegie Mus. Nat. Hist., 10:1–380.

DEWSBURY, D. A. 1981. An excercise in the prediction of monogamy in the field from laboratory data on 42 species of muroid rodents. The Biologist, 63:138–162.

———. 1982a. Pregnancy blockage following multiple-male copulation or exposure at the time of mating in deer mice (*Peromyscus maniculatus*). Behav. Ecol. Sociobiol., 11:37–42.

———. 1982b. Avoidance of incestuous breeding between siblings in two species of *Peromyscus* mice. Biol. Behav., 7:157–169.

DEWSBURY, D. A., and D. J. BAUMGARDNER. 1981. Studies of sperm competition in two species of muroid rodents. Behav. Ecol. Sociobiol., 9:121–133.

DICE, L. R., and W. E. HOWARD. 1951. Distance of dispersal by prairie deer mice from birthplaces to breeding sites. Contrib. Lab. Vert. Biol., Univ. Michigan, 50:1–15.

EISENBERG, J. F. 1968. Behavior Patterns. Pp. 451–495, in Biology of Peromyscus (Rodentia) (J. A. King, ed.). Spec. Publ., Amer. Soc. Mamm., 2:1–593.

ELWOOD, R. W., and M. C. OSTERMEYER. 1984. Does copulation inhibit infanticide in male rodents. Anim. Behav., 32:293–294.

FAIRBAIRN, D. J. 1977. The spring decline in deer mice: death or dispersal. Canadian J. Zool., 55:84–92.

———. 1978. Behavior of dispersing deer mice (Peromyscus maniculatus). Behav. Ecol. Sociobiol., 3:265–282.

FOLTZ, D. W. 1981. Genetic evidence for long term monogamy in a small rodent, Peromyscus polionotus. Amer. Nat., 117:665–675.

FORDHAM, R. A. 1971. Field populations of deermice with supplemental food. Ecology, 52:128–146.

GERZOFF, R. B. 1984. The effect of supplemental food on two sympatric Peromyscus species. Unpubl. M. S. thesis., Univ. Virginia, Charlottesville, 101 pp.

GLASER, H., and S. LUSTICK. 1975. Energetics and nesting behavior of the northern white-footed mouse, Peromyscus leucopus noveboracensis. Physiol. Zool., 48:105–113.

GLEASON, P. E., S. D. MICHAEL, and J. J. CHRISTIAN. 1980. Aggressive behavior during the reproductive cycle of female Peromyscus leucopus: effects of encounter site. Behav. Neural Biol., 29:506–511.

GOUNDIE, T. R. and S. H. VESSEY. 1986. Survival and dispersal of young white-footed mice born in nest boxes. J. Mamm., 67:53–60.

GRAVES, S., J. MALDONADO, and J. O. WOLFF. 1988. Use of ground and arboreal microhabitats by Peromyscus leucopus and Peromyscus maniculatus. Canadian J. Zool., 66:277–278.

HAIGH, G. R. 1983a. The effects of inbreeding and social factors on the reproduction of young female Peromyscus maniculatus bairdii. J. Mamm., 64:48–54.

———. 1983b. Reproductive inhibition and recovery in young female Peromyscus leucopus. J. Mamm., 64:706.

HAIGH, G. R., D. A. LOUNSBURY, and T. G. GORDON. 1985. Pheromone induced reproductive inhibition in young female Peromyscus leucopus. Biol. Reprod., 33:271–276.

HALPIN, Z. T. 1981. Adult-young interactions in island and mainland populations of the deermouse Peromyscus maniculatus. Oecologia, 51:419–425.

HANSEN, L., and G. O. BATZLI. 1978. The influence of food availability on the white-footed mouse: populations in isolated woodlots. Canadian J. Zool., 56:2530–2541.

HARLAND, R. M., P. J. BLANCHARD, and J. S. MILLAR. 1979. Demography of a population of Peromyscus leucopus. Canadian J. Zool., 57:323–328.

HEALEY, M. C. 1967. Aggression and self-regulation of population size in deermice. Ecology, 48:377–392.

HERMAN, T. B. 1984. Dispersion of insular Peromyscus maniculatus in coastal coniferous forest, British Columbia. Pp. 333–342, in Winter ecology of small mammals (J. F. Merritt, ed.). Spec. Publ. Carnegie Mus. Nat. Hist., 10:1–380.

HILL, J. L. 1974. *Peromyscus*: effects of early pairing on reproduction. Science, 186:1042–1044.

HOWARD, W. E. 1949. Dispersal, amount of inbreeding, and longevity in a local population of prairie deermice on the George Reserve, southern Michigan. Contrib. Lab. Vert. Biol., Univ. Michigan, 43:1–50.

HUCK, U. W., R. L. SOLTIS, and C. B. COOPERSMITH. 1982. Infanticide in male laboratory mice: effects of social status prior experience, and basis for discrimination between related and unrelated young. Anim. Behav., 30: 1158–1165.

KLEIMAN, D. G. 1977. Monogamy in mammals. Quart. Rev. Biol., 52:39–69.

KLEIN, H. G., and J. N. LAYNE. 1978. Nesting behavior in four species of mice. J. Mamm., 59:103–108.

KROHNE, D. T., B. A. DUBBS, and R. BACCUS. 1984. An analysis of dispersal in an unmanipulated population of *Peromyscus leucopus*. Amer. Midland Nat., 112:146–156.

LABOV, J. B. 1980. Factors influencing infanticidal behavior in wild male house mice (*Mus musculus*). Behav. Ecol. Sociobiol., 6:297–303.

LABOV, J. B., U. W. HUCK, R. W. ELWOOD, and R. J. BROOKS. 1985. Current problems in the study of infanticidal behavior of rodents. Quart. Rev. Biol., 60:1–20.

LAYNE, J. N. 1969. Nest-building behavior of three species of deer mice, *Peromyscus*. Behaviour, 35:288–303.

LOMBARDO, D. L., and C. R. TERMAN. 1980. The influence of the social environment on sexual maturation of female deermice (*Peromyscus maniculatus bairdii*). Res. Population Ecol., 22:93–100.

MADISON, D.M. 1977. Movements and habitat use among interacting *Peromyscus leucopus* as revealed by radiotelemetry. Canadian Field-Nat., 91:273–381.

MADISON, D. M., J. P. HILL, and P. E. GLEASON. 1984. Seasonality in the nesting behavior of *Peromyscus leucopus*. Amer. Midland Nat., 112:201–204.

MALLORY, F. F., and R. J. BROOKS. 1978. Infanticide and other reproductive strategies in the collard lemming (*Dicrostonyx groenlandicus*). Nature, 273: 144–146.

MERKT, J. R. 1981. An experimental study of habitat selection by the deer mouse, *Peromyscus maniculatus*, on Mandarte Island, B. C. Canadian J. Zool., 59: 589–597.

MERRITT, J. F. 1984. Growth and seasonal thermogenesis of *Peromyscus maniculatus* inhabiting the Appalachians and Rocky Mountains of North America. Pp. 201–213, *in* Winter ecology of small mammals (J. F. Merrit, ed.). Spec. Publ. Carnegie Mus. Nat. Hist., 10:1–380.

MERRITT, R. B., and B. J. WU. 1975. On the quantification of promiscuity (or "*Promyscus*" *maniculatus*?). Evolution, 29:575–578.

MESERVE, P. L. 1977. Three-dimensional home ranges of cricetid rodents. J. Mamm., 58:549–558.

METZGAR, L. H. 1971. Behavioral population regulation in the woodmouse, *Peromyscus leucopus*. Amer. Midland Nat., 86:434–448.

———. 1972. The measurement of home range shape. J. Wildl. Manag., 36: 643–645.

————. 1973a. Exploratory and feeding home ranges in *Peromyscus*. J. Mamm., 54:760–763.

————. 1973b. Home range shape and activity in *Peromyscus leucopus*. J. Mamm., 54:383–390.

————. 1973c. A comparison of trap- and track-revealed home ranges in *Peromyscus*. J. Mamm., 54:513–515.

————. 1979. Dispersion patterns in a *Peromyscus* population. J. Mamm., 60:129–145.

————. 1980. Dispersion and numbers in *Peromyscus* populations. Amer. Midland Nat., 103:26–31.

MIHOK, S. 1979. Behavioral structure and demography of subarctic *Clethrionomys gapperi* and *Peromyscus maniculatus*. Canadian J. Zool., 57:1520–1535.

MILLER, D. H., and L. L. GETZ. 1977. Comparisons of population dynamics of *Peromyscus* and *Clethrionomys* in New England. J. Mamm., 58:1–16.

MINEAU, P., and D. MADISON. 1977. Radiotracking of *Peromyscus leucopus*. Canadian J. Zool., 55:465–468.

MUMFORD, R. E., and J. O. WHITAKER, JR. 1982. Mammals of Indiana. Indiana Univ. Press, Bloomington, 537 pp.

MYTON, B. 1974. Utilization of space by *Peromyscus leucopus* and other small mammals. Ecology, 55:277–290.

NADEAU, J. H., R. T. LOMBARDI, and R. H. TAMARIN. 1981. Population structure and dispersal of *Peromyscus leucopus* on Muskeget Island. Canadian J. Zool., 59:793–799.

NICHOLSON, A. J. 1941. The homes and social habits of the wood-mouse (*Peromyscus leucopus noveboracensis*) in southern Michigan. Amer. Midland Nat., 25:196–223.

NOVAK, J. M. 1983. Multiple captures of *Peromyscus leucopus*: social behavior in a small rodent. J. Mamm., 64:710–713.

PETTIGREW, B. G., and R. M. F. S. SADLEIR. 1974. The ecology of the deer mouse, *Peromyscus maniculatus*, in a coastal coniferous forest. I. Population dynamics. Canadian J. Zool., 52:107–188.

ROWLEY, M. H., and J. J. CHRISTIAN. 1976. Intraspecific aggression of *Peromyscus leucopus*. Behav. Biol., 17:249–253.

SADLEIR, R. M. F. S. 1965. The relationship between agonistic behavior and population changes in the deer mouse, *Peromyscus maniculatus* (Wagner). J. Anim. Ecol., 34:331–352.

SAVIDGE, I. R. 1974. Social factors in dispersal of deer mice (*Peromyscus maniculatus*) from their natal site. Amer. Midland Nat., 91:395–405.

SEXTON, O. J., J. F. DOUGLASS, R. R. BLOYE, and A. PESCE. 1982. Thirteen-fold change in population size of *Peromyscus leucopus*. Canadian J. Zool., 60:2224–2225.

SHEPPE, W. 1966. Social behavior of the deer mouse, *Peromyscus leucopus*, in the laboratory. Wasmann J. Biol., 24:49–65.

————. 1967. The effect of livetrapping on the movements of *Peromyscus*. Amer. Midland Nat., 78:471–480.

SKRYJA, D. D. 1978. Reproductive inhibition in female cactus mice. J. Mamm., 59:543–550.

SMITH, D. A., and S. W. SPELLER. 1970. The distribution and behavior of *Peromyscus maniculatus gracilis* and *Peromyscus leucopus noveboracensis* (Rodentia:

Cricetidae) in a southern Ontario woodlot. Canadian J. Zool., 48: 1187–1198.

STAH, C. H.   1980.   Vertical nesting distribution of two species of *Peromyscus* under experimental conditions.   J. Mamm., 61:141–143.

STEBBINS, L. L.   1978.   Some aspects of overwintering in *Peromyscus maniculatus*. Canadian J. Zool., 56:386–390.

STICKEL, L. F.   1968.   Home range and travels.   Pp. 373–411, *in* Biology of *Peromyscus* (Rodentia) (J. A. King, ed.).   Spec. Publ., Amer. Soc. Mamm., 2:1–593.

SULLIVAN, T. P.   1977.   Demography and dispersal in island and mainland populations of the deer mouse, *Peromyscus* .   Ecology, 58:964–978.

TADLOCK, C. C., and H. G. KLEIN.   1979.   Nesting and food storage behavior of *Peromyscus maniculatus gracilis* and *P. leucopus noveboracensis*.   Canadian J. Zool., 93:239–242.

TAITT, M. J.   1981.   The effect of extra food on small rodent populations: I. Deermice (*Peromyscus maniculatus*).   J. Anim. Ecol., 50:111–124.

TERMAN, C. R.   1980.   Social factors influencing delayed reproductive maturation in prairie deermice (*Peromyscus maniculatus bairdii*).   J. Mamm., 61:219–223.

———.   1984.   Sexual maturation of male and female white-footed mice (*Peromyscus leucopus noveboracensis*): influence of urine and physical contact with adults.   J. Mamm., 65:97–102.

THOMAS, D., and C. R. TERMAN.   1975.   The effects of differential prenatal and postnatal social environments on sexual maturation of young prairie deermice (*Peromyscus maniculatus bairdii*).   Anim. Behav., 23:241–248.

TRUDEAU, A. M., G. R. HAIGH, and S. H. VESSEY.   1980.   Use of nest boxes to study behavioral ecology of white-footed mice (*Peromyscus leucopus*).   Ohio J. Sci., 80:90.

VAN HORNE, B.   1981.   Niches of adult and juvenile deer mice (*Peromyscus maniculatus*) in seral stages of coniferous forest.   Ecology, 63:992-1003.

VESTAL, B. M., and J. J. HELLACK.   1978.   Comparison of neighbor recognition in two species of deermice (*Peromyscus*).   J. Mamm., 59:339–346.

VICKERY, W. L., and J. S. MILLAR.   1984.   The energetics of huddling by endotherms.   Oikos, 43:88–93.

VOGT, F. D., and G. R. LYNCH.   1982.   Influence of ambient temperature, nest availability, huddling, and daily torpor on energy expenditure in the white-footed mouse *Peromyscus leucopus*.   Physiol. Zool., 55:56–63.

WHITSETT, J. M., L. E. GRAY, and G. M. BEDIZ.   1979.   Gonadal hormones and aggression toward juvenile conspecifics in prairie deer mice.   Behav. Ecol. Sociobiol., 6:165–168.

WOLFE, J. L.   1970.   Experiments on nest-building behavior in *Peromyscus* (Rodentia: Cricetinae).   Anim. Behav., 18:613–615.

WOLFF, J. O.   1985a.   The effects of food, density and interspecific interference on home range size in *Peromyscus leucopus* and *P. maniculatus*.   Canadian J. Zool., 63:2657–2662.

———.   1985b.   Comparative population ecology of *Peromyscus leucopus* and *P. maniculatus*.   Canadian J. Zool., 63:1548–1555.

———.   1985c.   Maternal aggression as a deterrent to infanticide in two species of *Peromyscus*.   Anim. Behav., 33:117–123.

———.   1986a.   The effects of food on midsummer demography of white-footed mice, *Peromyscus leucopus*.   Canadian J. Zool., 64:855–858.

———. 1986*b*. Infanticide in white-footed mice, *Peromyscus leucopus*. Anim. Behav., 34:1568.

———. 1986*c*. Life history strategies in white-footed mice, *Peromyscus leucopus*. Virginia J. Sci., 37:209–220.

WOLFF, J. O., and D. M. CICIRELLO. 1989. Field evidence for sexual selection and resource competition infanticide in white-footed mice. Anim. Behav., (in press).

WOLFF, J. O., and D. S. DURR. 1986. Winter nesting behavior of *Peromyscus leucopus* and *Peromyscus maniculatus*. J. Mamm., 67:409–411.

WOLFF, J. O., and B. HURLBUTT. 1982. Day refuges of *Peromyscus leucopus* and *Peromyscus maniculatus*. J. Mamm., 63:666–668.

WOLFF, J. O., M. H. FREEBERG, and R. D. DUESER. 1983. Interspecific territoriality in two sympatric species of *Peromyscus* (Rodentia: Cricetidae). Behav. Ecol. Sociobiol., 12:237–242.

WOLFF, J. O., and K. I. LUNDY. 1985. Intra-familial dispersion patterns in white-footed mice, *Peromyscus leucopus*. Behav. Ecol. Sociobiol., 17:381–384.

WOLFF, J. O., K. I. LUNDY, and R. BACCUS. 1988. Dispersal, inbreeding avoidance and reproductive success of white-footed mice. Anim. Behav., 36: 456–465.

XIA, X., and J. S. MILLAR. 1988. Paternal behavior by *Peromyscus leucopus* in enclosures. Canadian J. Zool., 66:1184–1187.

PEROMYSCUS AND *APODEMUS*: PATTERNS OF SIMILARITY IN ECOLOGICAL EQUIVALENTS

W. I. Montgomery

Abstract.—The aim of this paper is to review the population biology and community relationships of *Peromyscus* and *Apodemus* species and to assess the extent to which they may be considered ecological equivalents in the Nearctic and Palearctic, respectively. Methodological inconsistencies, lack of information and environmental heterogeneity preclude firm conclusions. Nevertheless, there are similarities among species in the two genera with regards to the seasonal control of reproduction, population dynamics and regulation, trophic relationships, and place in rodent communities. *Peromyscus* species, however, exploit a wider range of environments than *Apodemus*. The former have radiated widely in arid and tropical regions of the Nearctic. These southern species have markedly lower litter size than northern populations of *Peromyscus* and *Apodemus*. *Peromyscus* species are found in harsher conditions in the northern part of their range than *Apodemus*. This facility is associated with greater use of torpor, which may be more efficient in *Peromyscus* than *Apodemus*, and marked seasonal adaptation of metabolic thermoregulation. Sexual dimorphism and testes weight are greater in *Apodemus* than *Peromyscus* species. Grassland and woodland populations of *Apodemus* have greater production efficiencies than their North American counterparts. This may be due to the slightly larger litter size of the former genus. Rodent communities of Europe consist of fewer species than temperate and boreal communities in North America, and *Apodemus* species are less numerous in grassland habitats than *Peromyscus*. There is a greater prevalence of coexistence of congeners among *Apodemus* species than among northern populations of *Peromyscus*. Both *Apodemus* and *Peromyscus* are commonly found in association with *Clethrionomys* and *Microtus* with which they may compete for space. These interactions may be aggressive and their outcome dependent on habitat. There are clear intergeneric differences in activity and food between *Apodemus* and *Peromyscus* species and rodent species with which they share habitats. Intrageneric interaction may also involve competition though this is ameliorated by differences in population biology, behavior, and microhabitat selection. It is concluded that although *Peromyscus* and *Apodemus* species are broadly similar, each species is unique and caution should be employed in comparing species with species. It is more appropriate to compare processes where mechanisms and their results are alike, than to make general comparisons between species or genera.

Biologists have long been aware of morphological, ecological, and behavioral similarities of distantly related species and communities on different continents. Many striking examples are cited in ecology texts; see, for example, Whitaker's (1975) discussion of "Biome-types" and May's (1981) essay on patterns in multispecies communities. Organisms that have evolved independently but occupy similar niches in different zoogeographical regions, are sometimes called "ecological equivalents." Pianka (1978) suggests that convergent evolution occurs where species occupy·either relatively simple communities with predictable biotic interactions or extreme environments where selective forces are particularly strong. Pianka, however, urges caution: convergence may be read into a situation by overemphasizing superficial similarities without a comprehensive appreciation of "the inevitable dissimilarities between pairs of supposed ecological equivalents."

Mice of the murid genus *Apodemus*, of the Palearctic and northern Oriental regions, and the cricetid genus *Peromyscus*, of the Nearctic and northern Neotropical regions (Fig. 1), are regarded as ecological equivalents throughout the extensive literature concerning these genera. The use of Muridae and Cricetidae follows Walker (1975). The Family Cricetidae is known from the Oligocene of North America, Europe, and Asia, whereas the Muridae emerged late in the Miocene of Europe and Asia (Walker, 1975). North America and Europe separated in the middle of the Eocene, around 45 to 50 million years before present, and it is possible that the murid and New World cricetid lineages became distinct at this early date. This is supported by Brownell's (1983) estimation for the divergence of cricetids and murids as 38.5 to 58 million years before present based on DNA/DNA hybridization studies of muroid rodents. There is little doubt that *Peromyscus* and *Apodemus* lineages have been separate for a considerable period. Nor can there be much doubt that Modern representatives of these genera have similar appearances and, superficially at least, similar lifestyles. Representatives of *Peromyscus* and *Apodemus* are found in almost every habitat within their respective geographical ranges, are non-hibernators, are often numerically dominant in rodent communities, are subject to marked changes in abundance within a year but do not exhibit cyclical population dynamics between years, feed on a wide variety of plant and animal material, and are, in turn, a major component of the diets of avian and mammalian carnivores. Behaviorally, *Apodemus* and *Peromyscus* species may differ more within each genus than between genera. Representatives of both genera are nocturnal, use extensive burrow systems and subterranean nests, may be adept climbers, and may occupy home ranges which overlap among conspecifics.

The aim of this report is to examine the extent to which *Peromyscus* and *Apodemus* are similar in their population and community ecology. Data will be drawn from those areas and species that are relatively well researched. The contention that these genera are ecological equivalents in the New and Old Worlds, respectively, will be evaluated critically. It will become apparent that although there are many similarities between *Apodemus* and *Peromyscus* species, there are also differences that dictate caution in making general comparisons between genera. It is also clear that *Peromyscus* species are physiologically more diverse and live under a greater variety of conditions than *Apodemus* species.

A Taxonomic Note.—Hall (1981) lists some 59 species of *Peromyscus* in 12 subgeneric or species groups; Hooper's (1968) prediction that this genus would be resolved into 40 species has not been fulfilled despite the methodological advances of recent years (Carleton, this volume). The systematic complexity of *Peromyscus* is underlined by the 67 named subspecies of *P. maniculatus* alone (Hall, 1981). This contrasts markedly with the taxonomy of *Apodemus*; Corbet (1978) lists just 13 species, including two of doubtful status. Zimmermann (1962) suggests two subgeneric groups, *Sylvaemus* in the western and *Alsomys* in the eastern parts of the generic range, but again the validity of this division is in doubt although it is clear that the western species do not occur in the east. Neithammer and Krapp (1978) note 32 subspecies of *A. sylvaticus* but many of these are unlikely to withstand close scrutiny (Berry, 1985; Corbet, 1978). This review is necessarily selective and concentrates on the better known and more widespread species. These principal players and their approximate geographical distributions are listed in Table 1. For the purposes of this review, Hall's (1981) treatment of classification of *Peromyscus* is followed throughout.

ECOPHYSIOLOGY

Peromyscus and *Apodemus* species experience wide variation in climatic conditions; marked seasonal changes in temperature and precipitation occur in most parts of the generic ranges. Such parameters may limit the distribution of small granivores (Getz, 1968). MacMillen (1983) and MacMillen and Garland (this volume) reviewed water regulation in *Peromyscus* and described inter- and intraspecific variability in the ability to withstand water deprivation as determined by weight loss. *Peromyscus* from arid habitats suffer less weight loss during water deprivation than those from wetter habitats. *P. crinitus*, in particular, approaches the heteromyids' facility for water conservation. There are no comparable data for *Apodemus*, but there is intrageneric variation in (*ad libitum*) water consumption and tolerance to water shortage. Schropfer (1974) suggests that *A. sylvaticus* consumes more water when available but is better able to resist dehydration than *A. flavicollis*. These *Apode-*

TABLE 1.—*Taxonomic status and geographical distribution of the major species of* Apodemus *(L.) and* Peromyscus *(Gloger) referred to in this review (after Corbet, 1978 and Hall, 1981).*

Species	Approximate Geographical Range
Peromyscus eremicus Cactus mouse	S California, Baja, Nevada, Arizona, New Mexico, W Texas, N Mexico
P. californicus California mouse	Central and S California, N Baja
P. polionotus Oldfield mouse	Alabama, Georgia, South Carolina, N Florida
P. maniculatus Deer mouse	Widespread in North America, absent from some E and SE states, predominately non-coastal Mexico
P. leucopus White-footed mouse	Central and NE North America south of the Great Lakes, predomi- nantly eastern states of Mexico
P. gossypinus Cotton mouse	Southern states of United States as far west as E Texas
P. crinitus Canyon mouse	Nevada, Oregon, Arizona, Utah, SW Idaho, SE California
P. boylii Brush mouse	Central and S California, Utah, Arizona, New Mexico, widespread in non-coastal Central America
P. truei Piñon mouse	California, Nevada, Utah, Arizona, New Mexico, non-coastal states throughout Mexico
P. floridanus Florida mouse	Florida except S peninsula and northern counties
Apodemus flavicollis Yellow-necked mouse	NW Europe through Russia to the Urals, south to Italy, the Balkans, Asia Minor, Palestine
A. sylvaticus Wood mouse	Widespread in W Palearctic, east to Altai Mts, Nepal, Afghanistan, Iran, Palestine, NW Africa
A. agrarius Striped field mouse	West Germany, N Italy through E Europe and W Siberia to Lake Baikal, E Asia including Korea and China
A. microps Herb field mouse	Eastern Europe
A. mystacinus	SE Europe and offshore islands, Asia Minor, Iraq
A. argenteus Small Japanese field mouse	All four main islands of Japan
A. speciosus Large Japanese field mouse	All four main islands of Japan

mus species apparently consume less water than *Peromyscus* species (Table 2). Indeed, rates of water consumption in *Apodemus* are lower than some for *Peromyscus* from drier habitats. This may reflect a fundamental difference in the water regulatory capacity of the cricetids and murids. Hansson (1971) presented some data suggesting that total water intake among species of the latter family is generally low.

Torpor, occurring spontaneously (especially in northern populations) or induced by food deprivation, water shortage, or low ambient oxygen tension, has been established in at least nine species of *Peromyscus* from different climatic zones of North America (Hill, 1983). Typically, torpor is of short duration, lasting for only part of a day although it may recur over successive days, and mice recover to indulge in normal nocturnal activity. Torpor may also be in-

TABLE 2.—*Water consumption of* Peromyscus *and* Apodemus *species where water was provided* ad libitum.

Species	Rainfall (cm/year)	Mean body weight (g)	Daily water consumption (cc/g/day)	Reference
P. polionotus	100	12.4	0.30	Glenn, 1970
P. gossypinus	100	25.7	0.19	Glenn, 1970
P. floridanus	100	33.7	0.17	Glenn, 1970
P. maniculatus bairdii	50–100	18.6	0.16	Lindeborg, 1952
P. maniculatus gracilis	50–100	21.5	0.12	Lindeborg, 1952
P. leucopus noveboracensis	50–100	21.6	0.12	Lindeborg, 1952
P. californicus	25–50	34.3	0.13	MacMillen, 1964
P. eremicus	25–50	18.9	0.13	MacMillen, 1964
P. maniculatus gambelii	25–50	20.4	0.12	MacMillen, 1964
P. leucopus tornillo	25–50	29.4	0.06	Lindeborg, 1952
A. flavicollis	50–100	45.4	0.09	Schropfer, 1974
A. sylvaticus	50–100	29.5	0.11	Schropfer, 1974

duced in *Apodemus*; however, there is only anecdotal evidence that torpor occurs naturally in this genus (Wolton, 1983). Food deprivation and low ambient temperatures lead to a lowering of body temperature in *A. sylvaticus* and *A. flavicollis* (Cygan, 1984; Walton and Andrews, 1981). Like *Peromyscus*, *Apodemus* species show considerable individual variation in the tendency to enter torpor and the extent to which body temperature falls. However, the lowest reversible body temperature of 23°C recorded for *A. sylvaticus* is somewhat higher than the lower limit of 13°C for *P. maniculatus* (Morhardt, 1970). *Apodemus agrarius* does not seem capable of recovering from a body temperature of 24°C (Visinescu, 1970). These admittedly scant data suggest that *Apodemus* do not utilize torpidity for saving energy (Hill, 1983) or maintaining water balance (Mac-Millen, 1983) to the same extent as *Peromyscus* species.

Basal metabolic rates (BMR) of *A. sylvaticus*, *A. flavicollis*, and *A. agrarius* approximate those of *Peromyscus* species from temperate or mesic habitats (Table 3). In both genera, estimates of BMR among mice from such habitats are above those expected on the

TABLE 3.—*Basal metabolic rates of* Peromyscus *and* Apodemus *measured at rest or under anesthesia at 30–35°C, in most cases during the summer.*

Species	Body weight (g)	Basal metabolic rate (cc O_2/g/h)	Deviation from expected (%)[1]	Reference
P. leucopus	23.3	2.5	+53.9	Lynch, 1973
P. polionotus	14.5	2.2	+19.2	Mason, 1974
P. maniculatus bairdii	16.3	3.3	+84.5	Mason, 1974
P. maniculatus gambelii	19.1	1.7	+0.6	McNab and Morrison, 1963
P. californicus	45.5	1.0	−24.0	McNab and Morrison, 1963
P. crinitus	23.1	0.9	−44.7	McNab, 1968
P. eremicus	17.4	1.6	−8.9	MacMillen, 1965
A. sylvaticus	22.0	2.6	+57.7	Visinescu, 1967
A. flavicollis	25.9	2.9	+83.8	Cygan, 1984
A. agrarius	21.2	3.2	+89.1	Gorecki, 1969

[1] Expected oxygen consumption determined using the Brody (1945) equation: BMR $(O_2/h) = 3.8\ W^{0.73}$, where W is the body weight in grams.

basis of the Brody (1945) equation. There are no data for *Apodemus* from climatically more extreme regions of the Palearctic for comparison to the depressed BMR of semidesert or montane *Peromyscus* (McNab and Morrison, 1963). Similarly, little is known of the thermogenic and insulative properties of *Apodemus* in general. There is a seasonal fat cycle and some increase in the density and depth of fur in winter, at least in *A. flavicollis* (Grodzinski, 1985), and acclimation to low, winter temperatures through increased heart weight is apparent in *A. argenteus* (Sakai, 1976). Behavioral thermoregulation may be important in *A. flavicollis* and *A. agrarius* (Fedyk, 1971; Tertil, 1972) with a decrease of up to 45 percent in the rate of oxygen consumption among mice in small groups. Insulative nest material may also reduce oxygen demand significantly (Gorecki, 1969). In these respects *Apodemus* closely parallel temperate forms of *Peromyscus* such as *P. leucopus* (Hill, 1983).

Within the limitations of the evidence, there is a wider range of physiological adaptations among *Peromyscus* than among the *Apodemus* species. This is reflected in their respective geographical distributions (Fig. 1). Zoogeographical comparisons between land masses differing in size, shape, and many physical characteristics are only possible using an absolute criterion. Here, Lieth's distribution of terrestrial primary productivity (Cox and Moore, 1985), which encompasses precipitation and temperature, is used as a baseline for comparison between the distributions of *Apodemus* and *Peromyscus*. *Peromyscus* penetrates further north and south relative

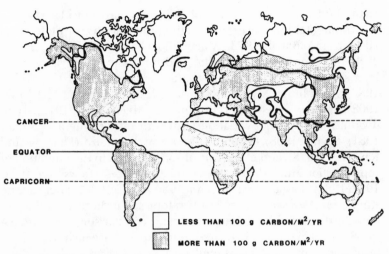

FIG. 1.—Distribution of *Apodemus* (Corbet, 1978) and *Peromyscus* (Hall, 1981) (heavy lines) superimposed on world distribution of plant productivity (after Leith *in* Cox and Moore, 1985).

to plant production than *Apodemus*. The latter, but for isolated populations of *A. sylvaticus* in Iceland and *A. penninsulae* in eastern Siberia, is absent from much of the northern half of the Palearctic. In North America, *P. maniculatus* is found well into the boreal zone. To the south, *Apodemus* species are limited by hot deserts in the west and tropical conditions in the east, although *A. agrarius* is found in Taiwan and parts of southeastern China, and there is an isolated record of *A. sylvaticus* in Qatar on the Arabian Peninsula (Corbet, 1984). Many forms of *Peromyscus* are, by contrast, found in the deserts of southwestern North America, whereas *P. flavidus* and *P. pirrensis*, among others, inhabit rain forests in Panama and Colombia (Baker, 1968). *Apodemus* species do not enter the corresponding desert regions of the Palearctic. The southern extension of the range of *A. sylvaticus* in central Asia closely follows the limits of a region, having slightly greater rainfall than neighboring desert or high plateau areas.

REPRODUCTION

The murids and the cricetids are, in general, characterized by moderate litter sizes, postpartum estrus with several litters per season, altricial young, and rapid growth rates to achieve early reproductive maturity, with young often breeding in their year of birth. Studies on reproduction of *Peromyscus* have elucidated 1) the environmental influences on breeding intensity (Gashwiler, 1979; Hansen and Batzli, 1978; Millar and Gyug, 1981; Taitt, 1981), 2) the interspecific, intraspecific, and intrapopulation variation in litter size and other life history characteristics (Layne, 1968; Millar, 1982, this volume; Myers and Master, 1983; Smith and McGinnis, 1968), 3) the energetics of reproduction (Millar, 1975, 1978), 4) individual and seasonal variation in growth rates (Gyug and Millar, 1981; Layne, 1968), 5) the evolution of litter size (Fleming and Rauscher, 1978), and 6) the role of social factors in the inhibition of reproduction and individual reproductive success (Dewsbury, 1981, 1982, 1984; Skryja, 1978; Terman, 1980; Wolff, 1985). There are no studies on *Apodemus* on the last two topics, but there are sufficient data on the first four to permit cautious comparison with *Peromyscus*. Some data on closely related murid species are included where information on *Apodemus* is limited to a few species.

Seasonality of Reproduction

Northern populations of *Peromyscus* generally breed during the spring, summer, and autumn, although the start and duration of reproduction is variable between taxa, location, and years (Millar, this volume, Millar and Gyug, 1981; Millar *et al.*, 1979). Many southern forms breed throughout the year but with a lower intensity than northern, seasonal populations (Millar, this volume). Lackey (1976), for example, found that *P. leucopus castaneus* and *P. yucatanicus* from the Yucatan Peninsula of Mexico breed aseasonally. Southern populations of *P. maniculatus* may breed almost all year (Jameson, 1953). *Apodemus* species resemble northern *Peromyscus* in the timing and variability of breeding (Table 4). Data for *A. spec-*

TABLE 4.—*Approximate breeding seasons of seven species of* Apodemus.

Species	Approximate limits of breeding season	Reference
A. flavicollis	February/April–September/October	Corbet and Southern, 1977
A. sylvaticus	March/April–September/October	Corbet and Southern, 1977
A. mystacinus	January/March–August/October	Niethammer and Krapp, 1978
A. microps	March–August	Niethammer and Krapp, 1978
A. agrarius	April–September/October	Niethammer and Krapp, 1978
A. argenteus	March–September	Fujimaki, 1969
A. speciosus	September–May	Yoshida, 1971

iosus in southern Japan are for only one year and may be exceptional. In some studies (Gibson and Delany, 1984; Montgomery, 1980*a*) certain females bred weeks or even months before the bulk of the population. This is also apparent in *P. maniculatus* where early breeding has been studied in detail by Fairbairn (1977*a*).

Four related factors may determine the periodicity of reproduction in both genera.

(1) Photoperiodism.—Day length is assumed to regulate seasonal breeding in vertebrates (Clarke, 1981). The role of photoperiodism has been experimentally verified in laboratory investigations on the effect of day length on reproductive condition. Clarke (1985) summarized data suggesting that sexual maturation is enhanced among male and female *A. sylvaticus* maintained under summer light. Similar data are available for *Peromyscus* though there may be inter- and intraspecific as well as individual variation in the response of reproductive condition to light regime (Millar, this volume; Price, 1966; Whitaker, 1940).

(2) Temperature.—Millar and Gyug (1981) confirmed the positive relationship between the start of breeding and spring temperatures in *P. leucopus* and *P. maniculatus* noted by earlier authors (Fuller, 1969; Sadlier, 1974). Sadleir *et al.* (1973) suggested that *P. maniculatus* are unable to meet the energy costs of lactation when temperatures are low; temperature fluctuation between years, therefore, accounts for the variation in the date breeding commences. The cost to a female *P. maniculatus* in rearing five young may be as high as an extra 134 percent over energy requirements while not breeding (Millar, 1979). Campbell (1974) estimated that a female *A. sylvaticus* needs to increase energy intake by 63 percent to rear a litter of 3.5 young. Clarke (1985) presented an experimental analysis that suggests that temperature has a direct effect on maturation of both sexes of *A. sylvaticus*. This has not been investigated in field populations, although there is evidence that low temperatures at night depress surface activity and therefore time spent foraging (Wolton, 1983). Low temperatures might prevent female *Apodemus* from obtaining sufficient food to reproduce.

(3) Food availability.—Several studies indicate that increased food supply enhances birth rate in populations of *P. leucopus* (Bendell, 1959; Hansen and Batzli, 1978) and *P. maniculatus* (Fordham, 1971; Gashwiler, 1979; Gilbert and Krebs, 1981; Taitt, 1981). Reproductive output is increased by prolongation of the breeding season rather than increased litter size or number of litters per female. *Apodemus* species respond in a similar fashion to natural changes in

tree seed production (Adamczewska, 1961; Gurnell, 1981; Hansson, 1971; Louarne and Schmidt, 1972; Smyth, 1966). Flowerdew (1973) demonstrated that supplementary food may advance the breeding season of *A. sylvaticus* by at least two to three weeks.

(4) Population size.—The breeding seasons of *A. sylvaticus* and *A. flavicollis* end more abruptly at higher densities (Adamczewska, 1961; Bobek, 1971; Watts, 1969). Montgomery (1981) presents some experimental evidence that breeding of both species commences earlier at reduced densities. Recent data for *A. sylvaticus* suggest that female reproductive condition is remarkably sensitive to female numbers at the start of the breeding season (Fig. 2). A similar suppression of reproduction occurs in at least four species of *Peromyscus* (Canham, 1969; Catlett and Brown, 1961; Skryja, 1978; Terman, 1968, 1973).

In summary the available evidence indicates that the timing of reproduction among temperate and boreal populations of *Peromyscus* and *Apodemus* is determined by similar processes. However, there are insufficient data to verify that this conformity is maintained throughout their generic ranges.

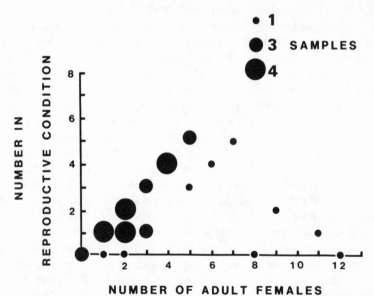

FIG. 2.—Number of breeding females in relation to total numbers of adult female *A. sylvaticus* on 40 grids in June. Each grid consisted of 50 Longworth traps set in 0.36 hectares of continuous mixed deciduous and coniferous forest in County Down, Northern Ireland (Montgomery, unpubl. data).

Variation in Litter Size and Weight

There is considerable variation in the number of young per litter in both genera. Litter size in mammals is associated with maternal size, parity, season, population density, and environmental parameters such as altitude and latitude. These factors are interrelated, and the manner in which they affect litter size is obscure. For example, Baker (1930) reported a slight increase and then a decrease in embryo number in wild *A. sylvaticus* as the breeding season progressed. This might reflect increasing frequency of multigravida females, which have larger litters than primagravida females, or seasonal variation in prenatal mortality. The former effect is known in field populations of *P. maniculatus* (Millar, 1982) and *A. sylvaticus* (Montgomery, 1982) and in laboratory populations, where there is also a decline in embryo number with extreme parity (Clarke, 1985; Drickamer and Vestal, 1973). Parity effects may be distinguished from possible growth effects that do not seem to influence litter size in *A. sylvaticus* (Montgomery, 1982) or *P. maniculatus* (Millar, 1982). Prenatal mortality is documented for both genera; preimplantation loss may be 18 percent and 30 percent in laboratory *P. maniculatus* and *P. polionotus*, respectively, whereas postimplantation loss is lower at around seven to thirteen percent (Liu, 1953). Dewsbury and Baumgardner (1981) and Dewsbury (1982) provided evidence that some egg or blastocyst loss may occur after multiple-male copulation in captive *P. maniculatus*. Loss of embryos is rare in field-caught *Apodemus*: two to five percent of embryos in *A. sylvaticus* and *A. flavicollis* and as little as one percent in *A. agrarius* and *A. microps* (Judes, 1979a; Pelikan, 1967a). Egg or blastocyst mortality may be up to 35 percent in *A. sylvaticus* and 39 percent in *A. flavicollis* (Judes, 1979a). Clarke and Egan (1984) reported total egg wastage in the order of 20 percent for primagravida female *A. sylvaticus*, 10 percent for multigravida females, and 27 percent among females that have had nine or more litters. It is uncertain whether these data reflect real differences between species and genera or between field and laboratory conditions.

There is greater variation in the mean litter size (number of embryos) of field-caught samples of *Peromyscus* than *Apodemus*; the data also suggest that *Apodemus* species generally have larger litters (Fig. 3). This is due to the inclusion of southern and often larger *Peromyscus* species, such as *P. californicus, P. eremicus, P. truei, P. gossypinus,* and *P. yucatanicus*. Samples for the wider ranging but predominantly northern species, *P. maniculatus* and *P. leucopus,* and

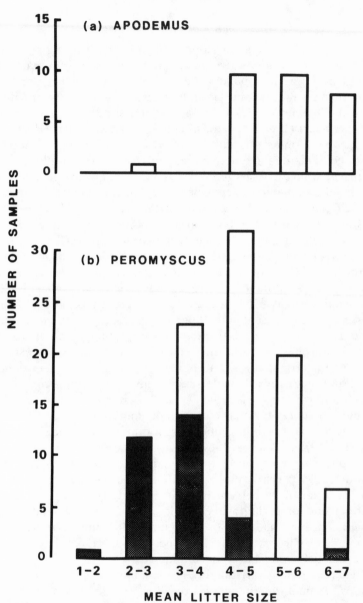

F<small>IG</small>. 3.—Frequency distribution of mean embryo number or litter size in 29 samples of seven *Apodemus* species and 95 samples of 20 *Peromyscus* species. Shaded portion of bars indicates samples from southern and southwestern *Peromyscus*. Data for *Apodemus* as in Table 5. Data for *Peromyscus* from Lackey (1976), Layne (1968), Millar (1982), Millar *et al.* (1979), Smith and McGinnis (1968), and Terman (1968). All data from samples taken from the field.

TABLE 5.—*Mean embryo number (field data) or litter size (laboratory data) recorded in seven* Apodemus *species.*

Location	n	Mean embryo number or litter size	Range	Reference
		A. mystacinus		
Caucasus	13	4.6	3−6	Miric, 1966
Yugoslavia	5	4.0		Miric, 1966
Bulgaria		4.6	2−9	Miric, 1966
Turkey	9	2.8	1−4	Felten *et al.*, 1973
		A. flavicollis		
Frankfurt	32	5.5	4−9	Niethammer and Krapp, 1978
Czechoslovakia	119	5.1	2−8	Pelikan, 1967*a*
Schleswig-Holstein	79	6.8	3−11	Judes, 1979*b*
		A. sylvaticus		
Oxford, England	62	5.0	3−7	Baker, 1930
Hertsfordshire, England	33	5.1		Jewell, 1966
Skomer Island	27	6.1		Jewell, 1966
St. Kilda Island	7	5.4		Jewell, 1966
Czechoslovakia	143	5.6	3−9	Pelikan, 1964
Austria	16	5.4	4−7	Steiner, 1968
Loire, France	82	4.5	2−9	Saint-Girons, 1966
Pyrenees	25	4.8	3−7	Sans-Coma and Gosalbez, 1976
Schleswig-Holstein	83	6.5	4−11	Judes, 1979*b*
N.E. Ireland	52	4.9		Montgomery, 1982
N.E. Ireland	59	4.9		Montgomery, 1982
S. Sweden (laboratory)		4.2		Eriksson, 1980
S. Sweden (field)		5.7		Clarke, 1985
S. Sweden (field)		6.9		Clarke, 1985
Oxford, England (laboratory)		5.0		Clarke, 1985
Surrey, England (laboratory)		5.3		Gurnell and Rennolls, 1983
		A. microps		
Czechoslovakia	84	6.3	3−10	Pelikan, 1964
Austria	6	4.8	4−6	Steiner, 1968
Bulgaria	73	5.9	3−8	Straka, 1966
		A. agrarius		
Czechoslovakia	79	6.6		Pelikan, 1965
Poland	32	6.7		Cerny, 1962
East Germany	30	6.0		Stein, 1955
European Turkey	10	5.1		Kahmann, 1961
		A. argenteus		
Sapporo, Japan	36	4.0		Fujimaki, 1969
		A. speciosus		
Fukuoka, Japan	18	4.3	2−7	Yoshida, 1971

Apodemus species have similar mean litter sizes, despite the greater variation in body size of the latter. There are no clear interspecific or interpopulation trends in litter size of *Apodemus* (Table 5). Production of offspring is determined not only by litter size but also by the length and intensity of the breeding season. These parameters may be measured as the proportion of females pregnant at any given time. Potential production of young is a function of body size in *Peromyscus* where fewer offspring are produced by larger species and subspecies. Smith and McGinnis (1968) attributed this relationship to the greater longevity and slower growth rates of larger rodents. No such relationship is apparent in data for *Apodemus* (Fig. 4).

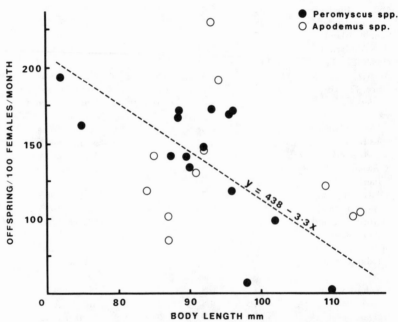

FIG. 4.—Relationship between number of offspring per 100 females per month and body length in eight *Peromyscus* species (15 samples) and seven *Apodemus* species (12 samples). Data for *Peromyscus* are from Smith and McGinnis (1968). Data for *Apodemus* are from Corbet and Southern (1977), Fairley (1965), Fujimaki (1969), Judes (1979b), Montgomery (1982), Niethammer and Krapp (1978), Ota (1968), Pelikan (1964, 1967a, 1967b), and Yoshida (1971) and follow criteria outlined in Smith and McGinnis (1968). The regression line indicates the relationship between production of offspring and size in *Peromyscus* and is drawn after Smith and McGinnis (1968).

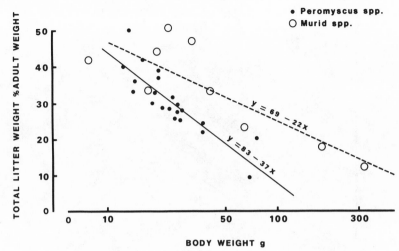

FIG. 5.—Litter weight (mean litter size times mean neonatal weight) expressed as percentage of adult weight, plotted against adult weight in 12 *Peromyscus* and nine murid species. Data for murids from Niethammar and Krapp (1978).

Neonatal weight of *Peromyscus*, which varies between 1.3 and 4.9 grams, is positively related to maternal body size but negatively related to litter size (Layne, 1968). Known birth weights of *Apodemus* range from 1.6 to 3.0 grams (Neithammer and Krapp, 1978) with larger species having heavier young just as in *Peromyscus* (Layne, 1968). Litter weight is a measure of maternal reproductive effort; this is expressed as the percentage of adult weight to control for size effects in comparisons within and between genera. Proportional reproductive effort declines with increasing body size in *Peromyscus* and nine murid species (Fig. 5). This trend is sharper among the former so that as size increases, reproductive effort is proportionally greater among the murids than *Peromyscus* of the same size. Within the size range of *A. sylvaticus, A. agrarius, A. flavicollis,* and *A. mystacinus,* most *Peromyscus* representatives are southern species with smaller litter sizes. This latitudinal bias confounds the comparison of reproductive effort and body size. However, relative litter weight is broadly similar in *P. maniculatus* and *P. leucopus* and the four *Apodemus* species despite the slightly lower weight range of the former.

Growth, Sexual Dimorphism, and Sex Ratios

Pre- and immediately postweaning growth of young *Peromyscus* and *Apodemus* has been studied under laboratory conditions. Al-

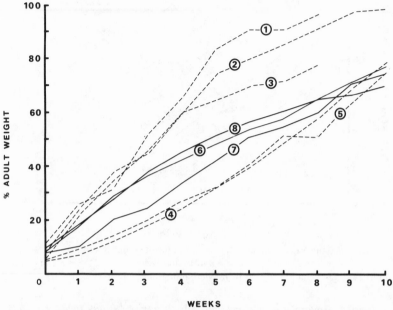

Fig. 6.—Growth of *Peromyscus* and *Apodemus* expressed as percentage of adult weight over time from birth. 1. *P. maniculatus bairdii*. 2. *P. maniculatus gambelii*. 3. *P. polionotus*. 4. *P. thomasi*. 5. *P. megalops* (data after Layne, 1968). 6. *A. agrarius* (Adamczewska-Andrzejewska, 1971). 7. *A. sylvaticus* (Flowerdew, 1972). 8. *A. flavicollis* (Bobek, 1969).

though nestling and juvenile growth rates may vary with sex (Dewsbury *et al.*, 1980), most data consider the sexes together over the early weeks of life. Small *Peromyscus* species grow more rapidly than larger ones (Layne, 1968; Millar, this volume). *Apodemus* species gain weight less quickly than *Peromyscus* of comparable size (Figs. 6 and 7). However, data are few and these apparent generic differences may be due to small sample size, methodological inconsistencies, or interpopulation variation. Growth in natural populations is complicated by variation between years, seasons, and localities. Gyug and Millar (1981), for example, demonstrated that after 40 days of age, *P. maniculatus* more or less cease growing during the non-breeding season, whereas those that have the potential to reproduce in the year of their birth continue to grow to reach adult weight. Similarly, growth rates of juvenile *A. agrarius* and *A. sylvaticus* are lower in autumn than earlier in the breeding season (Adamczewska-Andrzejewska, 1973; Green, 1979).

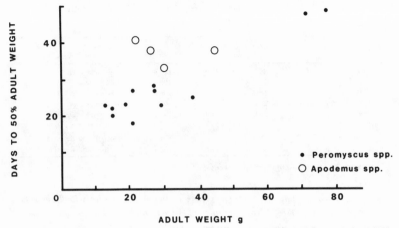

Fɪɢ. 7.—Age at which half of adult weight is attained in relation to adult weight in 13 species and subspecies of *Peromyscus* and four species of *Apodemus*. Data from the sources given in Figure 6 with the addition of data for *A. mystacinus* from Dieterlen (1965).

There are marked seasonal cycles in the average weight of male and female *Apodemus* species. These are due to a dynamic age structure, growth, and development of the reproductive and accessory organs (Adamczewska-Andrzejewska, 1973; Fujimaki, 1969; Montgomery, 1980a; Pelikan, 1967b; Yoshida, 1971). In all cases weight gain at the beginning of the breeding season commenced earlier in overwintered males than females. Such cycles are poorly documented for *Peromyscus* species, although it is apparent they occur in *P. maniculatus* and perhaps other temperate or boreal species (Fordham, 1971; Sadleir, 1970).

One consequence of annual cycles in adult body weight is the constantly changing relationship between male and female weight. Sexual dimorphism must be carefully defined or studied under controlled conditions (Dewsbury et al., 1980). Dewsbury et al. (1980) weighed male and female *Peromyscus* from mixed sex litters at the age of 90 days (Table 6). An alternative approach, using field data, is to measure dimorphism as the greatest ratio between mean male and female weights where pregnant females are excluded from the calculation (Table 7). In *Apodemus* sexual dimorphism is greatest among overwintered adults at the onset of breeding and increases with species size, albeit problematically, inasmuch as dimorphism is greatest in *A. microps*. Despite the paucity of comparable data,

TABLE 6.—*Mean weights of laboratory-reared male and female* Peromyscus *and* Apodemus *at 90 days of age.*

Species	Male wt. (g)	Female wt. (g)	M/F × 100	Reference
P. gossypinus	29.1	25.0	116.4	Dewsbury *et al.*, 1980
P. leucopus	18.9	17.2	109.9	Dewsbury *et al.*, 1980
P. maniculatus bairdii	16.2	15.2	106.6	Dewsbury *et al.*, 1980
P. maniculatus blandus	22.3	21.1	105.7	Dewsbury *et al.*, 1980
P. eremicus	24.6	23.9	102.9	Dewsbury *et al.*, 1980
P. polionotus	13.5	13.7	98.5	Dewsbury *et al.*, 1980
A. sylvaticus [1]	19.0	14.5	131.0	Gurnell and Rennolls, 1983
A. sylvaticus [1]	20.1	13.3	151.1	Gurnell and Rennolls, 1983

[1] Data are approximations from growth curves.

TABLE 7.—*Sexual dimorphism based on average weights of mature mice, with the exception of pregnant females, at times when the difference in male and female weight was greatest.*

Species	Season	Male wt. (g)	Female wt. (g)	M/F × 100	Reference
A. agrarius	spring	29.0	24.0	120.8	Adamczewska-Andrzejewska, 1973
A. flavicollis	winter	32.4	26.1	124.1	Pelikan, 1967*b*
A. sylvaticus	spring	22.8	19.2	118.8	Pelikan, 1967*b*
A. microps	spring	17.1	12.9	132.6	Pelikan, 1967*b*
A. mystacinus	spring	42.7	33.6	127.1	Niethammar and Krapp, 1978
A. argenteus	spring	15.0	12.3	120.0	Fujimaki, 1969
A. speciosus	winter	51.9	40.3	128.8	Yoshida, 1971
P. maniculatus austerus	spring	15.7	14.8	106.1	Fordham, 1971
P. maniculatus austerus	spring	16.0	14.0	114.3	Sadleir, 1970
P. maniculatus austerus	spring	17.8	16.1	110.6	Fairbairn, 1978*b*
P. leucopus	spring	23.1	23.2	99.6	Madison, 1977
P. leucopus	all year	19.8	20.1	96.2	Adler and Tamarin, 1984

trends are consistent; unlike *Peromyscus*, all *Apodemus* species are dimorphic and the difference between male and female body weight is greater in the Old World genus. Monomorphism has been associated with monogamy in mammals (Dewsbury *et al.*, 1980; Kleiman, 1977) but there are insufficient data on body size and social structure in the field to warrant even tentative conclusions on comparative social organization of *Peromyscus* and *Apodemus* species. The more marked sexual dimorphism of *Apodemus* compared to *Peromyscus* species is associated with greater testes weight in the former genus. Mean paired testes weight of *P. maniculatus* and *P. leucopus* rarely exceeds 0.5 gram (Bradley and Terman, 1981; Cornish and Bradshaw, 1978; Sadleir, 1974), whereas that of *A. sylvaticus, A. flavi-*

collis, and *A. agrarius* is frequently greater than 0.7 gram and may exceed 1.0 gram (Baker, 1930; Huminski, 1968; Steiner, 1968). The significance of the correlation between testis size and sexual dimorphism is not clear.

Adult sex ratios are biased towards males in virtually all studies of *Peromyscus* and *Apodemus* (Fig. 8). The proportion of males is highly variable within and between species and this aspect of *Peromyscus* and *Apodemus* biology is poorly understood. This bias is usually attributed to the method of collection rather than a real biological phenomenon. Juvenile sex ratios approach unity (Beer and MacLeod, 1966; Haitlinger, 1962; Kaufman and Kaufman, 1982; Montgomery, 1980a). Terman (1968) and Canham (1970) presented data suggesting that the sex ratio of neonatal *P. maniculatus* is also biased towards males. As far as is known, *Apodemus* species do not show this bias.

Reproductive patterns are consistent among the *Apodemus* species for which there are data. They differ little in time of breeding, litter size, and growth. There is greater variation among *Peromyscus* species, particularly southern and southwestern species. Northern populations and species of *Peromyscus* have much in common in their reproductive biology with *Apodemus*. In particular, the manner in which seasonal breeding is controlled is very similar between the two genera. Differences in growth rates, reproductive effort in relation to body size, offspring production, relative testes size, and sexual dimorphism merit closer scrutiny.

POPULATIONS IN TIME AND SPACE

The control and regulation of the distribution and abundance of *Peromyscus* and *Apodemus* species is one of the most intensively studied areas of small mammal ecology. Yet, the bulk of our knowledge is restricted to *P. maniculatus* and *P. leucopus* in the New World and *A. sylvaticus* and *A. flavicollis* in the Old. As a consequence, this review will deal mainly with these species.

Multi-annual Variation in Numbers

Flowerdew (1985) presented data for *A. sylvaticus* over 36 years at Wytham Woods, near Oxford, 14 years at Lathkill Dale, and 16 years at Woodchester Park. Although numbers varied markedly, the data give no indication of the three- to four-year cycles evident among some rodents (Krebs and Myers, 1974). Wendland (1975, 1981), who analyzed small mammal presence among remains in owl

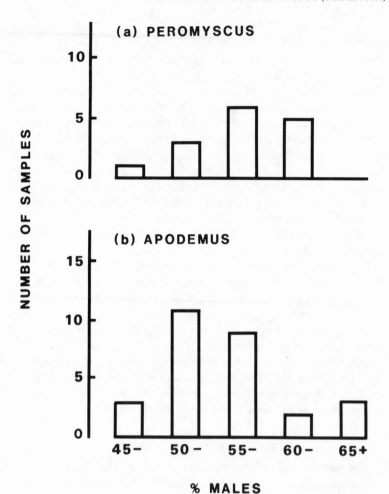

FIG. 8.—Frequency distribution of the proportion of males in samples of more than 100 adults for (a) *Peromyscus* and (b) *Apodemus*. Data for *Peromyscus* from Beer and MacLeod (1966), Bigler and Jenkins (1975), Canham (1970), Kaufman and Kaufman (1982), Martell (1983), Miller and Getz (1977a), Sadleir (1965), Sullivan (1979a), Terman (1968) and Van Horne (1981). Data for *Apodemus* from Adamczewska-Andrzejewska (1973), Baker (1930), Fujimaki (1969), Haitlinger (1969), Hansson (1971), Hsia (1962), Montgomery (1980a), Pelikan (1970), Steiner (1968) and Yoshida (1971).

pellets in Germany over a period of 21 years, suggested three-year cycles in the abundance of *A. flavicollis*. However, this might have resulted from the cyclic variation in numbers of alternative prey species, principally *Microtus arvalis*. Andrzejewski and Wroclawek (1961) reported a mass occurrence of *A. agrarius* in Poland, but there is no evidence of cyclical changes in the abundance of this species, or the more common *A. flavicollis*, in central and eastern Europe. Similarly, cyclic changes in population size do not occur in long-term studies of *Peromyscus* (Grant, 1976). Terman (1968), while presenting evidence that "*Peromyscus* exhibit relatively low numerical levels," noted several instances of population eruption; these (Catlett and Brown, 1961; Hoffmann, 1955) and later examples of unusually high or low abundance (Herman and Scott, 1984; Sexton *et al.*, 1982) are relatively rare, supporting Terman's contention that population size in *Peromyscus* is subject to "sensitive and effective controlling mechanisms." The key to population changes in *Apodemus* and *Peromyscus* lies in the processes that determine fluctuations between seasons.

Annual Cycles in Population Size

All *Apodemus* and northern populations of *Peromyscus* exhibit seasonal changes in population and size. There is a dearth of studies on southern populations of *Peromyscus* and *Apodemus*, but those that exist show some seasonality in abundance, although the amplitude of annual cycles is not as great as in northern populations; population size of *P. eremicus*, for example, changes little with season although numbers are lowest during the first quarter of the year (M'Closkey, 1972). Population size of *P. boylii* reaches a peak in December following a trough in August (Brown, 1964a). Meserve (1974) reported an increase in numbers of *P. californicus* during spring and early summer. Numbers of *P. polionotus* and *P. gossypinus* increase throughout the latter quarter of the year to reach a peak around January or February (Briese and Smith, 1974; Davenport, 1964; Smith, 1971; Smith and Vriese, 1979). The dynamics of the latter two species resemble those of populations of *A. argenteus* and *A. speciosus* in southern Japan (Doi and Iwamoto, 1973). These Japanese mice breed from November to the following spring with the maximum population increase from January to March (Fig. 9). This differs from the year-round reproduction of *P. gossypinus* and *P. polionotus* in the southern part of North America (Millar, this vol-

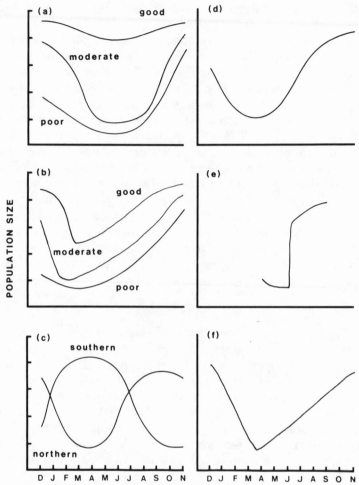

Fig. 9.—Idealized patterns of annual population dynamics of (a) *Apodemus sylvaticus* and (b) *A. flavicollis* following good, moderate, and poor seed crops, (c) northern- and southern-type dynamics of *A. speciosus* or *A. argenteus,* (d) *P. maniculatus* in coastal rain forest of British Columbia, (e) *P. maniculatus* at study sites 55° N or more, and (f) woodland populations of *P. leucopus.* Based on original sources, data given in Table 7 and Flowerdew (1985).

TABLE 8.—*Percentage of number of seasons sampled where population size of* Apodemus sylvaticus, A. flavicollis, Peromyscus maniculatus *and* P. leucopus *increased* (I), *remained stable* (S), *or decreased* (D)[1].

Species	Winter				Spring				Summer				Autumn			
	N	I	S	D	N	I	S	D	N	I	S	D	N	I	S	D
A. sylvaticus	66	14	41	45	67	4	36	60	64	31	64	5	66	88	6	6
A. flavicollis	19	16	16	68	20	40	45	15	19	63	32	5	21	67	9	24
P. maniculatus	30	7	29	73	48	23	40	37	72	51	38	11	43	49	21	28
P. leucopus	33	0	33	67	30	40	27	33	39	44	28	28	40	48	17	35
P. maniculatus (Brit. Col.)	27	4	22	74	29	17	41	41	37	43	41	16	36	50	28	22
P. maniculatus at 55°N or more	—	—	—	—	9	33	44	23	17	82	18	0	—	—	—	—
P. leucopus woodland only	25	0	23	72	19	53	37	10	26	46	15	38	28	39	25	36

[1] Based on trapline or grid trapping at intervals of approximately one month. Data for *Apodemus* species from Bergstedt (1965), Bobek (1973), Crawley (1969), Elton *et al.* (1931), Evans (1942), Fairley (1967), Fairley and Jones (1976), Flowerdew (1972), Gibson and Delany (1984), Green (1979), Gurnell (1978, 1981), Jensen (1975), Judes (1979*b*), Kikkawa (1964), Leigh Brown (1977*a*), Louarne and Schmidt (1972), Mermod (1969), Miller (1958), Montgomery (1980*a*, unpublished), Smal and Fairley (1982), and Watts (1969). Data for *Peromyscus* species from Adler and Tamarin (1984), Batzli (1977), Beer and McLeod (1966), Bendell (1959), Blair (1940, 1951), Fairbairn (1977*b*), Fuller (1969), Gilbert and Krebs (1981), Hansen and Batzli (1978, 1979), Howard (1949), Hunter *et al.* (1972), Martell (1983), Metzgar (1979), Middleton and Merriam (1981), Mihok (1979), Miller and Getz (1977*a*), Morris (1970), Myton (1974), Pettigrew and Sadleir (1974), Sadleir (1965, 1970), Snyder (1956), Sullivan (1977, 1979*a*, 1980), Sullivan and Sullivan (1982), Taitt (1981), and Van Horne (1981).

ume), suggesting that there is no fundamental similarity in the population dynamics of southern *Apodemus* and *Peromyscus* species. Data on the population dynamics of *A. sylvaticus, A. flavicollis, P. maniculatus,* and *P. leucopus* are summarized in Table 8. Population trends have been analyzed in each of four seasons; numbers either increase, remain stable, or decrease during a season. This crude compilation of data from different geographical locations and habitats and derived by differing techniques, affords only a preliminary basis for discussion of intraspecific and intra- and intergeneric variation in annual population cycles. Idealized annual cycles based on the data in Table 8 and original sources are illustrated in Figure 9. Numbers of both *Apodemus* species rarely rise during winter but while population size of *A. flavicollis* frequently decreases during this period, numbers of *A. sylvaticus* remain stable or decrease only slightly. By contrast the latter species suffers a major period of population decline in most springs while the former maintains numbers or even commences a period of population growth. Summer is a period of population growth in *A. flavicollis,* but the dynamics of *A. sylvaticus* are characterized by a period of numerical stability despite reproduction among overwintered mice. The major period of population growth in *A. sylvaticus* is the autumn. These differences in annual population dynamics in *A. sylvaticus* and *A. flavicollis* have been discussed at length elsewhere (Montgomery, 1980a, 1985).

There appears to be less variation in the dynamics of *P. maniculatus* and *P. leucopus*, especially when data for all studies are lumped. Greater intrageneric variation is apparent if data for *P. leucopus* are purged of studies conducted in non-woodland habitats and studies on *P. maniculatus* are restricted to those from British Columbia where habitat is chiefly coastal, coniferous rain forest (Table 8). Woodland populations of *P. leucopus* tend to increase during spring and summer. Those of *P. maniculatus* in British Columbia remain stable or decrease in spring and remain stable or increase in summer. (A summer period of stability is not evident, however, in northern populations of *P. maniculatus*; Table 8.) The data suggest that there is a greater tendency for population growth during the early part of the breeding season in *P. leucopus* than in *P. maniculatus*, whose numbers frequently remain static despite active reproduction. In these respects, the dynamics of *P. leucopus* and *P. maniculatus* parallel the dynamics of *A. flavicollis* and *A. sylvaticus*, respectively. During autumn, however, numbers of both *Peromyscus* species tend to decline to a much greater extent than in the two *Apodemus* species, and during the winter, both *Peromyscus* species are similar to *A. flavicollis* in that populations usually decrease. It is evident, therefore, that no two species have identical annual cycles of abundance (Fig. 9; Table 8), and thus, it would be more legitimate to compare processes affecting particular population events than to make more general comparisons between species or between genera. For example, the process and basis of spring decline and summer stability may be similar in *A. sylvaticus* and *P. maniculatus*, although more common in the former. Conversely, the manner in which numbers change and are limited throughout autumn and winter may be quite different in these frequently compared species.

Spatial Distribution

The spatial distribution of surface activity of mice is frequently inferred from trapping data. Individual range data, in turn, may form the basis of assertions concerning the social behavior and structure of wild populations. Methodology and the results of such analyses are extremely variable. Techniques that offer a more complete record of surface movements of unrestrained mice, such as feces-marking and tracking using radio transmitters or radioisotopes, are applied less widely. Wolton and Flowerdew (1985) reviewed this methodological tangle, concluding that valid comparisons may be made on a relative rather than absolute basis. This principle is adhered to throughout the following discussion of various aspects of spatial activity.

(1) Population dispersion in *Apodemus* and *Peromyscus*.—Population dispersion is rarely considered in studies of small mammals despite the numerous examples of aggregated distributions among natural populations (Taylor *et al.*, 1978). Contagious distributions are evident in field populations of *A. sylvaticus* and *A. flavicollis* (Montgomery, 1980*b*; Montgomery and Bell, 1981) and *P. maniculatus* (Metzgar, 1979). There is some evidence, however, that dispersion of female *A. sylvaticus* and *P. maniculatus* may approach regularity under certain conditions (Metzgar, 1979; Montgomery, 1980*b*).

(2) Resident and dispersing individuals.—Most trap-revealed movements of *Apodemus* and *Peromyscus* are modest in length, although there may be occasional, longer excursions, particularly across habitat boundaries (Blair, 1943; Crawley, 1970; Gashwiler, 1971; Wolton, 1985). Longer movements of several hundred meters or more, where individuals do not return to their original haunts, also occur and suggest that dispersal in *Apodemus* and *Peromyscus* species may be common, particularly among juveniles (Stickel, 1968; Watts, 1970). Other individuals, which constitute a resident group, may be caught many times at the same location (Montgomery, 1980*a*; Sadleir, 1965).

There is some evidence that resident and dispersing individuals differ with respect to age, sex, and condition, but the influence of these factors varies between species and with season. In spring and summer populations of A. *sylvaticus*, the sex ratio among first-time captives may be even more strongly male-biased than in established conspecifics (Montgomery, 1980*a*). First-time captives of *A. sylvaticus* and *A. flavicollis* may weigh less than residents during the non-breeding period (Montgomery, 1980*a*). Dispersing mice of the latter species suffer more trap mortality than resident individuals (Andrzejewski and Wroclawek, 1961). Fairbairn (1977*b*, 1978*a*) working on *P. maniculatus* in a coniferous forest in British Columbia, reported that lightweight, nonbreeding males disperse in spring and summer, juveniles and breeding males at the end of the breeding season, and lightweight males and females during the winter. The situation differs in *P. leucopus*; dispersing males may not differ in weight or reproductive condition from resident males, but dispersing females are lighter and more likely to be in reproductive condition than resident females (Nadeau *et al.*, 1981). Male *P. leucopus* may predominate among dispersing individuals (Harland *et al.*, 1979).

Further evidence that dispersers may be distinguished from residents is found in electrophoretic analyses of proteins (Fairbairn, 1978*a*); exploratory behavior of *P. polionotus* may (Garten, 1976) or may not (Blackwell and Ramsey, 1972) be related to heterozygosity. Several authors have demonstrated that males in small laboratory groups of *Apodemus* are aggressive towards introduced strangers (Montgomery and Gurnell, 1985). This is also evident in laboratory studies of *Peromyscus* species in which unrelated neighbors and unfamiliar kin are discriminated from unrelated strangers (Ayer and Whitsett, 1980; Grau, 1982; Healey, 1967; Sadleir, 1965; Vestal and Hellack, 1978). Ayer and Whitsett (1980) found that female *P. maniculatus bairdii* were particularly aggressive towards strange juveniles.

(3) Variation in home range area with age, sex, population density, habitat quality, and season.—These factors have been reviewed by Stickel (1968) and Wolton and Flowerdew (1985) for *Peromyscus* and *Apodemus*, respectively. Adult home ranges are larger than those of juveniles. Ranges of males in both genera are larger than those of female conspecifics, although male and female home range size is similar in some studies on *Peromyscus* (Stickel, 1968). Population density and habitat quality and their effects, if any, on home range area are difficult to distinguish in the field. Stickel (1968) and Wolton and Flowerdew (1985) reviewed a number of studies that indicate no consistent relationship between home range size and population density, although habitat differences may confound any valid conclusions. Some studies, such as those on *A. sylvaticus* in yew and oak woods in Ireland (Smal and Fairley, 1982) and *P. leucopus* in upland and lowland woods in Maryland (Stickel, 1960; Stickel and Warbach, 1960), suggest that home range area is smaller when food supply is good and density is high. Supplementary food leads to a decrease in male and female home range area in *P. maniculatus* during winter (Taitt, 1981). Seasonal effects are better understood; home ranges of male *Apodemus* increase sharply in spring, as females become sexually receptive, and decrease again in autumn (Miller, 1958; Randolph, 1977; Wolton and Flowerdew, 1985). Female home range area remains relatively constant throughout the year (Randolph, 1977), although measurement of range length indicates contrary results (Crawley, 1969; Green, 1979). Metzgar (1979) described increases in male and female range size at the be-

ginning of the breeding season in *P. maniculatus*, and Stickel (1968) reviewed evidence for similar changes in populations of *P. leucopus* and *P. polionotus*.

(4) Home range overlap, exclusivity, and social behavior.— Territoriality in small, nocturnal species is difficult to establish through direct observation under field conditions. However, spatial distribution of surface activity, elucidated by trapping or tracking, may provide some insight into the manner in which spatial division operates within a population. Social organization of *Apodemus* and *Peromyscus* is not well known, although there is some evidence from several species of territoriality in males, females, or both. In all species of both genera, male home ranges overlap with those of females. For example, Wolton (1985), radio tracking *A. sylvaticus* in deciduous woodland, found that adult male ranges overlapped with those of five to 10 females. Male and female *A. sylvaticus* may form bisexual pairs during the first part of the breeding season (Randolph, 1977) or a mature male may spend successive nights with one female (Garson, 1975). Loose associations between breeding males and females also occur in *Peromyscus* species (Metzgar, 1979; Mihok, 1979).

Home ranges of male conspecifics in *A. sylvaticus* and *A. flavicollis* usually overlap, although they may vary greatly in area (Montgomery, 1979; Randolph, 1977; Wolton, 1985; Wolton and Flowerdew, 1985). Brown (1966, 1969) and Cody (1982) suggested that in *A. sylvaticus*, dominant males have large home ranges that encompass those of females and subordinate males, and from which other dominant males are excluded. Similar polygynous groups were reported in *P. maniculatus* (Mihok, 1979). Male aggression is widely regarded as being of paramount importance in the spatial relationships and social structure of *Peromyscus* (Healey, 1967; Mihok, 1979; Sadleir, 1965). Male ranges in *P. maniculatus* and *P. leucopus*, overlap, at least to some degree, in some studies but not in others (Stickel, 1968). Mihok (1979), Nadeau *et al.* (1981), and Wolff (1985, this volume) suggested that social order and spatial distribution in *Peromyscus* is dictated by overall population density; at high densities mice become territorial, whereas at low densities ranges overlap more and mice simply avoid contact with same-sex conspecifics. Although there is no evidence that this density-dependent constraint on spatial dynamics operates on *A. sylvaticus* (Wolton and Flowerdew, 1985), male *A. sylvaticus* avoid other males within their overlapping home ranges (Brown, 1969; Garson, 1975; Montgomery and Gurnell, 1985).

Several studies on *P. leucopus* and to a lesser extent on *P. maniculatus*, indicate that females occupy exclusive territories, are aggressive towards intruders, or both (Harland *et al.*, 1979; Manville, 1949; Metzgar, 1971; Myton, 1974; Nadeau *et al.*, 1981; Redman and Sealander, 1958). The maintenance of territories in female rodents is associated with female aggression in defense of young against infanticide by adult conspecifics (Wolff, 1985, this volume). A polyandrous, hierarchical system occurs in *P. leucopus* where several males, one of which is dominant, are associated with one adult female (Metzgar, 1979; Myton, 1974; Stickel and Warbach, 1960). Recent evidence also suggests that female home ranges do not overlap in breeding populations of *A. sylvaticus* (Wolton, 1985). Earlier work on home range of *A. sylvaticus* (Brown, 1969) is compatible with Wolton's observations. There is anecdotal evidence that female *A. sylvaticus* are more aggressive before and after giving birth (Montgomery and Gurnell, 1985).

The absence of careful, experimental analyses of the effect of environmental factors on social structure in *Apodemus* and *Peromyscus* precludes a definitive assessment of mating systems in these rodents. Some home range data for both genera indicate promiscuity or polygyny associated with a hierarchical system among adult males. Promiscuity results in multiple paternity of litters in *P. maniculatus* (Birdsall and Nash, 1973). Other home range data suggest monogamy for at least part of the mating season. Foltz (1981) presented genetic evidence of obligate monogamy in *P. polionotus*. The application of genetic techniques may revolutionize comprehension of social behavior in *Apodemus* and *Peromyscus*.

Population Control and Regulation

Population size is determined by reproduction, mortality, and dispersal (immigration and emigration). Although many processes influence the abundance of small rodents, only those producing density dependent losses or gains have the capacity to regulate numbers within the limits set by the biotic potential of *Apodemus* and *Peromyscus* species. Control and regulation of population size are considered below. It is clear, however, that many of these processes are interrelated.

(1)   Habitat heterogeneity and climatic factors.—Environmental heterogeneity is rarely investigated in population studies of small mammals. It is evident that specific distributions in *Apodemus* and *Peromyscus* are constrained by physical and biological phenom-

ena (see above and below). *A. flavicollis*, for example, is less abundant and *A. sylvaticus* more abundant, in woodland edges and old-
fields in continental regions of sympatry (Hoffmeyer and Hansson,
1974; Jensen, 1984). In an area of continuous but heterogeneous
deciduous woodland and conifer plantations, the population density of *A. sylvaticus* in winter varied by a factor of 40, with 14 percent of this variation accounted for by the density of high cover
(Montgomery, unpubl. data). Numbers of this species in winter are
clearly much less stable than suggested by Watts (1969). Catastrophic changes in habitat structure or history also may have a direct impact on numbers; for example, numbers of *P. maniculatus*
sometimes increase on areas cleared of trees (Martell, 1983; Van
Horne, 1981) or decrease in areas of grasslands devastated by fire
(Crowner and Barrett, 1979). Other studies, notably the experimental analysis of Parmenter and MacMahon (1983), suggest that
even drastic disruption of a habitat has little effect on numbers of
P. maniculatus. Climatic factors may also influence abundance; recruitment of *P. maniculatus* in northern populations may be inhibited by long, cold, wet springs, but there is little evidence of a consistent relationship between winter severity and population change
(Fuller, 1969). Similarly, winter severity has little impact on over-
winter survival in *A. sylvaticus* (Flowerdew, 1985). Although habitat
disruption and adverse weather conditions may affect various aspects of the dynamics of *Apodemus* and *Peromyscus*, there is no evidence that they act in anything other than a density-independent
manner. Weather conditions may interact with food supply to further reduce overwinter survival following poor seed crops.

(2) Predation.—Avian and mammalian predators do not exert
any braking effect during population increases and only delay the
recovery phase in rodents exhibiting three- to four-year cycles in
abundance (Southern, 1979). The impact of predators on rodents
that do not cycle between years is less certain. There are few data
for *Peromyscus*, but the impact of predators on these and *Apodemus*
species is likely to be similar. In southern England, the weasel, *Mustela nivalis*, and the tawny owl, *Strix aluco*, are the principal predators of the small woodland rodents *A. sylvaticus* and *Clethrionomys
glareolus*. King (1980) estimated that weasels take up to 10 percent
of the population of *A. sylvaticus* per month. Southern and Lowe
(1982) were reluctant to estimate losses due to the tawny owl in the
same area of woodland but concluded that *A. sylvaticus* is either
more vulnerable to predation, because it frequents open habitat, or
is preferred over *C. glareolus* The latter species suffered up to 30

percent mortality due to tawny owl predation over a two-month period. Therefore loss from the population of *A. sylvaticus* must have been considerable. Predatory loss, however, may be quite variable; for example, weasels take few *Apodemus* where *Microtus* are available (Tapper, 1979) and readily switch to alternative prey items when rodents are scarce (Dunn, 1977). Population size of *P. maniculatus* was largely unaffected by the introduction of *Mustela frenata* (Sullivan and Sullivan, 1980).

The most persuasive evidence that predators have a regulatory role in the dynamics of *Apodemus* was provided by Erlinge *et al.* (1983), working on small rodents in oldfield habitats of southern Sweden. They calculated that losses to four avian and four mammalian predator species accounted for the entire production of *Apodemus* during the course of a year, with greatest predation rates in spring and early summer (about 20 to 30 percent a month) and least in autumn (10 percent) as the population increased. It is not clear whether these findings are typical of *Apodemus* living in grassland habitats.

Losses from populations of *A. sylvaticus* due to predation may contribute to declines in numbers but they do not prove that predation acts in a regulatory manner. Flowerdew (1985) suggested that predation is likely to act in an inversely density-dependent fashion whereby a relatively small loss is incurred at higher densities. Predatory loss explains neither numerical stability in the absence of reproduction nor changes in population size between years, because many predator populations remain relatively stable over many years (Southern, 1970). Stability within an annual population cycle and variability in numbers between years are characteristic of many populations of *Apodemus* and *Peromyscus* (Table 8).

(3) Food supply.—Food supply, principally tree seeds for woodland rodents, varies within and between years. Seed production may be correlated across wide geographical areas, but is unpredictable; at best, mast-producing, deciduous tree species produce a sizeable crop every other year (Harper, 1977). Studies on *A. sylvaticus* demonstrate that there is a direct relationship between food supply and peak population size and overwinter losses (Fig. 9; Flowerdew, 1985; Gurnell, 1981; Hansson, 1971; Watts, 1969). This is mediated by enhanced reproduction, immigration, and survival. An exceptional mast crop may affect numbers in the following summer and autumn (Hansson, 1971). The role of food in winter and spring population dynamics of *A. sylvaticus* was verified experimentally by Hansson (1971) and Flowerdew (1972). Tree

seed abundance apparently has a similar role in the dynamics of *A. flavicollis* (Fig. 9; Adamczewska, 1961; Bergstedt, 1965; Bobek, 1973; Hoffmeyer and Hansson, 1974).

There is also a positive relationship between seed abundance and numbers of *P. maniculatus, P. leucopus,* and *P. polionotus* (Bendell, 1959; Fordham, 1971; Gashwiler, 1965, 1979; Gilbert and Krebs, 1981; Hansen and Batzli, 1978, 1979; Smith, 1971; Taitt, 1981). Prolonged reproduction, enhanced immigration, and increased survival are involved in population growth, though not always all three. Despite the addition of food in experimental studies on *Peromyscus* species and *A. sylvaticus,* numbers still declined during spring months. This suggests that factors other than food supply affect the dynamics of these rodents (Flowerdew, 1972). Spatial and temporal variation in food availability account for only 20 to 30 percent of the variation in peak numbers of *A. sylvaticus* (Flowerdew, 1985; Montgomery, unpubl. data). Other authors believe that numbers of *Apodemus* and *Peromyscus* in woodlands are restricted more rigorously by the seasonal availability of food (Hansson, 1971; Taitt, 1981). This might well be so during poor seed years or in marginal habitats. Watts (1969), for example, suggested that *A. sylvaticus* population size is limited by food one year in three in southern England.

(4)   Intraspecific aggression.—The importance of intrinsic mechanisms of population regulation in *P. maniculatus* and *A. sylvaticus* has been recognized since the 1960s. Flowerdew (1974), Gurnell (1978), Healey (1967), Pettigrew and Sadleir (1974), Sadleir (1965), and Watts (1969), suggested that the population decline, frequently associated with the onset of breeding, and the low stable populations evident during summer, result from poor juvenile survival caused by male aggression. Spring declines are density-dependent (Pettigrew and Sadleir, 1974; Watts, 1969) so that after poor winter survival there is little, if any, decrease. Spring losses involve dispersal, especially of males, and mortality, especially of females (Fairbairn, 1977*b*). Juvenile recruitment during late summer and autumn is enhanced by the death of overwintered adults, the cessation of breeding, or a change in adult behavior, or both (Flowerdew, 1985). Gonadal hormones control male aggression towards juveniles (Whitsett *et al.,* 1979). It is now clear that the control of the density of females at the start of the breeding season and female aggression play as great a role as male aggression in determining spring and summer populations of *P. maniculatus* (Haplin, 1981; Taitt, 1981). This is indicated by a strong negative relationship

between juvenile survival and density of lactating females (Taitt, 1981). The latter effect has not been studied in *A. sylvaticus* populations, but it is possible that the exclusive ranges in females of this species (Wolton, 1985) are indicative of increased aggression and that females suppress recruitment throughout the early part of the breeding season.

Spring declines are uncommon in *A. flavicollis* and *P. leucopus* (Table 7). Montgomery (1985) suggested that intraspecific aggression does not operate in the same way in *A. flavicollis* and *A. sylvaticus*. Similarly, social interaction in *P. leucopus* may differ from that of *P. maniculatus* during the first half of the breeding season; Harland *et al.* (1979), Metzgar (1971), and Nadeau *et al.* (1981) found little evidence of male aggression in *P. leucopus*. It is unclear whether the lack of male aggression in these studies was due to low densities (see above) or reflected a real interspecific difference.

Although there is much in common in the control and regulation of the annual cycles of abundance in *Apodemus* and *Peromyscus*, the dynamics of each of the four better-known species seem unique. Habitat differences in the availability of tree seeds and alternative foods, and the manner in which these are utilized, may result in intra- and intergeneric differences in the timing of population declines in winter or spring (Table 8). Similarly, intra- and intergeneric differences in social behavior during the early part of the breeding season may produce differences in the tendency for numbers to remain stable or increase during spring and early summer. Nevertheless, there are basic similarities at key stages in the dynamics of *Apodemus* and *Peromyscus* species. This is particularly so in spring and summer populations of *P. maniculatus* and *A. sylvaticus*. Finally, electrophoretic analyses demonstrate that changes in population size in these species are associated with changes in gene frequency (Leigh Brown, 1977*a*; Smith *et al.*, 1978). Polymorphism in *A. sylvaticus*, for example, may be associated with differential survival during food shortage (Leigh Brown, 1977*b*). Further research is required to establish whether cyclical changes in genotype frequency drive, or are driven by, cycles in population size.

TROPHIC RELATIONSHIPS

Among the essential features of the biology of any species are what it consumes and what consumes it. Here, the food, predators, and parasites of *Peromyscus* and *Apodemus* are reviewed to assess the similarity of their respective food webs.

TABLE 9.—*Foods of* Peromyscus *and* Apodemus *species. Crosses indicate relative frequency (+ = rare, ++ = moderately common, +++ = very common) of major food classes in the stomachs or feces of wild caught mice.*

Species	Reference[1]	Seed	Fruit	Green plant	Animal	Seasonality Yes/No
P. maniculatus	1	+++		+	+	?Y
	2	++		+++	+++	?
	3	+++	++	+	+	Y
	4	+++			++	N
	5	+++		++	++	Y
	6	++		+	+++	Y
P. leucopus	7	+++			+++	Y
	4	+++			++	N
	5	+++		+	++	Y
	8		++		+++	?
P. californicus	9	+++	+	++	+	?
	3	++	++	+	+	Y
P. eremicus	3	++	++	+	+	Y
P. truei	9	++	+	++	+	?
P. boylii	4	+++	+		++	N
A. sylvaticus	10	+++	++	+	++	Y
	11	+++	+	+	++	Y
	12	++			+++	Y
	13	+++			+	Y
A. flavicollis	12	++			+++	Y
	14	+++		+	+	Y
	15	+++	++	+	++	Y
A. agrarius	16	+++			+++	Y

[1](1) Vaughan (1974), (2) Kritzman (1974), (3) Meserve (1976), (4) Brown (1964*b*), (5) Whitaker (1966), (6) Flake (1973), (7) Batzli (1977), (8) M'Closkey and Fieldwick (1975), (9) Merritt (1974), (10) Hansson (1971), (11) Watts (1968), (12) Holisova and Obrtel (1980), (13) Holisova and Obrtel (1977), (14) Zemanek (1972), (15) Drozdz (1966), (16) Holisova (1967).

Food

All species in both genera primarily eat seeds and animal material (Table 9). *Peromyscus* may take a little more green plant material and fruit than *Apodemus*, but this difference is marginal. Within each genus there is little consistent variation among species. Coexisting *P. maniculatus, P. leucopus,* and *P. boylii,* for example, have very similar feeding habits in the Missouri Ozarks (Brown, 1964*b*). There is, perhaps, more variation in diet between habitats; for example, *A. sylvaticus* living in spruce forests, take relatively more insects than those in reed swamps (Holisova and Obrtel, 1977, 1980). There may also be substantial changes in diet between years (Montgomery, 1982). All studies on *Apodemus* reveal seasonal changes in diet. Seasonality in diet is also evident among most *Peromyscus* species, but there are studies, particularly of southern populations, that indicate little or no annual dietary cycle. In general *Peromyscus* and *Apodemus* are granivores with omnivorous tendencies that reflect highly opportunistic feeding habits.

Predators

Erlinge *et al.* (1983), in reviewing the impact of predators on small rodents in southern Sweden, designated certain predators as rodent specialists and others as facultative. Both groups of predators account for substantial losses from populations of *A. sylvaticus* (see above). Small or medium-sized mammalian carnivores and nocturnal or crepuscular avian predators are the major predators of *Apodemus* and *Peromyscus*. Other losses may be due to larger mammalian carnivores, diurnal raptors, or reptiles and amphibians, but these are probably localized and rare. There is insufficient published data to permit a comparison of predators throughout the ranges of *Apodemus* and *Peromyscus*, so data for northern and central populations only are summarized in Table 10. Further, there is little indication of the relative losses owing to each predator; undoubtedly some will have greater impact on their prey than others but the impact of each may be subject to temporal and spatial constraints. The apparent similarity among predators of *Apodemus* and *Peromyscus* is striking. Some predators, such as *Mustela erminea* and *Asio flammeus*, are found both in the Nearctic and the Palearctic. Others, although of less importance as predators of small rodents, such as *Taxidea taxus* in the Nearctic and *Meles meles* in the Palearctic, have similar biology.

Parasites

Parasites, often neglected in population studies of small mammals, constitute an important part of the economy of rodent communities. Parasite burdens may be substantial and their effects on the individual host and host population are potentially of great ecological significance. *Peromyscus* and *Apodemus* harbor many kinds of endo- and ectoparasites (Healing and Nowell, 1985; Whitaker, 1968). Parasite species differ between host genera and species and between populations of the same host. As one would expect, *P. maniculatus* and *P. leucopus* share many parasite genera, although the index of similarity varies substantially among the major parasite groups, being low among mites and cestodes and high in the nematodes and chiggers (Table 11). Levels of similarity between *A.*

TABLE 10.—*Principal mammalian and avian predators of* Peromyscus *and* Apodemus *in northern and central parts of their ranges (after various sources).*

Peromyscus species	*Apodemus* species
Mammalian predators	
Didelphis virginiana	
Virginia opossum	
Mustela nivalis (= *rixosa*)	*M. nivalis*
Least weasel	Weasel
M. frenata	*M. erminea*
Long-tailed weasel	Stoat
M. erminea	*M. putorius*
Ermine	Polecat
M. vison	*M. vison*
Mink	Mink
Taxidea taxus	*Meles meles*
Badger	Badger
Spilogale putorius	*Genetta genetta*
Eastern spotted skunk	Genet
Mephitis mephitis	
Striped skunk	
Felis lynx	*Felis lynx*
Lynx	Lynx
F. rufus	*F. silvestris*
Bobcat	Wild cat
F. catus	*F. catus*
Domestic cat	Domestic cat
Vulpes vulpes	*Vulpes vulpes*
Red fox	Fox
Avian predators	
Tyto alba	*Tyto alba*
Barn owl	Barn owl
Otus asio	*Asio otus*
Screech owl	Long-eared owl
A. flammeus	*A. flammeus*
Short-eared owl	Short-eared owl
	Strix aluco
	Tawny owl

sylvaticus and the two *Peromyscus* species are lower than in the former comparison. Nevertheless, there is broad similarity in the genera of parasites in the New and Old World rodents. Nematode genera, for example, of *A. sylvaticus* and *P. leucopus* are almost as similar as those of *P. leucopus* and *P. maniculatus* (Table 11).

The similarity of food, predators, and parasites between *Peromyscus* and *Apodemus* species suggest that the trophic relationships of these mice are similar. Closer examination, however, indicates

TABLE 11.—*Indices*[1] *of similarity among parasitic genera infecting* P. maniculatus, P. leucopus *and* A. sylvaticus.

Parasite group	P. maniculatus P. leucopus	P. maniculatus A. sylvaticus	P. leucopus A. sylvaticus
trematodes	57 (7)	20 (10)	18 (11)
cestodes	46 (13)	35 (17)	20 (10)
nematodes	67 (18)	58 (24)	64 (22)
acanthocephalans	0 (1)	67 (3)	0 (2)
ticks	80 (5)	50 (4)	67 (3)
fleas	64 (53)	33 (48)	48 (29)
mites	44 (54)	22 (63)	26 (39)
chiggers	80 (25)	0 (13)	0 (12)
lice	67 (6)	50 (4)	50 (4)

[1] Similarity is calculated as $100 \times (2W/(A+B))$, where W is the number of genera in common, and A and B are the total number of genera in species one and two, respectively. Numbers in brackets are the sum of A and B. Data for *Peromyscus* spp. are from Whitaker (1968) and for *A. sylvaticus* from Montgomery (1982) and additional references.

inter- and intrageneric and interpopulation differences in feeding, predation, and parasitic burdens. Despite these differences, it is evident that food webs involving *Apodemus* and *Peromyscus* have similar themes; only the characters differ.

ECOLOGICAL ENERGETICS

Grodzinski and French (1983) reviewed the relationship between total energy utilization and net production in small mammal communities. Average production efficiency (production/respiratory loss) for granivores, omnivores, and herbivores is 2.4 percent, 2.7 percent, and 3.5 percent, respectively. Data for *Peromyscus* and *Apodemus* clearly indicate that production efficiency is markedly greater in the latter (Table 12). This may reflect differences between ecosystems in as much as three of the four years of data for *Peromyscus* are from oldfield or desert habitats, whereas most (19 out of 23 years) of the data for *Apodemus* are from woodland studies. Production efficiency is, however, very consistent among the *Peromyscus* species and is, if anything, lower in the only study on woodland populations. Furthermore, production efficiency of *Apodemus* species in grasslands is higher than in *Peromyscus* (Table 12). An alternative explanation may lie in the less dominant position of *Peromyscus* species in southern, grassland rodent communities (Grant and Birney, 1979; Grant *et al.*, 1982; see below) and their more conservative breeding habits. Southern populations characteristically have small litters, breed aseasonally, or have a less intensive breeding season. This contrasts with the larger litters recorded for north-

TABLE 12.—*Comparison of production and production efficiency of*
Peromyscus *and* Apodemus.

Species	Productivity (Kj/ha/yr)	Production efficiency (production/respiratory loss)
P. eremicus	340	1.75
P. polionotus	5020	1.95
P. leucopus	8138	1.97
Peromyscus spp.	419	1.61
A. sylvaticus[1]	2535	4.54
A. sylvaticus	2044	5.82
A. flavicollis	3327	2.83
A. flavicollis	2675	3.41
A. flavicollis	650	3.79
A. flavicollis	1117	4.10

[1] Replicate data for *A. sylvaticus* and *A. flavicollis* are averages for different studies. All data from Grodzinski and French (1983).

TABLE 13.—*Comparative energy strategies*[1] *in* Apodemus *species
and northern populations of* Peromyscus.

Characteristic	Peromyscus	Apodemus
Food digestibility and assimilation	87–88%	81–96%
Weight loss or cessation of growth	Cessation of growth	Cessation of growth
Reduced activity and food intake	Yes	Reduced activity
Seasonal fat cycle	Yes	Yes
Seasonal cycle in coat thermoinsulation	Yes	Yes
Seasonal adaptation of metabolic thermoregulation	Yes	? No
Torpor	Great importance	Yes
Behavioral thermoregulation	Yes	Yes

[1] Based heavily on Grodzinski (1985); additional material from Baar and Fleharty (1976), Hill (1983), Schreiber (1979), and Stebbins (1978).

ern populations of *Apodemus* (Table 5). Reproduction undoubtedly plays the major role in productivity of small mammals. Humphreys (1984) suggested that there is a strong positive relationship between production efficiency and litter size, and this may be sufficient to explain the observed discrepancies in the production efficiencies of *Apodemus* and *Peromyscus*.

Winter is a critical period for small mammals in temperate and boreal regions, and several energy-saving adaptations are observed in northern populations of *Peromyscus* and *Apodemus* (Table 13). Individuals of both genera exploit more digestible foods than at other seasons, cease growing in winter, lay down fat, increase the thermo-insulation properties of their pelage, and conserve energy through huddling together. *Peromyscus* also reduce food intake (Stebbins, 1978), increase the facility for nonshivering thermogenesis (Hill, 1983), and exhibit marked seasonality in daily energy requirements (Baar and Fleharty, 1976). These latter characteristics are shared with other members of the Cricetidae (Grodzinski, 1985) and may partially explain the more northerly distribution of *Peromyscus* than *Apodemus* (Fig. 1).

The Impact of *Apodemus* and *Peromyscus* in Temperate Habitats

The proportion of primary production available to populations of *Peromyscus* and *Apodemus* is relatively small; however, habitat influences may be important. Grodzinski *et al.* (1970), for example, estimated that only 2.4 percent of net primary production was available to a population of *A. flavicollis* in a deciduous forest. Odum *et al.* (1962) estimated that seven percent of net primary production is available to granivores in oldfields, whereas Chew and Chew (1970) estimated that 16.85 percent of primary productivity is available to *Peromyscus* in semidesert. The proportion of available primary production that is consumed by *Peromyscus* and *Apodemus* is also variable but usually low. In grasslands, small mammals may consume 12 percent of available food (Odum *et al.*, 1962), but generally consumption is closer to five percent in grassland and desert populations (Hayward and Phillipson, 1979). Forest small mammals account for less than three percent of available energy (Hayward and Phillipson, 1979), but Smal and Fairley (1981) calculated that *A. sylvaticus* consumed 12 to 15 percent of available food in an oak wood and a yew wood in southern Ireland. Therefore, it is difficult to make generalizations concerning the role of *Apodemus* and *Peromyscus* in energy flow through different ecosystems. Populations of both genera may, at certain times and in certain habitats, make a significant contribution to energy transfer.

Apodemus and *Peromyscus* are significant "predators" of seeds and seedlings and may have a direct impact on regeneration of long-lived species. For example, *A. flavicollis* may destroy up to 78 percent of beech seedlings in a deciduous forest; losses of seeds and seedlings depend on plant species and rodent density (Golley *et al.*, 1975). Seed predation by *P. maniculatus* and other rodents may prevent the reforestation of cutover forest lands (Sullivan, 1979*a*,

1979*b*). In grassland, seed losses attributable to *P. maniculatus* may play a significant role in the distribution and abundance of herbs (Mittlebach and Gross, 1984). The impact of rodents on survival of seeds may be responsible for the irregular production of seed in mast-producing tree species (for example, Silvertown, 1980); abnormally heavy crops interspersed by one or more poor seasons help ensure that rodent populations are insufficient to remove all seeds before some can germinate and grow beyond vulnerability to rodents. Under certain circumstances mice may prevent regeneration and thereby alter the course of succession in a plant community.

PEROMYSCUS AND APODEMUS IN RODENT COMMUNITIES

Many rodent families are represented in the rodent communities of the Palearctic and Nearctic (Table 14). The Sciuridae and Zapodidae are found in both regions. The Geomyidae and Heteromyidae contain fossorial and bipedal forms that dominate rodent communities in arid regions of the Nearctic. In the Palearctic, these are paralleled by the Spalacidae and Dipodidae, respectively. The Cricetidae are found in both regions but, apart from the voles and lemmings of northern and temperate habitats, different tribes are found in the Palearctic and Nearctic. In the Palearctic, the Myospalacinae, Gerbillinae, and the hamsters among the Cricetinae,

TABLE 14.—*Nearctic and Palearctic families of rodents.*

Rodent family	Nearctic	Palearctic	Common name
Sciuridae	+	+	Squirrels, ground squirrels, chipmunks
Geomyidae	+		Pocket gophers
Heteromyidae	+		Pocket mice, kangaroo mice and rats
Cricetidae			
Myospalacinae		+	Mole rats
Microtinae	+	+	Voles and lemmings
Gerbillinae		+	Gerbils and jirds
Cricetinae O.W.		+	Hamsters
Cricetinae N.W.	+		New World mice and rats
Spalacidae		+	Mole rat
Muridae	+ (introduced)	+	Old World rats and mice
Myoxidae		+	Dormice
Seleviniidae		+	Desert dormice
Zapodidae	+	+	Jumping mice
Dipodidae		+	Jerboas

are forms of drier habitats. In the New World, the Cricetinae are represented by mice and rats of the genera *Peromyscus, Reithrodontomys, Ochrotomys, Sigmodon, Neotoma,* and so on, which are found in many kinds of habitats. The Muridae are primarily an Old World family, but such has been the influence of man that *Rattus* and *Mus* are now found in many habitats throughout the New World. The species composition of rodent communities in which *Apodemus* or *Peromyscus* are represented, is variable, not only between the New World and the Old but also between climatological zones and habitats within each region. Consequently specific differences in the use of space and other resources, and interactions between species vary from place to place and, indeed, from time to time.

Species Composition and Relative Abundance

It is not possible to make comparisons of Old and New World rodent communities for all habitats in which *Apodemus* or *Peromyscus* are found. There are few available data, other than occasional species records, for much of the southern and eastern parts of the range of *Apodemus*. It is likely that the rodent communities in these "remote" arid areas of the Palearctic are dominated by jerboas and gerbils and are in many ways comparable to those dominated by the heteromyids in arid North America (Kotler, 1984a). A further constraint is inadequate documentation of rodent community composition; Asher and Thomas (1985) indicated that single surveys may omit species or give aberrant diversity indices. They emphasized the variable nature of community structure over time and space. The number of small mammal species captured over an 11-year study in deciduous forests in Quebec, for example, ranged from three to eight in any one year (Grant, 1976). The present survey, therefore, draws data from studies conducted over at least several months in habitats that allow meaningful comparison between the New and Old Worlds.

Numbers of species of small, terrestrial rodents tend to increase towards the Southwest of North America. This is apparent among communities that contain *Peromyscus*, where samples from southern locations frequently contain eight or more species (Fig. 10); one habitat contained 14 species, including *P. eremicus, P. maniculatus,* and *P. leucopus* (Whitford, 1976). Most northern samples from North America consist of between five and eight rodent species. Comparable parts of the range of *Apodemus* in central and northern Europe have fewer species; usually there are only three or four

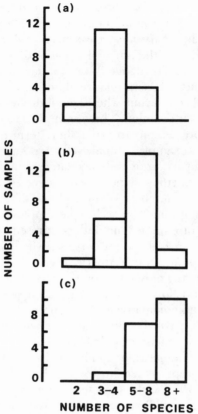

FIG. 10.—Numbers of species of small, terrestrial, or semi-arboreal rodents in (a) northern samples (≥40° N) containing *Apodemus,* (b) northern (≥35° N) samples containing *Peromyscus* and (c) southern (≤34° N) samples containing *Peromyscus;* samples from all habitat types throughout continental area only. Data for *Peromyscus* from Adler and Tamarin (1984), Anthony *et al.* (1981), Asher and Thomas (1985), Bigler and Jenkins (1975), Blair (1940, 1943), Briese and Smith (1974), Canham (1970), Chew and Chew (1970), Daly *et al.* (1980), Dueser and Shugart (1979), Fleherty and Nevo (1983), Fuller (1969), Garland and Bradley (1984), Hallett (1982), Hallett *et al.* (1983), Halvorson (1982), Hansen and Batzli (1978), Howard (1960), Kirkland and Griffin (1974), Kitchings and Levy (1981), Kotler (1984a), Martell (1983), M'Closkey (1972), Metzgar (1979), Mihok (1979), Myton (1974), Parmenter and MacMahon (1983), Platt (1968), Price and Waser (1984), Reynolds (1980), Rickart (1977), Smith *et al.* (1974), Whitford (1976), Wirtz and Pearson (1960), and Yahner (1983). Data for *Apodemus* from Bergstedt (1965), Flousek *et al.* (1985), Grodzinski, *et al.* (1966), Hansson (1968), Jensen (1975), Larsson *et al.* (1973), Louarne *et al.* (1970), Mermod (1969), Muller (1972), Paspaleva and Andreescu (1975), Pelikan and Nesvadbova (1979), Spitzenberger and Steiner (1967), Vlcek (1984), Yalden *et al.* (1973), and Zejda and Holisova (1970).

small, terrestrial or semi-arboreal rodents. In general *Peromyscus* species inhabit rodent communities that are richer in species than those inhabited by *Apodemus* species; this is evident even in central and northern parts of the range of *Peromyscus*.

Community diversity depends on species numbers and their relative abundance. Herein, relative abundance is estimated using a scale of zero through four where one indicates present but rare and four extremely common. This was applied to samples collected over two or more seasons to avoid short term changes in abundance due to, for example, annual cycles in population size. A fully quantitative comparison of rodent relative abundance is not possible because of methodological inconsistencies among researchers. For example, numbers of individuals of rarer species frequently are not given. The technique employed here probably over-represents rare species and under-represents the most numerous species. Data are collated for northern woodland and grassland habitats of Europe and North America (Fig. 11). *Apodemus* and *Peromyscus* species are the most common rodents of woodlands and forests in Europe and North America (Fig. 11). *Apodemus* but not *Peromyscus* are less common in grasslands than woodlands. Cricetids are the most common cohabitants with *Apodemus*; in woodland, the most common cricetid associate is *C. glareolus* but in grassland, *Microtus* species are numerically dominant. In North America, these genera also are represented in woodland and grassland, respectively, although *Microtus* does not seem to be so numerically dominant in the presence of *Peromyscus* species as in the presence of *Apodemus* species. The relationships among *Apodemus*, *Microtus*, and *Clethrionomys* in Europe and *Peromyscus*, *Microtus*, and *Clethrionomys* in North America provide a most striking example of similarity in the rodent communities of Europe and North America. The major difference between rodent communities of the two continents is the presence of small ground-living squirrels, particularly of the genus *Tamias* in North American samples. *Apodemus* come into contact with ground-living squirrels only in Asian parts of their range (Corbet, 1978).

Two or more species of *Peromyscus*, in the sense of Hall (1981), or *Apodemus* are sometimes found sharing the same macrohabitat. The maximum number of congeners in a single habitat is three in both genera (Fig. 12). Some studies report four species of *Peromyscus* (Baker, 1968; Smith, 1981) living in sympatry, but three seems to be the largest number of congeneric species persisting in one

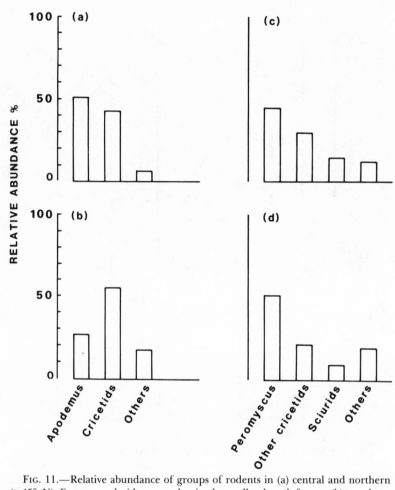

FIG. 11.—Relative abundance of groups of rodents in (a) central and northern (≥45° N) European deciduous and mixed woodland and forest, (b) northern (≥45° N) European oldfields and grassland, (c) northern (≥35° N) deciduous woodland of North America, and (d) northern (≥40° N) oldfields and grassland of North America. Relative abundance is based on a scale of 0–4 where 1 indicates present but rare and 4 very common. Percent relative abundance is the proportion of scores totalled for all studies for the given taxa. Data from studies listed in the legend of Figure 10.

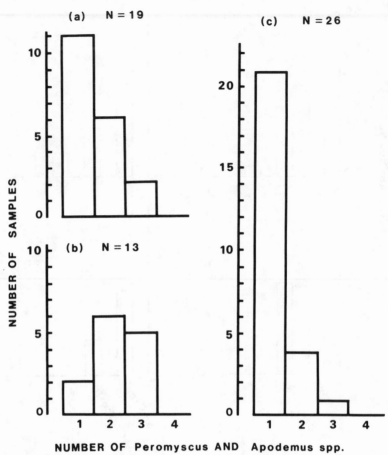

NUMBER OF Peromyscus AND Apodemus spp.

FIG. 12.—Number of *Apodemus* and *Peromyscus* species sharing macrohabitats in sympatry in rodent communities of (a) continental Eurasia, (b) southern (≤34° N) continental N. America and (c) northern (≥35° N) continental N. America. Data from studies listed in the legend of Figure 10.

habitat type. Few southwestern samples contain less than two *Peromyscus* species, but northern communities usually have just one. There is just one species, *P. maniculatus*, throughout much of Canada (Baker, 1968). Central and northern European communities frequently have two *Apodemus* species. Many parts of western Europe, however, have just one *Apodemus* species, *A. sylvaticus*; like *P. maniculatus*, *A. sylvaticus* occupies a wide range of habitats.

Rodent communities may change over time as a result of successional changes in the habitat following some catastrophic disruption, such as clearcutting or fire, or gradual change in climate, habitat, or both. For example, Crowner and Barrett (1979) found that *Microtus* and *Mus* disappeared from grassland, whereas numbers of *P. maniculatus* were reduced by half after burning. Clearing and burning woodland also results in a decline in species richness; although, *P. maniculatus* survives and may even flourish following catastrophic events (Gashwiler, 1970). This perhaps reflects the facility of *Peromyscus* species to exploit open grassland-type habitats and woodlands (Fig. 11). *A. sylvaticus*, however, may suffer in disrupted habitats; Venables and Venables (1971) and Remmert (1980) describe declines in abundance of *A. sylvaticus* over approximately 10 years after the planting of conifers. Ferns (1979), conversely, reported an increasing dominance of this species in a developing, three- to six-year-old larch plantation close to more mature woodland. Decline in the numbers and relative abundance of *A. flavicollis* at one locality in southwest England over 10 years was associated with disease of a major tree species and habitat destruction (Montgomery, 1985). *A. flavicollis* was present outside its current range in England and Wales during Roman times (1800 years before present), and its present restricted distribution in Britain may result from the removal of mature forest (Yalden, 1984). Similarly, progressive climatic shifts in the American Southwest over 200 to 300 years have been associated with local decreases in rodent diversity (Holbrook, 1980).

Division of Resources Within Communities

The most important resources for small rodents are food and space. The manner in which these are exploited determines not only species composition but also relative abundance within any particular habitat. Inter- and intrageneric differences in diet and use of space will be considered in turn. *Apodemus* and *Peromyscus* are commonly found in association with *Clethrionomys*, *Microtus*, and *Mus musculus* in woodland, grassland, and agricultural habitats, respectively. Particular emphasis will be given to resource exploitation in these genera.

(1) Food.—There is relatively little intrageneric variation in the diet of *Apodemus* and *Peromyscus* (Table 9). Coexisting populations of *Apodemus* and *Peromyscus* have similar diets (Hamilton, 1941; Holisova and Obrtel, 1980; Jameson, 1952; Sviridenko, 1940; Whitaker, 1966). Trophic niche overlap between *A. sylvaticus* and

A. flavicollis, for example, maybe as high as 0.85 (Holisova and Obrtel, 1980). In European habitats, *Clethrionomys* and *Microtus* species are less dependent on seeds and animal food than *Apodemus*. *Clethrionomys* have a very diverse diet, including seeds, animal material, fungi, fruit, and the vegetative parts of plants (Holisova and Obrtel, 1980; Watts, 1968). *Microtus* feed almost exclusively on green plant material with little in common with the diet of *Apodemus* (Evans, 1973; Holisova and Obrtel, 1980). Hamilton (1941) indicated that North American *Clethrionomys* are also more catholic in their tastes than *Peromyscus*, eating a wide range of animal and green plant material, seeds, fruit, fungi, and roots. North American *Microtus* are, like European species, specialists on the exposed tissues of herbaceous plants and grass (M'Closkey and Fieldwick, 1975; Vaughan, 1974).

Mus musculus is primarily a granivore although animal food and a variety of other items are eaten. There is considerable but not total overlap in food between *Mus* and *Peromyscus* (Whitaker, 1966) and, probably, between *Mus* and *Apodemus*. *Peromyscus* species in rodent communities of the short grasslands and desert have more general diets than the coexisting heteromyid species. There may be considerable overlap in the size of seeds taken by heteromyids and *Peromyscus* (Brown and Lieberman, 1973), but heteromyids concentrate on seeds or invertebrates or, alternatively, their diets are less seasonal than *Peromyscus* (Flake, 1973; Kritzman, 1974; Meserve, 1974).

Dietary overlap between rodents and other taxa is poorly documented. It is possible that *Apodemus* and *Peromyscus* in all habitats are dependent on seeds also eaten by a wide range of insects, birds, and other mammals. Brown and Davidson (1977), for example, have highlighted the large degree of dietary overlap between granivorous rodents and ants in desert environments. *Apodemus* and *Clethrionomys* share an acorn crop with rabbits, deer, badgers, wild boar, pigeons, and corvids of all kinds (Corbet, 1974). During certain seasons, or in certain habitats, there may be moderate to large dietary overlap between *Peromyscus* and *Apodemus* and insectivorous shrews of the genera *Blarina* and *Sorex* (Churchfield, 1984; Zegers and Ha, 1981).

(2) Space.—Intergeneric, differential use of space usually occurs in communities in which *Peromyscus* are present. This is evident in woodland (Dueser and Hallett, 1980; Dueser and Shugart, 1978, 1979; Holbrook, 1979a; Kirkland and Griffin, 1974; M'Closkey, 1975; Mihok, 1979; Morris, 1984; Seagle, 1985; Zegers and Ha,

1981), grassland (Hallett *et al.*, 1983; M'Closkey and Lajoie, 1975; Whitaker, 1967), arid shrub and scrub (Meserve, 1976; Price and Waser, 1984), and desert environments (Brown and Lieberman, 1973; Hallett, 1982). Often, at least one *Peromyscus* species is more general in its use of space than coexisting species (Brown and Lieberman, 1973; Dueser and Hallett, 1980; Dueser and Shugart, 1979; Hallett *et al.*, 1983; Holbrook, 1979*b*; Kirkland and Griffin, 1974). Sympatric populations of *Peromyscus* tend to have complementary spatial distributions (Hallett, 1982; Holbrook, 1978, 1979*b*; M'Closkey and Lajoie, 1975; Parren and Capen, 1985; Price and Waser, 1984; Smith and Speller, 1970; Whitaker, 1967). Several authors report that *Peromyscus* may make differential use of space vertically, as well as, or instead of, horizontally (Holbrook, 1979*a*; Meserve, 1976; Smith and Speller, 1970). Sexual (Bowers and Smith, 1979; Morris, 1984) and age (Van Horne, 1982) effects among conspecific *Peromyscus* species may further complicate the division of space in North American rodent communities.

Interspecific spatial relationships in Eurasian rodent communities are similar to those observed among their North American counterparts. In a number of investigations sympatric populations of *Apodemus, Microtus, Clethrionomys*, and *Mus* make differential use of shared habitats (Chelkowska, 1969; Doi *et al.*, 1978; Fairley and Jones, 1976; Flowerdew *et al.*, 1977; Krylov, 1975; Loy and Boitani, 1984; Miyao *et al.*, 1974; Montgomery, 1985; Otsu, 1969; Southern and Lowe, 1968). Other studies, however, reveal little spatial division between *Apodemus* and *Clethrionomys* (Gliwicz, 1981; Jonge and Dienske, 1979). Flowerdew *et al.* (1977), Montgomery (1985), and Southern and Lowe (1968) found that *Apodemus*, in particular *A. sylvaticus*, made more general use of space than *Clethrionomys* and *Microtus*, which favor areas of dense low-level cover. In farmland, *Mus* are associated with farm buildings, whereas *Apodemus* species are found in hedgerows, fields, and woods (Pelikan and Nesvadbova, 1979; Vlcek, 1984); this spatial separation may change with the seasons (Montgomery and Dowie, unpubl. data). Differential use of space has also been demonstrated among coexisting population of *Apodemus* (Abramsky, 1981; Doi and Iwamoto, 1982; Gliwicz, 1981; Hoffmeyer and Hansson, 1974; Montgomery, 1980*b*, 1981; Otsu, 1970; Yoshida, 1970).

Habitat use by different species of *Apodemus* is, in general, poorly documented, possibly because of the wide range of environments that these rodents inhabit. However, Corbet (1966) indicated that in continental Europe *A. sylvaticus* is found in almost every habitat type, whereas *A. flavicollis* is more characteristic of mature woodland, and *A. microps* and *A. agrarius* are associated with grassland, woodland edges, and cultivated land. There is some evidence of vertical stratification of activity in woodland habitats shared by populations of *Apodemus* (Hoffmeyer, 1976, 1983; Imaizuma, 1978; Montgomery 1980c). It is clear, however, that all *Apodemus* are better climbers and more arboreal in habit than *Clethrionomys* species (Holisova, 1969; Montgomery, 1980c).

Differential timing of surface activity is, in some cases, associated with relatively high levels of spatial overlap. *Microtus, Clethrionomys* and *Mus* are less strictly nocturnal than *Apodemus* or *Peromyscus* (Corbet and Southern, 1977; Falls, 1968). Common ground may also be exploited at different times by cohabiting *Apodemus* species (Montgomery, 1985).

Division of space and food in rodent communities are inextricably linked; different foods may be sought in the same place and vice versa. Drickamer (1976) and Miller and Getz (1977b) indicated the direct effect of diet on use of space in *Peromyscus* species. Narrow habitat selection may be predetermined by a diet set by genotype and early experience with particular food types. A broader, more flexible diet may permit exploitation of a wider range of habitats. The use of space by individual *A. sylvaticus* may also be subject to genetic and learned responses to food; for example, these mice exhibit a distinct preference for familiar food items in laboratory experiments (Partridge, 1981).

Periods of activity of several nocturnal and diurnal species frequently coincide. It is apparent, however, that spatial division is virtually universal among populations constituting rodent communities. There are two explanations for this consistent spatial organization in animal communities.

(1) Specific adaptations to environmental heterogeneity.—Differential adaptations to particular habitat features, such as the distribution of food, nesting sites, and refuges from predators, may promote spatial division. Some *Peromyscus* species, for example, are better adapted climbers than others; this may be reflected in longer tails (Horner, 1954), development of plantar tubercles (Layne, 1970), and larger brains (Lemen, 1980) among semi-arboreal species. Vertical stratification of activity is common in habitats with two or more *Peromyscus* species (Barry *et al.*, 1984; Holbrook, 1979a). Some *Apodemus* species are also more arboreal than others (Abramsky, 1981; Hoffmeyer, 1973; Montgomery, 1980c), and there is some evidence that more arboreal species, such as *A. flavicollis* and

A. mystacinus, have proportionately longer tails than less arboreal species, for example, *A. sylvaticus* (Larina, 1964; Zatsepina, 1960). Vertical separation of nesting sites is also apparent in *Peromyscus* (Smith and Speller, 1970; Stah, 1980; Wolff and Hurlbutt, 1982) and *Apodemus* species (Hoffmeyer, 1973) although competitive interactions may affect the latter (see below). Differential selection of food and foraging sites also results in spatial separation of *Peromyscus* species (Drickamer, 1970; Harris, 1984; Miller and Getz, 1977*b*). This may be embodied in differences in mandibular morphology where woodland types have a deeper jaw with longer diastema than grassland mice (Holbrook, 1982). Interspecific differences in morphology associated with feeding are unknown in *Apodemus*, and the evidence concerning differential food selection is equivocal. In two studies, *A. sylvaticus* and *A. flavicollis* selected food in accord with their habitat or geographical background (Hoffmeyer, 1976; Pfeiffer and Neithammer, 1972). Hoffmeyer (1976) found a clear dietary difference between *A. flavicollis* from woodland and *A. sylvaticus* from grassland habitats. Laboratory-reared offspring of these species retained some element of the parental choice of diet. Genetic and early experience effects on habitat selection also are evident in grassland and woodland subspecies of *P. maniculatus* (Harris, 1952; Wecker, 1963). Similar data on *Apodemus* are lacking.

Risk of predation may enhance spatial division; less vulnerable species exploit more open habitat patches, whereas vulnerable ones are associated with cover (Kotler, 1984*a*, 1984*b*). *P. maniculatus*, for example, is a vulnerable species in desert rodent communities and is restricted to the vicinity of bushes, whereas *Dipodomys deserti*, a large, bipedal species, exploits more open spaces (Kotler, 1984*a*). *P. maniculatus* increases in abundance if distances between bushes are reduced artificially (Thompson, 1982). Price (1984) suggested, however, that foraging economics are more important in determining microhabitat use in desert and scrub rodents than risk of predation. Similar data are not available for *Apodemus*. However, it seems unlikely that spatial division of habitats shared by *A. sylvaticus* and *C. glareolus*, both hunted by tawny owls, is produced by differential risk of predation. *A. sylvaticus* occupy more open habitat but appear more vulnerable to predation than *C. glareolus* (Southern and Lowe, 1968, 1982).

Differences in behavior, particularly social behavior, may result in quite different patterns in the use of space by different species. Populations consisting of family groups may have a very different distribution than one made up of monogamous pairs. Interspecific differences in behavior are well-known in *Peromyscus* (Eisenberg, 1962, 1968; King *et al.*, 1968). Hoffmeyer (1983) suggested that *A. flavicollis* and *A. sylvaticus* differ with respect to levels of antagonistic behavior within each species, the former being more tolerant of conspecifics than the latter. Spatial distribution may be determined by intraspecific interactions but result in some degree of spatial division between species. Differences in social behavior may also promote different annual cycles of population size so that large numbers of two congeneric species seldom occur in the same habitat at the same time (Montgomery 1980*a*, 1985).

The most convincing evidence that the manner in which different rodent species respond to habitat discontinuities is responsible for spatial division in rodent communities, is the effect of habitat changes on distribution and abundance. Artificial or natural changes in habitat quality result in changes in relative abundance and, possibly, species composition (Montgomery, 1985; Parmenter and MacMahon, 1983; Price and Waser, 1984; Thompson, 1982).

(2) Competitive interactions between species.—Competitive exclusion on whatever scale has long been considered the major determinant of community organization. Yet, evidence for this contention has been far from unequivocal. There are two main kinds of evidence: inverse distribution and abundance patterns between supposed competitors and the results of experimental manipulations in the field. Regression analyses of the number of captures of species A on the number of captures of species B, C, D, and so on, fall into the former category. Such analyses suggest that *P. maniculatus* and *P. leucopus* are competitively weak, habitat generalists (Dueser and Hallett, 1980; Hallett *et al.*, 1983), whereas *P. eremicus* is a competitively strong, habitat specialist (Hallett, 1982). Abramsky (1981) reported a similar analysis on *A. sylvaticus* and *A. mystacinus*; in this case, however, distributions of both species were dependent on habitat structure rather than on any interspecific effect.

Experimental removal of unenclosed or enclosed populations has provided strong evidence of interspecific competition between *Peromyscus* and *Microtus* species (Abramsky *et al.*, 1979; Grant, 1970*a*, 1971; Morris, 1972; Redfield *et al.*, 1977), *Dipodomys* species (Munger and Brown, 1981), *Mus musculus* (Sheppe, 1967), and *Neotoma* species (Holbrook, 1979*b*). Data for interactions between *Apodemus* species and other genera are few. Population size of *A. agrarius* increased following the removal of *A. flavicollis* and *C. glareolus* (Gliwicz, 1981), with the intrageneric effect likely to have been greater than the intergeneric interaction. Direct aggression of *A. flavicollis* towards *C. glareolus* may be important in determining the distribution of the latter (Andrzejewski and Olszewski, 1963). In the presence of *Apodemus*, *C. glareolus* becomes more diurnal and, on the whole, less aggressive (Bergstedt, 1965; Brown, 1956; Greenwood, 1978). *Apodemus sylvaticus* avoid *Mus musculus* in staged encounters (Montgomery, unpublished data), and there is circumstantial evidence of competition in the field in that the former species underwent microhabitat expansion following the extinction of *Mus musculus* on the island of St. Kilda (Delany, 1970).

Experimental evidence for competition between *Peromyscus* species is scarce. Sheppe (1961) reported the invasion of a ravine by *P. maniculatus* after the removal of a population of *P. oreas*. Similarly, Holbrook (1979*b*) found that the removal of *P. boylii* resulted in an expansion of the microhabitat of *P. maniculatus*. Laboratory encounters suggest that interspecific interactions between *Peromyscus* species are usually aggressive, or one species avoids the other, or both (Brown, 1964*b*; Llewellyn, 1980; Smith, 1965). Sympatric populations of *P. leucopus* and *P. maniculatus* may exhibit interspecific territoriality where resident individuals are dominant over intruders regardless of their species (Wolff *et al.*, 1983). This may be indicative of the high degree of ecological and morphological similarity of these species. Interspecific territoriality is not apparent among *Apodemus* species; Montgomery (1979) found that in woodland, home ranges of *A. sylvaticus* and *A. flavicollis* largely overlapped, although these species are negatively associated and have differential affinities for low and medium-level cover (Montgomery, 1980*b*). Although there is little overt aggression, *A. sylvaticus* avoid *A. flavicollis* in laboratory encounters (Hoffmeyer, 1973; Montgomery, 1978). A removal experiment, conducted without enclosures, however, suggested that the presence of either species inhibits reproduction in the other such that each comes into reproductive condition earlier in isolation. There was also some encroachment by *A. sylvaticus* and *A. flavicollis* into parts of the habitat in which congenerics were no longer present (Montgomery, 1981).

There are clear similarities in the composition of rodent communities in the Palearctic and Nearctic. Division of resources, principally food and space, follow similar lines in these remote rodent communities. *Apodemus* and *Peromyscus* species have different diets from cohabiting genera and, often, are active at different times. However, diet and activity patterns are broadly similar within each genus. Spatial separation occurs between congeneric *Peromyscus* and *Apodemus* species in shared habitats. Both environmental adaptations and interspecific competition are involved in maintaining microspatial separation of species in communities with *Apodemus* and *Peromyscus*.

PEROMYSCUS AND APODEMUS ON ISLANDS

Numbers of species of terrestrial mammals decrease on smaller, more remote islands (Lomolino, 1984). *Apodemus* and *Peromyscus* species are known from a large number of coastal islands around the fringe of their continental ranges. Isolated representatives of both genera occur on islands and are the most frequent of all small rodents on islands. For example, *P. maniculatus* occurs on 50 islands off the west coast of Canada, *Microtus* species on 12, and *C. gapperi* on just one (Grant, 1970b). Similarly, *A. sylvaticus* is present on 24 British islands, *Microtus* species on 13, and *C. glareolus* on five (Grant, 1970b). On these islands *Microtus* and *Clethrionomys* seem to exclude one another. (Note that *Microtus* may be more successful island inhabitants than *Peromyscus* on other island groups; Mehlhop and Lynch, 1978). Grant (1970b) suggests that occupancy is more likely to be by a generalist, such as *Apodemus* and *Peromyscus*, than a specialist. This is supported by Crowell and Pimm's (1976) experimental introduction of *P. maniculatus* and *C. gapperi* to islands with *Microtus* only; this resulted in *C. gapperi* displacing *P. maniculatus* from woodland habitats and *P. maniculatus* ousting *Microtus* from shrubby habitats. *C. gapperi* may also be a poorer disperser than *P. maniculatus* (Crowell, 1973). Habitat diversity also plays a vital role in determining the constitution of the rodent faunas of islands (Dueser and Brown, 1980; McPherson and Krull, 1972).

The characteristics of island populations of small rodents may differ markedly from mainland populations (Gliwicz, 1980). High densities, depressed dispersal, decreased reproduction, poor survival of young, good survival of adults, and altered spatial distribution, may be evident in island-living rodents. Studies of island populations of *P. maniculatus* (Haplin, 1981; Sullivan, 1977) and *A. sylvaticus* (Delany, 1970; Fullagar *et al.*, 1963; Jewell, 1966) suggest that these species exhibit altered population characteristics on islands, although the effect appears to be greater in the *Peromyscus* than the *Apodemus* species. Island populations of both genera display gigantism; Lawlor (1982) noted that 16 of 29 island populations of *Peromyscus* are larger than mainland forms; similarly, 17 out of 27 island populations of *A. sylvaticus* around Scotland and Ireland are larger than mainland mice (Berry, 1969; Kelly *et al.*,

1982). Lawlor (1982) suggested that gigantism in island populations occurs in generalist species as a response to low resource levels and absence of competitors and predators. Regardless of the underlying mechanism, this morphological response to island life emphasizes the high degree of similarity in the population and community ecology of *Apodemus* and *Peromyscus*.

Discussion and Conclusions

Methodological inconsistencies plague much of mammal ecology. Often it is impossible to make valid comparisons between studies simply because of the manner in which they were conducted. In this review, data have been extracted from the literature in as basic a form as possible or only when methods are reasonably alike: for example, using only data on embryo numbers as a basis for comparison for litter size. Much more information would be available for comparative studies if some standardization were accepted. For example, data concerning mammal species present at a given study site are often lacking. This obstructs the construction of large data sets for the analysis of community structure. Clear indications of the sex and the age composition of samples also are often lacking in morphological records, rendering much of the effort useless. Nevertheless, sufficient information is available to allow some preliminary conclusions concerning similarities in the biology of *Apodemus* and *Peromyscus*.

It is clear that southern populations and species of *Peromyscus* differ in many aspects of their population and community biology from northern populations. In truth, *P. maniculatus* and *P. leucopus* are atypical of the genus whose center of speciation and diversity is in Mexico. The situation among *Apodemus* species is less clear, simply because of the lack of data for southern populations. This reflects the northern and western distribution of research interest, rather than a lack of southern populations. More data for central and southern populations of both genera are required. Comparison of northern populations of *Peromyscus* and *Apodemus* reveal many points of similarity in the factors controlling the onset and termination of reproduction, litter size, noncyclic population dynamics, population regulation, trophic position, impact in temperate habitats, and community structure. The two genera have much in common.

There are, however, differences in the biology of species within and between genera. This is clear in the dynamics and regulation of populations of *A. sylvaticus* and *A. flavicollis* and, possibly, *P. ma niculatus* and *P. leucopus*. *Peromyscus* seem to make greater use of torpor and have a distinct cycle in thermogenesis capability. These and other characteristics allow mice of this genus to exploit more extreme subarctic habitats than *Apodemus*. The existence of such northern populations, and the diversity of *Peromyscus* in Central America and southern North America, indicate that this genus is the most adaptable of all small rodents. However, it remains to be seen whether *Peromyscus* species could exploit areas from which *Apodemus* species are absent in the Palearctic. For example, would any *Peromyscus* survive the rigors of the Central Asian Plateau?

This review confirms that unrelated species living in similar but geographically remote environments adapt similarly to their surroundings. Nevertheless, species have unique attributes, which prohibit all-enveloping comparisons between ecological equivalents. It is more appropriate to make comparisons between processes in population and community ecology than to attempt to establish species or genera as ecologically equivalent. *Apodemus* and *Peromyscus* have much in common; students of the biology of *Apodemus* and *Peromyscus* have even more.

Acknowledgments

I wish to thank John Flowerdew, John Gurnell, Derek Yalden, Gordon L. Kirkland, Jr., James N. Layne, and Sally Montgomery for their assistance in the preparation of this review. I am also indebted to Elizabeth Purdy for her tolerance in typing the manuscript.

This review is dedicated to the late Professor Wladyslaw Grodzinski.

Literature Cited

Abramsky, Z.   1981.   Habitat relationships and competition in two Mediterranean *Apodemus* spp.   Oikos, 36:219–225.

Abramsky, Z., M. I. Dyer, and P. D. Harrison.   1979.   Competition among small mammals in experimentally perturbed areas of the shortgrass prairie. Ecology, 60:530–536.

Adamczewska, K. A.   1961.   Intensity of reproduction of the *Apodemus flavicollis* (Melchior, 1834) during the period 1954–1959.   Acta Theriol., 5:1–21.

Adamczewska-Andrzejewska, K. A.   1971.   Methods of age determination in *Apodemus agrarius* (Pallas, 1771).   Ann. Zool. Fennici, 8:68–71.

———.   1973.   Growth variations and age criteria in *Apodemus agrarius* (Pallas, 1771).   Acta Theriol., 19:353–394.

Adler, G. H., and R. H. Tamarin.   1984.   Demography and reproduction in island and mainland white-footed mice (*Peromyscus leucopus*) in southeastern Massachusetts.   Canadian J. Zool., 62:58–64.

Andrzejewska, R., and J. Olszewski.   1963.   Social behavior and interspecific relations in *Apodemus flavicollis* (Melchior, 1834) and *Clethrionomys glareolus* (Schreber, 1780).   Acta Theriol., 9:155–168.

Andrzejewska, R., and H. Wroclawek.   1961.   Mass occurrence of *Apodemus agrarius* (Pallas, 1771) and variations in the numbers of associated Muridae. Acta Theriol., 13:173–184.

ANTHONY, R. G., L. J. NILES, AND J. D. SPRING. 1981. Small-mammal associations in forested and old-field habitats—a quantitative comparison. Ecology, 62:955–963.

ASHER, S. O., AND V. G. THOMAS. 1985. Analysis of temporal variation in the diversity of a small mammal community. Canadian J. Zool., 63:1106–1109.

AYER, M. L., AND J. M. WHITSETT. 1980. Aggressive behavior of female prairie deer mice in laboratory populations. Anim. Behav., 28:763–771.

BAAR, S. L., AND E. D. FLEHARTY. 1976. A model of the daily energy budget and energy flow through a population of the white-footed mouse. Acta Theriol., 21:179–193.

BAKER, J. R. 1930. The breeding season in the British wild mice. Proc. Zool. Soc. London, 1930:113–127.

BAKER, R. H. 1968. Habitats and distribution. Pp. 98–126, in Biology of *Peromyscus* (Rodentia) (J. A. King, ed.). Spec. Publ., Amer. Soc. Mamm., 2:1–593.

BARRY, R. E., JR., M. A. BOTJE, AND L. B. GRANTHAM. 1984. Vertical stratification of *Peromyscus leucopus* and *P. maniculatus* in southwestern Virginia. J. Mamm., 65:145–148.

BATZLI, G. O. 1977. Population dynamics of the white-footed mouse in floodplain and upland forest. Amer. Midland Nat., 97:18–32.

BEER, J. R., AND C. F. MACLEOD. 1966. Seasonal population changes in the prairie deer mouse. Amer. Midland Nat., 76:277–289.

BENDELL, J. F. 1959. Food as a control of a population of white-footed mice, *Peromyscus leucopus noveboracensis* (Fischer). Canadian J. Zool., 37:173–209.

BERGSTEDT, B. 1965. Distribution, reproduction, growth and dynamics of the rodent species *Clethrionomys glareolus* (Schreber), *Apodemus flavicollis* (Melchior), and *Apodemus sylvaticus* (Linne) in southern Sweden. Oikos, 16:132–160.

BERRY, R. J. 1969. History in the evolution of *Apodemus sylvaticus* (Mammalia) at one edge of its range. J. Zool. (London), 159:311–328.

———. 1985. Evolutionary and ecological genetics of the bank vole and wood mouse. Symp. Zool. Soc. London, 55:1–32.

BIGLER, W. J., AND J. H. JENKINS. 1975. Population characteristics of *Peromyscus gossypinus* and *Sigmodon hispidus* in tropical hammocks of South Florida. J. Mamm., 56:633–644.

BIRDSALL, D. A., AND D. NASH. 1973. Occurrence of successful multiple insemination of females in natural populations of deer mice (*Peromyscus maniculatus*). Evolution, 27:106–110.

BLACKWELL, T. L., AND P. R. RAMSEY. 1972. Exploratory activity and lack of genotypic correlates in *Peromyscus polionotus*. J. Mamm., 53:401–403.

BLAIR, W. F. 1940. A study of deer-mouse populations in southern Michigan. Amer. Midland Nat., 24:273–305.

———. 1943. Populations of the deer-mouse and associated small mammals in the mesquite association of southern New Mexico. Contrib. Lab. Vert. Biol., Univ. Michigan, 21:1–40.

———. 1951. Population structure, social behavior, and environmental relations in a natural population of the beach mouse (*Peromyscus polionotus leucocephalus*). Contrib. Lab. Vert. Biol., Univ. Michigan, 48:1–47.

BOBEK, B. 1969. Survival, turnover and production of small rodents in a beech forest. Acta Theriol., 14:191–210.

————. 1971. Influence of population density upon rodent production in a deciduous forest. Ann. Zool. Fennici, 8:137–144.

————. 1973. Net production of small rodents in a deciduous forest. Acta Theriol., 18:403–434.

BOWERS, M. A., AND H. D. SMITH. 1979. Differential habitat utilization by sexes of the deermouse, *Peromyscus maniculatus*. Ecology, 60:869–875.

BRADLEY, E. L., AND C. R. TERMAN. 1981. Serum testosterone concentrations in male prairie deermice (*Peromyscus maniculatus bairdii*) from laboratory populations. J. Mamm., 62:811–814.

BRIESE, L. A., AND M. H. SMITH. 1974. Seasonal abundance and movement of nine species of small mammals. J. Mamm., 55:615–629.

BRODY, S. 1945. Bioenergetics and growth. Reinhold Publishing Co., New York, 1023 pp.

BROWN, J. H., AND D. W. DAVIDSON. 1977. Competition between seed-eating rodents and ants in desert ecosystems. Science, 196:880–882.

BROWN, J. H., AND G. A. LIEBERMAN. 1973. Resource utilization and coexistence of seed-eating desert rodents in sand dune habitats. Ecology, 54:788–797.

BROWN, L. E. 1956. Field experiments on the activity of small mammals, *Apodemus*, *Clethrionomys* and *Microtus*. Proc. Zool. Soc. London, 126:549–564.

————. 1966. Home range and movement of small mammals. Symp. Zool. Soc. London, 18:111–142.

————. 1969. Field experiments on the movements of *Apodemus sylvaticus* L. using trapping and tracking techniques. Oecologia, 2:198–222.

BROWN, L. N. 1964a. Dynamics in an ecologically isolated population of the brush mouse. J. Mamm., 45:436–442.

————. 1964b. Ecology of three species of *Peromyscus* from southern Missouri. J. Mamm., 45:189–202.

BROWNELL, E. 1983. DNA/DNA hybridization studies of muroid rodents: symmetry and rates of molecular evolution. Evolution, 37:1034–1051.

CAMPBELL, I. 1974. The bioenergetics of small mammals, particularly *Apodemus sylvaticus* (L.) in Wytham Woods, Oxfordshire. Unpubl. Ph.D. dissert., Univ. Oxford, 124 pp.

CANHAM, R. P. 1969. Early cessation of reproduction in an unusually abundant population of *Peromyscus maniculatus borealis*. Canadian Field-Nat., 83:279.

————. 1970. Sex ratios and survival in fluctuating population of the deer mouse, *Peromyscus maniculatus borealis*. Canadian J. Zool., 48:809–811.

CATLETT, R. H., AND R. Z. BROWN. 1961. Unusual abundance of *Peromyscus* at Gothic, Colorado. J. Mamm., 42:415.

CERNY, V. 1962. K nektorym otazkam romnozovani u mysice temnopase (*Apodemus agrarius* Pall.). Cas. Narod. Mus. (Nar.), 131:15–17.

CHELKOWSKA, H. 1969. Numbers of small rodents in five plant associations. Ekol. Pol. ser. A, 117:847–854.

CHEW, R. M., AND A. E. CHEW. 1970. Energy relationships of the mammals of a desert shrub (*Larrea tridentata*) community. Ecol. Monogr., 40:1–21.

CHURCHFIELD, S. 1984. Dietary separation in three species of shrew inhabiting water-cress beds. J. Zool., 204:211–228.

CLARKE, J. R. 1981. Physiological problems of seasonal breeding in eutherian mammals. Pp. 244–312, *in* Oxford reviews of reproductive biology. Vol. 3 (C. A. Finn, ed.). Clarendon Press, Oxford, 334 pp.

————. 1985. The reproductive biology of the bank vole (*Clethrionomys glareolus*) and the wood mouse (*Apodemus sylvaticus*). Symp. Zool. Soc. London, 55: 33–60.

CLARKE, J. R., AND E. A. EGAN. 1984. Wastage of ova: experimental studies in the field vole (*Microtus agrestis*), the bank vole (*Clethrionomys glareolus*) and the wood mouse (*Apodemus sylvaticus*). Acta Zool. Fennica, 171 : 141–144.

CODY, C. B. J. 1982. Studies on behavioural and territorial factors relating to the dynamics of woodland rodent populations. Unpubl. Ph.D. dissert., Univ. Oxford, 140 pp.

CORBET, G. B. 1966. The terrestrial mammals of western Europe. Foulis and Co. Ltd., London, 264 pp.

————. 1974. The importance of oak to mammals. Pp. 312–324, *in* The British oak: its history and natural history (M. G. Morris and F. H. Perrins, eds.). E. W. Classey, 376 pp.

————. 1978. The mammals of the Palaearctic region: a taxonomic review. British Museum (Natural History), London, 314 pp.

————. 1984. Supplement to The mammals of the Palaearctic region: a taxonomic review. British Museum (National History), London, 89 pp.

CORBET, G. B., AND H. N. SOUTHERN (EDS.). 1977. The handbook of British mammals. Second ed. Blackwell Sci. Pub., Oxford, 244 pp.

CORNISH, L. M., AND W. N. BRADSHAW. 1978. Patterns in twelve reproductive parameters for the white-footed mouse (*Peromyscus leucopus*). J. Mamm., 59:731–739.

COX, C. B., AND P. D. MOORE. 1985. Biogeography: an ecological and evolutionary approach. Blackwell Sci. Pub., Oxford, 244 pp.

CRAWLEY, M. C. 1969. Some population dynamics of the bank vole, *Clethrionomys glareolus* and the wood mouse, *Apodemus sylvaticus* in mixed woodland. J. Zool., 180:71–89.

————. 1970. Movements and home-ranges of *Clethrionomys glareolus* Schreber and *Apodemus sylvaticus* L. in north-east England. Oikos, 20:310–319.

CROWELL, K. L. 1973. Experimental zoogeography: introductions of mice to small islands. Amer. Nat., 107:535–558.

CROWELL, K. L., AND S. L. PIMM. 1976. Competition and niche shifts of mice introduced onto small islands. Oikos, 27:251–258.

CROWNER, A. W., AND G. W. BARRETT. 1979. Effects of fire on the small mammal component of an experimental grassland community. J. Mamm., 60: 803–813.

CYGAN, T. 1984. Seasonal changes in thermoregulation and maximum metabolism in *Apodemus flavicollis* (Melchior, 1834). Unpubl. M.S. thesis, Jagiellonian University, Cracow, 22 pp.

DALY, M., M. I. WILSON, AND P. BEHRENDS. 1980. Factors affecting rodent responses to odours of strangers with odour-baited traps. Behav. Ecol. Sociobiol., 6:323–329.

DAVENPORT, L. B. 1964. Structure of two *Peromyscus polionotus* populations in old-field ecosystems at the AEL Savannah River Plant. J. Mamm., 45:95–113.

DELANY, M. J. 1970. Variation and ecology of island populations of the long-tailed field mouse (*Apodemus sylvaticus* L.). Symp. Zool. Soc. London, 26:283–295.

DEWSBURY, D. A. 1981. Social dominance, copulatory behavior, and differential reproduction in deer mice (*Peromyscus maniculatus*). J. Comp. Physiol. Psychol., 95:880–895.

———. 1982. Pregnancy blockage following multiple-male copulation or exposure at the time of mating in deer mice, *Peromyscus maniculatus*. Behav. Ecol. Sociobiol., 11:37–42.

———. 1984. Aggression, copulation, and differential reproduction of deer mice (*Peromyscus maniculatus*) in a semi-natural enclosure. Behaviour, 91: 1–31.

DEWSBURY, D. A., AND D. J. BAUMGARDNER. 1981. Studies of sperm competition in two species of muroid rodents. Behav. Ecol. Sociobiol., 9:121–133.

DEWSBURY. D. A., D. J. BAUMGARDNER, R. L. EVANS, AND D. C. WEBSTER. 1980. Sexual dimorphism for body mass in 13 taxa of muroid rodents under laboratory conditions. J. Mamm., 61:146–149.

DIETERLEN, F. C. 1965. Von der Lebensweise und dem Verhalten der Felsenmaus, *Apodemus mystacinus* (Danford und Alston, 1877), nebst Beitragen zur vergleichenden Ethologie der Gattung *Apodemus*. Saugetierk. Mitt., 13: 152–161.

DOI, T., AND T. IWAMOTO. 1973. A preliminary study on the seasonal fluctuation of the population density of the Japanese forest mice, genus *Apodemus*, in natural forest in Kyushu. Rep. Ebino Biol. Lab., Kyushu Univ, 1:64–72.

———. 1982. Local distribution of two species of *Apodemus* in Kyushu. Res. Population Ecol., 24:110–122.

DOI, T., T. IWAMOTO, AND K. EGUCHI. 1978. Population and biomass of small mammals. Pp. 212–221, *in* Biological production in a warm-temperate evergreen oak forest of Japan (T. Kira, Y. Ono, and T. Hosokawa, eds.). Univ. of Tokyo Press, Tokyo, 289 pp.

DRICKAMER, L. C. 1970. Seed preferences in wild caught *Peromyscus maniculatus bairdii* and *Peromyscus leucopus noveboracensis*. J. Mamm., 51:191–194.

———. 1976. Hypotheses linking food habits and habitat selection in *Peromyscus*. J. Mamm., 57:763–766.

DRICKAMER, L. C., AND B. M. VESTAL. 1973. Patterns of reproduction in a laboratory colony of *Peromyscus*. J. Mamm., 54:523–528.

DROZDZ, A. 1966. Food habits and food supply of rodents in the beech forest. Acta Theriol., 11:363–364.

DUESER, R. D., AND W. C. BROWN. 1980. Ecological correlates of insular rodent diversity. Ecology, 61:50–56.

DUESER, R. D., AND J. G. HALLETT. 1980. Competition and habitat selection in a forest-floor small-mammal fauna. Oikos, 35:293–297.

DUESER, R. D., AND H. H. SHUGART, JR. 1978. Microhabitats in a forest-floor small mammal fauna. Ecology, 59:89–98.

———. 1979. Niche pattern in a forest-floor small mammal fauna. Ecology, 60:108–118.

DUNN, E. 1977. Predation by weasels *Mustela nivalis* on breeding tits (*Parus* spp.) in relation to the density of tits and rodents. J. Anim. Ecol., 46:633–654.

EISENBERG, J. F. 1962. Studies on the behavior of *Peromyscus maniculatus gambelii* and *Peromyscus californicus parasiticus*. Behaviour, 19:177–207.

———. 1968. Behavior patterns. Pp. 451–495, *in* Biology of *Peromyscus* (Rodentia) (J. A. King, ed.). Spec. Publ., Amer. Soc. Mamm., 2:1–593.

ELTON, C., E. B. FORD, J. R. BAKER, AND A. D. GARDNER. 1931. The health and parasites of a wild mouse population. Proc. Zool. Soc. London, 1931: 657–721.

ERIKSSON, M. 1980. Breeding in a laboratory colony of wood mice, *Apodemus sylvaticus*. Saugetierk. Mitt., 28:79–80.

ERLINGE, S., *ET AL.* 1983. Predation as a regulating factor on small rodent populations in southern Sweden. Oikos, 40:36–52.

EVANS, F. C. 1942. Studies of a small mammal population in Bagley Wood, Berkshire. J. Anim. Ecol., 11:182–197.

———. 1973. Seasonal variation in the body composition and nutrition of the vole *Microtus agrestis*. J. Anim. Ecol., 42:1–18.

FAIRBAIRN, D. J. 1977*a*. Why breed early? A study of reproductive tactics in *Peromyscus*. Canadian J. Zool., 55:862–871.

———. 1977*b*. The spring decline in deer mice: death or dispersal? Canadian J. Zool., 55:84–92.

———. 1978*a*. Behavior of dispersing deer mice (*Peromyscus maniculatus*). Behav. Ecol. Sociobiol., 3:265–282.

———. 1978*b*. Dispersal of deer mice, *Peromyscus maniculatus*: proximal causes and effects on fitness. Oecologia, 32:171–193.

FAIRLEY, J. S. 1965. Studies on the biology and present status of the long-tailed fieldmouse in Ireland. Unpubl. Ph.D. dissert., The Queen's Univ. of Belfast, 210 pp.

———. 1967. A woodland population of *Apodemus sylvaticus* (L.) at Seaforde, Co. Down. Proc. R. Ir. Acad., 65 B:407–424.

FAIRLEY, J. S., AND J. M. JONES. 1976. A woodland population of small rodents *Apodemus sylvaticus* (L.) and *Clethrionomys glareolus* (Schreber) at Adare, Co. Limerick. Proc. R. Ir. Acad., 76 B:323–336.

FALLS, J. B. 1968. Activity. Pp. 543–570, *in* Biology of *Peromyscus* (Rodentia), (J. A. King, ed.). Spec. Publ., Amer. Soc. Mamm., 2:1–593.

FEDYK, A. 1971. Social thermoregulation in *Apodemus flavicollis* (Melchior, 1834). Acta Theriol., 16:221–229.

FELTEN, H., F. SPITZENBERGER, AND G. STORCH. 1973. Zur Kleinsaugerfauna West Anatoliens. 2. Sneckenbergiana Biol., 54:227–290.

FERNS, P. N. 1979. Successional changes in the small mammal community of a young larch plantation in south west Britain. Mammalia, 43:439–452.

FLAKE, L. D. 1973. Food habits of four species of rodents on a short-grass prairie in Colorado. J. Mamm., 54:636–647.

FLEHARTY, E. D., AND K. W. NAVO. 1983. Irrigated cornfield as habitat for small mammals in the sandsage prairie region of western Kansas. J. Mamm., 64:367–379.

FLEMING, T. H., AND R. J. RAUSCHER. 1978. On the evolution of litter size in *Peromyscus leucopus*. Evolution, 32:45–55.

FLOUSEK, J., Z. FLOUSKOVA, AND K. TOMASOVA. 1985. To the knowledge of small mammals (Insectivora, Rodentia) in the Rodnei Mts. (Rumania). Vest. Czech. Spolec. Zool., 49:6–17.

FLOWERDEW, J. R. 1972. The effect of supplementary food on a population of wood mice (*Apodemus sylvaticus*). J. Anim. Ecol., 41:553–566.

———. 1973. The effect of natural and artificial changes in food supplies on breeding in woodland mice and voles. J. Reprod. Fert., Suppl., 19: 259–269.

————. 1974. Field and laboratory experiments on the social behaviour and population dynamics of the wood mouse (*Apodemus sylvaticus*). J. Anim. Ecol., 43:499–511.

————. 1985. The population dynamics of wood mice and yellow-necked mice. Symp. Zool. Soc. London, 55:315–338.

FLOWERDEW, J. R., S. J. G. HALL, AND J. C. CLEVEDON BROWN. 1977. Small rodents, their habitats, and the effects of flooding at Wicken Fen, Cambridgeshire. J. Zool. (London), 182:323–342.

FOLTZ, D. W. 1981. Genetic evidence for long-term monogamy in a small rodent, *Peromyscus polionotus*. Amer. Nat., 117:665–675.

FORDHAM, R. A. 1971. Field populations of deermice with supplemental food. Ecology, 52:138–146.

FUJIMAKI, Y. 1969. Reproductive activity in *Apodemus argenteus* Temmick. J. Mamm. Soc. Japan, 4:74–80.

FULLAGAR, P. J., P. A. JEWELL, R. M. LOCKLEY, AND I. W. ROWLANDS. 1963. The Skomer vole (*Clethrionomys glareolus skomerensis*) and long-tailed field mouse (*Apodemus sylvaticus*) on Skomer Island, Pembrokeshire in 1960. Proc. Zool. Soc. London, 140:295–314.

FULLER, W. A. 1969. Changes in numbers of three species of small rodent near Great Slave Lake, N.W.T. Canada, 1964–1967, and their significance for general population theory. Ann. Zool. Fennici, 6:113–144.

GARLAND, T., JR., AND W. G. BRADLEY. 1984. Effects of a highway on Mojave Desert rodent populations. Amer. Midland Nat., 111:47–56.

GARSON, P. J. 1975. Social interactions of wood mice *Apodemus sylvaticus* studied by direct observation in the wild. J. Zool., 177:496–500.

GARTEN, C. T. 1976. Relationships between aggressive behavior and genic heterozygosity in the oldfield mouse, *Peromyscus polionotus*. Evolution, 30:59–72.

GASHWILER, J. S. 1965. Tree seed abundance vs. deer mouse populations in Douglas-fir clearcuts. Proc. Amer. Soc. For., 1965:219–222.

————. 1970. Plant and mammal changes on a clearcut in west-central Oregon. Ecology, 51:1018–1026.

————. 1971. Deer mouse movement in forest habitat. Northwest Science, 45:163–170.

————. 1979. Deer mouse reproduction and its relationship to the tree seed crop. Amer. Midland Nat., 102:95–104.

GETZ, L. L. 1968. Relationship between ambient temperature and respiratory water loss of small mammals. Comp. Biochem. Physiol., 24:335–342.

GIBSON, D. ST. C., AND M. J. DELANY. 1984. The population ecology of small rodents in Pennine woodlands. J. Zool. (London), 203:63–85.

GILBERT, B. S., AND C. J. KREBS. 1981. Effects of extra food on *Peromyscus* and *Clethrionomys* populations in the southern Yukon. Oecologia, 51:326–331.

GLENN, M. E. 1970. Water relations in three species of deer mice (*Peromyscus*). Comp. Biochem. Physiol., 33:231–248.

GLIWICZ, J. 1980. Island populations of rodents: their organization and functioning. Biol. Rev., 55:109–138.

————. 1981. Competitive interactions within a forest rodent community in central Poland. Oikos, 37:353–362.

GOLLEY, F. B., L. RYSZKOWSKI, AND J. T. SOKUR. 1975. The role of small mammals in temperate forests, grassland and cultivated fields. Pp. 223–242, *in*

Small mammals: their productivity and population dynamics (F. B. Golley, K. Petrusewicz, and L. Ryszkowski, eds.). Cambridge University Press, Cambridge, 451 pp.

GORECKI, A. 1969. Metabolic rate and energy budget of the striped field mouse. Acta Theriol., 14: 181–190.

GRANT, P. R. 1970a. Experimental studies of competitive interaction in a two-species system. II. The behaviour of *Microtus*, *Peromyscus* and *Clethrionomys* species. Anim. Behav., 18:411–426.

———. 1970b. Colonization of islands by ecologically dissimilar species of mammals. Canadian J. Zool., 48:545–553.

———. 1971. Experimental studies of competitive interaction in a two-species system. III. *Microtus* and *Peromyscus* species in enclosures. J. Anim. Ecol., 40:323–335.

———. 1976. An 11-year study of small animal populations at Mont St. Hilaire, Quebec. Canadian J. Zool., 54:2156–2173.

GRANT, W. E., AND E. C. BIRNEY. 1979. Small mammal community structure in North American grasslands. J. Mamm., 60:23–36.

GRANT, W. E., E. C. BIRNEY, N. R. FRENCH, AND D. M. SWIFT. 1982. Structure and productivity of grassland small mammal communities related to grazing-induced changes in vegetative cover. J. Mamm., 63:248–260.

GRAU, H. J. 1982. Kin recognition in white-footed deermice (*Peromyscus leucopus*). Anim. Behav., 30:497–505.

GREEN, R. 1979. The ecology of wood mice (*Apodemus sylvaticus*) on arable farmland. J. Zool. (London), 188:357–377.

GREENWOOD, P. J. 1978. Timing of activity of the bank vole *Clethrionomys glareolus* and the wood mouse *Apodemus sylvaticus* in a deciduous woodland. Oikos, 31:123–127.

GRODZINSKI, W. 1985. Ecological energetics of bank voles and wood mice. Symp. Zool. Soc. London, 55:169–192.

GRODZINSKI, W., AND N. R. FRENCH. 1983. Production efficiency in small mammal populations. Oecologia, 56:41–49.

GRODZINSKI, W., B. BOBEK, A. DROZDZ, AND A. GORECKI. 1970. Energy flow through small rodent populations in a beech forest. Pp. 291–298, *in* Energy flow through small mammal populations (K. Petrusewicz and L. Ryszkowski, eds.). Inst. Ecol., Polish Acad. Sci., IBP, Small Mammal Group, Warsaw, 298 pp.

GRODZINSKI, W., Z. PUCEK, AND L. RYSZKOWSKI. 1966. Estimation of rodent numbers by means of prebaiting and intensive removal. Acta Theriol., 11: 297–314.

GURNELL, J. 1978. Seasonal changes in numbers and male behavioural interaction in a population of wood mice, *Apodemus sylvaticus*. J. Anim. Ecol., 47:741–755.

———. 1981. Woodland rodents and tree seed supplies. Pp. 1191–1214, *in* The worldwide furbearer conference proceedings, 3 vols. (J. A. Chapman and D. Pursley, eds.). R. R. Donnelly and Sons Co., Falls Church, Virginia, 2056 pp.

GURNELL, J., AND K. RENNOLLS. 1983. Growth in field and laboratory populations of wood mice (*Apodemus sylvaticus*). J. Zool., 200:355–365.

GYUG, L. W., AND J. S. MILLAR. 1981. Growth of seasonal generation in three natural populations of *Peromyscus*. Canadian J. Zool., 59:510–514.

HAITLINGER, R. 1962. Morphological variability in *Apodemus agrarius* (Pallas, 1771). Acta Theriol., 6:239–255.

———. 1969. Morphological variability of the Wroclaw population of *Apodemus sylvaticus*. Acta Theriol., 14:285–302.

HALL, E. R. 1981. The mammals of North America. John Wiley and Sons, New York, 2:vi+601–1181+90.

HALLETT, J. G. 1982. Habitat selection and the community matrix of a desert small-mammal fauna. Ecology, 63:1400–1410.

HALLETT, J. G., M. A. O'CONNELL, AND R. L. HONEYCUTT. 1983. Competition and habitat selection: test of a theory using small mammals. Oikos, 40:175–181.

HALVORSON, C. H. 1982. Rodent occurrence, habitat disturbance, and seed fall in a larch-fir forest. Ecology, 63:423–433.

HAMILTON, W. J. 1941. The food of small forest mammals in eastern United States. J. Mamm., 22:250–263.

HANSEN, L. P., AND G. O. BATZLI. 1978. The influence of food availability on the white-footed mouse: populations in isolated woodlots. Canadian J. Zool., 56:2530–2541.

———. 1979. Influence of supplemental food on local populations of *Peromyscus leucopus*. J. Mamm., 60:335–342.

HANSSON, L. 1968. Population densities of small mammals in open field habitats in south Sweden in 1964–1967. Oikos, 19:53–60.

———. 1971. Small rodent food, feeding and population dynamics. A comparison between granivorous and herbivorous species in Scandinavia. Oikos, 22:183–198.

HAPLIN, Z. T. 1981. Adult-young interactions in island and mainland populations of the deer mouse *Peromyscus maniculatus*. Oecologia, 51:419–425.

HARLAND, R. M., P. J. BLANCHER, AND J. S. MILLAR. 1979. Demography of a population of *Peromyscus leucopus*. Canadian J. Zool., 57:323–328.

HARPER, J. L. 1977. Population biology of plants. Academic Press, London, 892 pp.

HARRIS, J. H. 1984. An experimental analysis of desert rodent foraging ecology. Ecology, 65:1579–1584.

HARRIS, T. V. 1952. An experimental study of habitat selection by prairie and forest races of the deer mouse, *Peromyscus maniculatus*. Contrib. Lab. Vert. Biol., Univ. Michigan, 56:1–53.

HAYWARD, G. P., AND J. PHILLIPSON. 1979. Community structure and functional role of small mammals in ecosystems. Pp. 135–211, *in* Ecology of small mammals (D. M. Stoddart, ed.). Chapman and Hall, London, 386 pp.

HEALEY, M. C. 1967. Aggression and self-regulation of population size in deermice. Ecology, 48:377–392.

HEALING, T. D., AND F. NOWELL. 1985. Diseases and parasites of woodland rodent communities. Symp. Zool. Soc. London, 55:143–218.

HERMAN, T. B., AND F. W. SCOTT. 1984. An unusual decline in abundance of *Peromyscus maniculatus* in Nova Scotia. Canadian J. Zool., 62:175–178.

HILL, R. W. 1983. Thermal physiology and energetics of *Peromyscus*: ontogeny, body temperature, metabolism, insulation, and microclimatology. J. Mamm., 64:19–37.

HOFFMANN, R. S. 1955. A population high for *Peromyscus maniculatus*. J. Mamm., 36:571–572.

HOFFMEYER, I. 1973. Interaction and habitat selection in the mice *Apodemus flavicollis* and *A. sylvaticus*. Oikos, 24:108–116.

———. 1976. Experiments on the selection of food and foraging site by the mice *Apodemus sylvaticus* (Linne, 1758) and *A. flavicollis* (Melchior, 1834). Saugetierk. Mitt., 2:112–124.

———. 1983. Interspecific behavioural niche separation in wood mice (*Apodemus flavicollis* and *A. sylvaticus*) and scent marking relative to social dominance in bank voles (*Clethrionomys glareolus*). Unpubl. Ph.D. dissert., Univ. Lund, 116 pp.

HOFFMEYER, I., AND L. HANSSON. 1974. Variability in number and distribution of *Apodemus flavicollis* (Melch.) and *A. sylvaticus* (L.) in south Sweden. Z. Saugetierk., 39:15–23.

HOLBROOK, S. J. 1978. Habitat relationships and coexistence of four sympatric species of *Peromyscus* in northwestern New Mexico. J. Mamm., 59:18–26.

———. 1979a. Vegetational affinities, arboreal activity, and coexistence of three species of rodents. J. Mamm., 60:528–542.

———. 1979b. Habitat utilization, competitive interactions, and coexistence of three species of cricetine rodents in east-central Arizona. Ecology, 60:758–769.

———. 1980. Species diversity patterns in some present and prehistoric rodent communities. Oecologia, 44:355–367.

———. 1982. Ecological inferences from mandibular morphology of *Peromyscus maniculatus*. J. Mamm., 63:399–408.

HOLISOVA, V. 1967. The food of *Apodemus agrarius* (Pall.). Zool. Listy, 16:1–14.

———. 1969. Vertical movements of some small mammals in a forest. Zool. Listy, 18:121–141.

HOLISOVA, V., AND R. OBRTEL. 1977. Food resource allocation in four myomorph rodents coexisting in a reed swamp. Acta Sci. Nat. Brno, 11:1–35.

———. 1980. Food resource partitioning among four myomorph rodent populations coexisting in a spruce forest. Folia Zool., 29:193–207.

HOOPER, E. T. 1968. Classification. Pp. 27–74, *in* Biology of *Peromyscus* (Rodentia) (J. A. King, ed.). Spec. Publ., Amer. Soc. Mamm., 2:1–593.

HORNER, B. E. 1954. Arboreal adaptations of *Peromyscus*, with special reference to use of the tail. Contrib. Lab. Vert. Biol., Univ. Michigan, 61:1–84.

HOWARD, W. E. 1949. Dispersal, amount of inbreeding and longevity in a local population of prairie deermice on the George Reserve, southern Michigan. Contrib. Lab. Vert. Biol., Univ. Michigan, 43:1–51.

———. 1960. Innate and environmental dispersal of individual vertebrates. Amer. Midland Nat., 63:152–161.

HSIA, W. 1962. On the populations and home range of the greater wood mouse, *Apodemus speciosus penninsulae* Thomas. Acta Zool. Sinica, 13:171–182.

HUMINSKI, S. 1968. Maturation and seasonal variations of the testes and the male accessory glands in the field mouse, *Apodemus agrarius* (Pallas, 1771). Zool. Polon., 18:69–80.

HUMPHREYS, W. F. 1984. Production efficiency in small mammal populations. Oecologia, 62:85–90.

HUNTER, D. M., R. M. F. S. SADLEIR, AND J. M. WEBSTER. 1972. Studies on the ecology of cuterebrid parasitism in deermice. Canadian J. Zool., 50:25–29.

IMAIZUMA, Y.   1978.   Climbing behaviour of *Apodemus argenteus* and *Apodemus speciosus* (Rodentia, Muridae).   Appl. Ent. Zool., 13:304–307.

JAMESON, E. W.   1952.   Food of deer mice, *Peromyscus maniculatus* and *P. boylii*, in the northern Sierra Nevada, California.   J. Mamm., 33:50–60.

————.   1953.   Reproduction of the deer mice (*Peromyscus maniculatus* and *P. boylii*) in the Sierra Nevada, California.   J. Mamm. 34:44–58.

JENSEN, T. S.   1975.   Population estimations and population dynamics of two Danish forest rodent species.   Vidensk. Meddr. Dansk Naturn. Foren., 138:65–86.

————.   1984.   Habitat distribution, home range and movements of rodents in mature forest and reforestations.   Acta Zool. Fennica, 171:305–307.

JEWELL, P. A.   1966.   Breeding season and recruitment in some British mammals confined on small islands.   Pp. 99–116, *in* Comparative biology of reproduction in mammals (I. W. Rowlands, ed.). Academic Press, London, 559 pp.

JONGE, G. DE, AND H. DIENSKE.   1979.   Habitat and interspecific displacement of small mammals in the Netherlands.   Netherlands J. Zool., 29:177–214.

JUDES, U. VON.   1979a.   Untersuchungen zur Okologie der Waldmaus (*Apodemus sylvaticus* Linne, 1758) und der Gelbhalsmaus (*Apodemus flavicollis* Melchior, 1834) im Raum Kiel (Schleswig-Holstein). II. Pranatale Mortalitat. Z. Saugetierk., 44:185–195.

————.   1979b.   Untersuchungen zur Okologie der Waldmaus (*Apodemus sylvaticus* Linne, 1758) und der Gelbhalsmaus (*Apodemus flavicollis* Melchior, 1834) im Raum Kiel (Schleswig-Holstein). I. Populationsdichte, Gewichtveranderungen, Fortpflanzungs-Jahreszyklus, populationsbiologische Parameter. Z. Saugetierk., 44:81–95.

KAHMANN, H.   1961.   Beitrage zur Saugetierkunde der Turkei 2. Die Brandmaus in Thrakien (*Apodemus agrarius* Pallas, 1774) und die sudeuropaische Verbreitung der Art.   Rev. Fac. Sci. Univ. Istanbul. (B), 26:87–106.

KAUFMAN, D. W., AND G. A. KAUFMAN.   1982.   Sex ratio in natural populations of *Peromyscus leucopus*.   J. Mamm., 63:655–658.

KELLY, P. A., G. A. T. MAHON, AND J. S. FAIRLEY.   1982.   An analysis of morphological variation in the fieldmouse *Apodemus sylvaticus* (L.) on some Irish islands.   Proc. R. Ir. Acad., 82B:39–51.

KIKKAWA, J.   1964.   Movement, activity and distribution of small rodents *Clethrionomys glareolus* and *Apodemus sylvaticus* in woodland.   J. Anim. Ecol., 33:259–299.

KING, C. M.   1980.   The weasel *Mustela nivalis* and its prey in an English woodland.   J. Anim. Ecol., 49:127–159.

KING, J. A., E. O. PRICE, AND P. L. WEBER.   1968.   Behavioral comparisons within the genus *Peromyscus*.   Michigan Acad. Sci. Arts Lett., 53:113–136.

KIRKLAND, G. L., JR., AND R. J. GRIFFIN.   1974.   Microdistribution of small mammals at the coniferous-deciduous forest ecotone in northern New York.   J. Mamm., 55:417–427.

KITCHINGS, J. T., AND D. J. LEVY.   1981.   Habitat patterns in a small mammal community.   J. Mamm., 62:814–820.

KLEIMAN, D. G.   1977.   Monogamy in mammals.   Quart. Rev. Biol., 52:39–69.

KOTLER, B. P.   1984a.   Risk of predation and the structure of desert rodent communities.   Ecology, 65:689–701.

————.   1984b.   Harvesting rates and predatory risk in desert rodents; a comparison of two communities on different continents.   J. Mamm., 65:91–96.

KREBS, C. J., AND J. H. MYERS. 1974. Population cycles in small mammals. Adv. Ecol. Res., 8:267–399.

KRITZMAN, E. B. 1974. Ecological relationships of *Peromyscus maniculatus* and *Perognathus parvus* in eastern Washington. J. Mamm., 55:172–188.

KRYLOV, D. G. 1975. Tendency to grouping in spatial distribution of small mammals in a forest habitat. Ekol. Pol. Ser. A, 23:335–345.

LACKEY, J. A. 1976. Reproduction, growth and development in the Yucatan deer mouse, *Peromyscus yucatanicus*. J. Mamm., 57:638–655.

LARINA, N. I. 1964. Geographic change in biotopical adoption of the sylvan mouse-like rodents in European part of U.S.S.R. Bull. Mosk. Obskch. Ispytot. Prirods. Otd. Biol., 63:21–33.

LARSSON, T. B., L. HANSSON, AND E. NYHOLM. 1973. Winter reproduction in small rodents in Sweden. Oikos, 24:475–476.

LAWLOR, T. E. 1982. The evolution of body size in mammals: evidence from insular populations in Mexico. Amer. Nat., 119:54–72.

LAYNE, J. N. 1968. Ontogeny. Pp. 148–253, *in* Biology of *Peromyscus* (Rodentia) (J. A. King, ed.). Spec. Publ., Amer. Soc. Mamm., 2:1–593.

———. 1970. Climbing behavior of *Peromyscus floridanus* and *Peromyscus gossypinus*. J. Mamm., 51:580–591.

LEIGH BROWN, A. J. 1977*a*. Physiological correlates of an enzyme polymorphism. Nature, 269:803–804.

———. 1977*b*. Genetic changes in a population of fieldmice (*Apodemus sylvaticus*) during one winter. J. Zool., 182:281–289.

LEMEN, C. 1980. Relationship between relative brain size and climbing ability in *Peromyscus*. J. Mamm., 61:360–364.

LINDEBORG, R. G. 1952. Water requirements of certain rodents from xeric and mesic habitats. Contrib. Lab. Vert. Biol., Univ. Michigan, 58:1–32.

LIU, T. T. 1953. Prenatal mortality in *Peromyscus* with special reference to its bearing on reduced fertility in some interspecific and intersubspecific crosses. Contrib. Lab. Vert. Biol., Univ. Michigan, 60:1–32.

LLEWELLYN, J. B. 1980. Seasonal change in the aggressive behavior of *Peromyscus maniculatus* inhabiting a pinyon-juniper woodland in western Nevada. J. Mamm., 61:341–345.

LOMOLINO, M. V. 1984. Mammalian island biogeography: effects of area, isolation and vagility. Oecologia, 61:376–382.

LOUARNE, H. LE, AND A. SCHMITT. 1972. Relations observees entre la production de faines et la dynamique de population du mulot, *Apodemus sylvaticus* L. en foret de Fontainebleau. Ann. Sci. Forest., 30:205–214.

LOUARNE, H. LE, F. SPITZ, AND B. DASSONVILLE. 1970. Ecological distribution of small mammals in the forests of Briancon area (Hautes-Alpes). Ann. Zool. Ecol. Anim., 2:427–432.

LOY, A., AND L. BOITANI. 1984. The structural microhabitat, in the Mediterranean scrub environment, of two rodent species, *Apodemus sylvaticus* and *Mus musculus*. Suppl. Ric. Biol. Selvaggina., 9:143–160.

LYNCH, G. R. 1973. Seasonal changes in thermogenesis, organ weight, and body composition in the white-footed mouse, *Peromyscus leucopus*. Oecologia, 13:363–376.

MACMILLEN, R. E. 1964. Population ecology, water relations, and social behavior of a southern California semidesert rodent fauna. Univ. California Publ. Zool., 71:1–66.

————. 1965. Aestivation in the cactus mouse, *Peromyscus eremicus*. Comp. Biochem. Physiol., 16:227–248.

————. 1983. Water regulation in *Peromyscus*. J. Mamm., 64:38–47.

MADISON, D. M. 1977. Movements and habitat use among interacting *Peromyscus leucopus* as revealed by radiotelemetry. Canadian Field-Nat., 91:273–281.

MANVILLE, R. H. 1949. A study of small mammal populations in northern Michigan. Misc. Publ. Mus. Zool., Univ. Michigan, 73:1–83.

MARTELL, A. M. 1983. Demography of southern red-backed voles (*Clethrionomys gapperi*) and deer mice (*Peromyscus maniculatus*) after logging in north-central Ontario. Canadian J. Zool., 61:958–969.

MASON, E. B. 1974. Metabolic response of two species of *Peromyscus* raised in different thermal environments. Physiol. Zool., 47:68–74.

MAY, R. M. (ED.). 1981. Theoretical ecology: principles and applications. Second ed. Blackwell Sci. Pub., Oxford, 489 pp.

M'CLOSKEY, R. T. 1972. Temporal changes in populations and species diversity in a California rodent community. J. Mamm., 53:657–676.

————. 1975. Habitat succession and rodent distribution. J. Mamm., 56:950–955.

M'CLOSKEY, R. T., AND B. FIELDWICK. 1975. Ecological separation of sympatric rodents (*Peromyscus* and *Microtus*). J. Mamm., 56:119–129.

M'CLOSKEY, R. T., AND D. T. LAJOIE. 1975. Determinants of local distribution and abundance in white-footed mice. Ecology, 56:467–472.

McNAB, B. K. 1968. The influence of fat deposits on the basal rate of metabolism in desert homiotherms. Comp. Biochem. Physiol., 26:337–343.

McNAB, B. K., AND P. MORRISON. 1963. Body temperature and metabolism in subspecies of *Peromyscus* from arid and mesic environments. Ecol. Monogr., 33:63–82.

McPHERSON, A. D., AND J. N. KRULL. 1972. Island populations of small mammals and their affinities with vegetation type, island size and distance from mainland. Amer. Midland Nat., 88:384–392.

MEHLHOP, P., AND J. F. LYNCH. 1978. Population characteristics of *Peromyscus leucopus* introduced to islands inhabited by *Microtus pennsylvanicus*. Oikos, 31:17–26.

MERMOD, C. 1969. Ecologie et dynamique des populations de trois rongeurs sylvicoles. Mammalia, 33:1–57.

MERRITT, J. F. 1974. Factors influencing the local distribution of *Peromyscus californicus* in northern California. J. Mamm., 55:102–114.

MESERVE, P. L. 1974. Temporary occupancy of a coastal sage scrub community by a seasonal immigrant, the California mouse (*Peromyscus californicus*). J. Mamm., 55:836–840.

————. 1976. Habitat and resource utilization by rodents of a California coastal sage scrub community. J. Anim. Ecol., 45:647–666.

METZGAR, L. H. 1971. Behavioral population regulation in the woodmouse, *Peromyscus leucopus*. Amer. Midland Nat., 86:434–448.

————. 1979. Dispersion patterns in a *Peromyscus* population. J. Mamm., 60:129–145.

MIDDLETON, J., AND G. MERRIAM. 1981. Woodland mice in a farmland mosaic. J. Appl. Ecol., 18:703–710.

MIHOK, S. 1979. Behavioural structure and demography of subarctic *Clethrionomys gapperi* and *Peromyscus maniculatus*. Canadian J. Zool., 57:1520–1535.

MILLAR, J. S.   1975.   Tactics of energy partitioning in breeding *Peromyscus*. Canadian J. Zool., 53:967–976.

———.   1978.   Energetics of reproduction in *Peromyscus leucopus*: the cost of lactation. Ecology, 59:1055–1061.

———.   1979.   Energetics of lactation in *Peromyscus maniculatus*. Canadian J. Zool., 57:1015–1019.

———.   1982.   Life cycle characteristics of northern *Peromyscus maniculatus borealis*. Canadian J. Zool., 60:510–515.

MILLAR, J. S., AND L. W. GYUG.   1981.   Initiation of breeding by northern *Peromyscus* in relation to temperature. Canadian J. Zool., 59:1094–1098.

MILLAR, J. S., P. B. WILLE, AND S. L. IVERSON.   1979.   Breeding by *Peromyscus* in seasonal environments. Canadian J. Zool., 57:719–727.

MILLER, D. H., AND L. L. GETZ.   1977*a*.   Comparisons of population dynamics of *Peromyscus* and *Clethrionomys* in New England. J. Mamm., 58:1–16.

———.   1977*b*.   Factors influencing local distribution and species diversity of forest small mammals in New England. Canadian J. Zool., 55:806–814.

MILLER, R. S.   1958.   A study of a wood mouse population in Wytham Woods, Berkshire.   J. Mamm., 39:477–493.

MIRIC, D.   1966.   Die Felsenmaus (*Apodemus mystacinus* Danford und Alston, 1877—Rodentia, Mammalia) als Glied der Nagetierfauna Jugoslawiens. Z. Saugetierk., 31:417–440.

MITTLEBACH, G. G., AND K. L. GROSS.   1984.   Experimental studies of seed predation in old-fields. Oecologia, 65:7–13.

MIYAO, T., T. MOROZUMI, AND M. MOROZUMI.   1974.   Small mammals in Kirigamine and Shirakabako plateaus. J. Mamm. Soc. Japan, 6:33–38.

MONTGOMERY, S. S. J.   1982.   A field study on the biology of *Apodemus sylvaticus* (Rodentia: Muridae) and its helminth parasites. Unpubl. Ph.D. dissert., Queen's Univ. Belfast, 126 pp.

MONTGOMERY, W. I.   1978.   Intraspecific and interspecific interactions of *Apodemus sylvaticus* and *Apodemus flavicollis* under laboratory conditions. Anim. Behav., 26:1247–1254.

———.   1979.   Trap-revealed home range in sympatric populations of *Apodemus sylvaticus* and *A. flavicollis*.   J. Zool., 189:535–540.

———.   1980*a*.   Population structure and dynamics of sympatric *Apodemus* species (Rodentia: Muridae).   J. Zool., 192:351–377.

———.   1980*b*.   Spatial organisation in sympatric populations of *Apodemus sylvaticus* and *A. flavicollis* (Rodentia: Muridae).   J. Zool., 192:397–401.

———.   1980*c*.   The use of arboreal runways by woodland rodents, *Apodemus sylvaticus* (L.), *A. flavicollis* (Melchior) and *Clethrionomys glareolus* (Schreber). Mamm. Rev., 10:189–195.

———.   1981.   A removal experiment with sympatric populations of *Apodemus sylvaticus* (L.) and *A. flavicollis* (Melchior) (Rodentia: Muridae). Oecologia, 51:123–132.

———.   1985.   Interspecific competition and the comparative ecology of *Apodemus* species. Pp. 126–187, *in* Case studies in population biology (L. M. Cook, ed.). Manchester University Press, Manchester, 218 pp.

MONTGOMERY, W. I., AND D. V. BELL.   1981.   Dispersion of the woodmouse in deciduous woodland. Acta Theriol., 26:107–112.

MONTGOMERY, W. I. AND J. GURNELL.   1985.   The behaviour of *Apodemus*. Symp. Zool. Soc. London., 55:89–115.

MORHARDT, J. E.   1970.   Body temperature of white-footed mice (*Peromyscus* sp.) during daily torpor.   Comp. Biochem. Physiol., 33:423–439.

MORRIS, D. W.   1984.   Sexual differences in habitat use by small mammals: evolutionary strategy or reproductive constraint?   Oecologia, 65:51–57.

MORRIS, R. D.   1970.   The effects of endrin on *Microtus* and *Peromyscus*. I. Unenclosed field populations.   Canadian J. Zool., 48:695–708.

———.   1972.   The effects of endrin on *Microtus* and *Peromyscus*. II. Enclosed field populations.   Canadian J. Zool., 50:885–896.

MULLER, J. P.   1972.   Die Verteilung der Kleinsauger auf die Lebensraume an einem nordhang im Churer Rheintal.   Z. Saugetierk., 37:257–286.

MUNGER, J. C. AND J. H. BROWN.   1981.   Competition in desert rodents: an experiment with semipermeable exclosures.   Science, 211:510–512.

MYERS, P., AND L. L. MASTER.   1983.   Reproduction by *Peromyscus maniculatus*: size and compromise.   J. Mamm., 64:1–18.

MYTON, B.   1974.   Utilization of space by *Peromyscus leucopus* and other small mammals.   Ecology, 55:277–290.

NADEAU, J. H., R. T. LOMBARDI, AND R. H. TAMARIN.   1981.   Population structure and dispersal of *Peromyscus leucopus* on Muskeget Island.   Canadian J. Zool., 59:793–799.

NIETHAMMER, J., AND P. KRAPP (EDS.).   1978.   Handbuch der Saugetiere Europas.   Akademische Verlagsgesellschaft, Wiesbaden, 1:1–476.

ODUM, E. P., C. E. CONNELL, AND L. B. DAVENPORT.   1962.   Population energy flow of three primary consumer components of old-field ecosystems.   Ecology, 43:88–96.

OTA, K.   1968.   Studies on the ecological distribution of murid rodents in Hokkaido.   Hokkaido Univ. Coll. Exp. For. Res. Bull., 26:223–295.

OTSU, S.   1969.   On wood mice (voles) in Yamagata Prefecture. I. The distribution of wood mice in mountainous districts.   Japan J. Appl. Ent. Zool., 13:5–8.

———.   1970.   On wood mice (voles) in Yamagata Prefecture. II. The distribution of wood mice according to different types of the forests.   Japan J. Appl. Ent. Zool., 14:85–88.

PARMENTER, R. R., AND J. A. MACMAHON.   1983.   Factors determining the abundance and distribution of rodents in a shrub-steppe ecosystem: the role of shrubs.   Oecologia, 59:145–156.

PARREN, S. G., AND D. E. CAPEN.   1985.   Local distribution and coexistence of two species of *Peromyscus* in Vermont.   J. Mamm., 66:36–44.

PARTRIDGE, L.   1981.   Increased preferences for familiar foods in small mammals.   Anim. Behav., 29:211–216.

PASPALEVA, M., AND I. ANDREESCU.   1975.   Contributions a la connaisance de la dynamique des populations de micromammiferes (Ord. Rodentia) de la region Sinaia-Roumanie.   Tran. du Mus. d'Hist. Nat. Gr. Antipa., 16:283–293.

PELIKAN, J.   1964.   Vergleich einiger populationsdynamischer faktoren bei *Apodemus sylvaticus* (L.) und *A. microps* Kr. et Ros.   Z. Saugetierk., 29:242–251.

———.   1965.   Reproduction, population structure and elimination of males in *Apodemus agrarius* (Pall.).   Zool. Listy, 14:317–332.

———.   1967a.   Analysis of three population dynamical factors in *Apodemus flavicollis* (Melch.).   Z. Saugetierk., 31:31–37.

———.   1967b.   Variability of body weight in three *Apodemus* species.   Zool. Listy, 16:199–220.

————. 1970. Sex ratio in three *Apodemus* species. Zool. Listy, 19:23–34.

PELIKAN, J., AND J. NESVADBOVA. 1979. Small mammal communities in farms and surrounding fields. Folia Zool., 28:209–217.

PETTIGREW, B. G., AND R. M. F. S. SADLEIR. 1974. The ecology of the deer mouse *Peromyscus maniculatus* in a coastal coniferous forest. I. Population dynamics. Canadian J. Zool., 52:107–118.

PFEIFFER, VON H., AND J. NIETHAMMER. 1972. Versuche zur Nahrungswahl von Wald- und Gelbhalsmaus (*Apodemus sylvaticus* und *A. flavicollis*). Z. Saugetierk., 37:57–65.

PIANKA, E. R. 1978. Evolutionary ecology. Second ed. Harper and Row, New York, 397 pp.

PLATT, A. P. 1968. Differential trap mortality as a measure of stress during times of population increase and decrease. J. Mamm., 49:331–335.

PRICE, E. O. 1966. Influence of light on reproduction of *Peromyscus maniculatus gracilis*. J. Mamm., 47:343–344.

PRICE, M. V. 1984. Microhabitat use in rodent communities: predator avoidance or foraging economics. Netherlands J. Zool., 34:63–80.

PRICE, M. V., AND N. M. WASER. 1984. On the relative abundance of species: postfire changes in coastal sage scrub rodent community. Ecology, 65: 1161–1169.

RANDOLPH, S. E. 1977. Changing spatial relationships in a population of *Apodemus sylvaticus* with the onset of breeding. J. Anim. Ecol., 46:653–676.

REDFIELD, J. A., C. J. KREBS, AND M. J. TAITT. 1977. Competition between *Peromyscus maniculatus* and *Microtus townsendii* in grasslands of coastal British Columbia. J. Anim. Ecol., 46:607–616.

REDMAN, J. P. AND J. A. SELANDER. 1958. Home ranges of deer mice in southern Arkansas. J. Mamm., 39:390–395.

REMMERT, H. 1980. Ecology: a text book. Springer-Verlag, Berlin, 289 pp.

REYNOLDS, T. D. 1980. Effects of some different land management practices on small mammals. J. Mamm., 61:558–561.

RICKART, E. A. 1977. Reproduction, growth and development in two species of cloud forest *Peromyscus* from southern Mexico. Occas. Pap. Mus. Nat. Hist., Univ. Kansas, 67:1–22.

SADLEIR, R. M. F. S. 1965. The relationship between agonistic behaviour and population changes in the deermouse, *Peromyscus maniculatus* (Wagner). J. Anim. Ecol., 34:331–352.

————. 1970. Population dynamics and breeding of the deermouse (*Peromyscus maniculatus*) on Burnaby Mountain, British Columbia. Syesis, 3:67–74.

————. 1974. The ecology of the deer mouse *Peromyscus maniculatus* in a coastal coniferous forest. II. Reproduction. Canadian J. Zool., 52:119–131.

SADLEIR, R. M. F. S., K. D. CASPERSON, AND J. HARLING. 1973. Intake and requirements of energy and protein for the breeding of wild deermice, *Peromyscus maniculatus*. J. Reprod. Fert. Suppl., 19:237–252.

SAINT GIRONS, M. -C. 1966. Le rhythme circadien de'l activite chez les mammiferes holartiques. Mem. Mus. Nat. Hist. Nat. (Paris), Zool. N.S., 40: 102–187.

SAKAI, A. 1976. Seasonal changes in heart weights of wood mice, *Apodemus argenteus*. J. Mamm. Soc. Japan, 6:224–230.

SANS-COMA, V., AND J. GOSALBEZ. 1976. Sobre la reproducion de *Apodemus sylvaticus* L., 1758 en el nordeste iberico. Miscelanea Zoologica, 3:227–233.

SCHREIBER, R. K. 1979. Coefficients of digestibility and caloric diet of rodents in the northern Great Basin. J. Mamm., 60:416–420.

SCHROPFER, R. VON. 1974. Comparative ecological studies on water requirement of *Apodemus tauricus* (Pallas, 1811) and *Apodemus sylvaticus* (Linne, 1758) (Rodentia, Muridae). Zool. Jb. Syst., 101:236–248.

SEAGLE, S. W. 1985. Patterns of small mammal microhabitat utilization in cedar glade and deciduous forest habitats. J. Mamm., 60:22–35.

SEXTON, O. J., J. F. DOUGLASS, R. R. BLOYE, AND A. PESCE. 1982. Thirteen-fold change in population size of *Peromyscus leucopus*. Canadian J. Zool., 60: 2224–2225.

SHEPPE, W., JR. 1961. Systematic and ecological relations of *Peromyscus oreas* and *P. maniculatus*. Proc. Amer. Phil. Soc., 105:421–446.

———. 1967. Habitat restriction by competitive exclusion in the mice *Peromyscus* and *Mus*. Canadian Field-Nat., 81:81–98.

SILVERTOWN, J. W. 1980. The evolutionary ecology of mast seeding in trees. Biol. J. Linn. Soc., 14:235–250.

SKRYJA, D. D. 1978. Reproductive inhibition in female cactus mice (*Peromyscus eremicus*). J. Mamm., 59:543–550.

SMAL, C. M., AND J. S. FAIRLEY. 1981. Energy consumption of small rodent populations in two Irish woodland ecosystems. Acta Theriol., 26:449–458.

———. 1982. The dynamics and regulation of small rodent population in the woodland ecosystems of Killarny, Ireland. J. Zool., 196:1–30.

SMITH, A. T., AND J. M. VRIESE. 1979. Population structure of everglades rodents: responses to a patchy environment. J. Mamm., 60:778–794.

SMITH, D. A., AND S. W. SPELLER. 1970. The distribution and behavior of *Peromyscus maniculatus gracilis* and *P. leucopus noveboracensis* (Rodentia: Cricetidae) in a southeastern Ontario woodlot. Canadian J. Zool., 48:1187–1199.

SMITH, M. F. 1981. Relationships between genetic variability and niche dimensions among coexisting species of *Peromyscus*. J. Mamm., 62:273–285.

SMITH, M. H. 1965. Behavioral discrimination shown by allopatric and sympatric males of *Peromyscus eremicus* and *Peromyscus californicus*, between females of the same species. Evolution, 19:430–435.

———. 1971. Food as a limiting factor in the population ecology of *Peromyscus polionotus* (Wagner). Ann. Zool. Fennici, 8:109–112.

SMITH, M. H., J. B. GENTRY, AND J. PINDER. 1974. Annual fluctuations in small mammal population in an eastern hardwood forest. J. Mamm., 55: 231–234.

SMITH, M. H., M. N. MANLOVE, AND J. JOULE. 1978. Spatial and temporal dynamics of genetic organization of small mammal populations. Pp. 99–113, *in* Populations of small mammals under natural conditions (D. Snyder, ed.). Spec. Publ. Pymatuning Lab. Ecol., Univ. Pittsburgh, 5:1–237.

SMITH, M. H., AND J. T. MCGINNIS. 1968. Relationships of latitude, altitude, and body size to litter size and mean annual production of offspring in *Peromyscus*. Res. Popul. Ecol., 10:115–126.

SMYTH, M. 1966. Winter breeding in woodland mice, *Apodemus sylvaticus*, and voles, *Clethrionomys glareolus* and *Microtus agrestis*, near Oxford. J. Anim. Ecol., 35:471–485.

SNYDER, D. P. 1956. Survival rates, longevity, and population fluctuations in the white-footed mouse, *Peromyscus leucopus* in southeastern Michigan. Misc. Publ. Mus. Zool., Univ. Michigan, 95:1–33.

SOUTHERN, H. N.   1970.   The natural control of a population of tawny owls (*Strix aluco*).   J. Zool., 162:197–285.

———.   1979.   The stability and instability of small mammal populations. Pp. 103–134, *in* Ecology of small mammals (D. M. Stoddart, ed.). Chapman and Hall, London, 386 pp.

SOUTHERN, H. N. AND V. P. W. LOWE.   1968.   The pattern of distribution of prey and predation in tawny owl territories.   J. Anim. Ecol., 37:75–97.

———.   1982.   Predation by tawny owls (*Strix aluco*) on bank voles (*Clethrionomys glareolus*) and wood mice (*Apodemus sylvaticus*).   J. Zool., 198:83–102.

SPITZENBERGER, F., AND H. M. STEINER.   1967.   Die Okologie der Insectivora und Rodentia (Mammalia) der Stockerauer Donau-Auen (Niederosterreich). Bonn. Zool. Beitr., 18:258–296.

STAH, C. D.   1980.   Vertical nesting distribution of two species of *Peromyscus* under experimental conditions.   J. Mamm., 61:141–143.

STEBBINS, L. L.   1978.   Some aspects of overwintering in *Peromyscus maniculatus*. Canadian J. Zool., 56:386–390.

STEIN, G. H. W.   1955.   Die Kleinsauger ostdeutscher Ackerflachen.   Z. Saugetierk., 20:89–113.

STEINER, H. M.   1968.   Untersuchungen uber die variabilitat und Bionomie der Gattung *Apodemus* (Muridae, Mammalia) der Donau-Auen von Stockerau (Niederosterreich).   Z. Wiss. Zool., 177:1–96.

STICKEL, L. F.   1960.   *Peromyscus* ranges at high and low population densities.   J. Mamm., 41:433–441.

———.   1968.   Home range and travels. Pp. 373–411, *in* Biology of *Peromyscus* (Rodentia) (J. A. King, ed.). Spec. Publ., Amer. Soc. Mamm., 2:1–593.

STICKEL, L. F., AND O. WARBACH.   1960.   Small-mammal populations in a Maryland woodlot, 1949–1954.   Ecology, 41:269–286.

STRAKA, F.   1966.   Zur Bionomie von *Apodemus microps* Krat. und Ros. in Bulgarien.   Zool. Listy, 15:97–104.

SULLIVAN, T. P.   1977.   Demography and dispersal in island and mainland populations of the deermouse, *Peromyscus maniculatus*.   Ecology, 58:964–978.

———.   1979*a*.   Repopulation of clear-cut habitat and conifer seed predation by deer mice.   J. Wildl. Manag., 43:861–871.

———.   1979*b*.   The use of alternative foods to reduce conifer seed predation by the deer mouse, (*Peromyscus maniculatus*).   J. Appl. Ecol., 16:475–495.

———.   1980.   Comparative demography of *Peromyscus maniculatus* and *Microtus oregoni* populations after logging and burning of coastal forest habitats. Canadian J. Zool., 58:2252–2259.

SULLIVAN, T. P., AND D. S. SULLIVAN.   1980.   The use of weasels for natural control of mouse and vole populations.   Oecologia, 47:125–129.

———.   1982.   Responses of small mammal populations to a forest herbicide application in a 20-year-old conifer plantation.   J. Appl. Ecol., 19:95–106.

SVIRIDENKO, P. A.   1940.   The nutrition of mouse-like rodents and their significance in the problem of reafforestation.   Zool. Zhur., 19:680–703.

TAITT, M. J.   1981.   The effect of extra food on small rodent populations: 1. Deer mice (*Peromyscus maniculatus*).   J. Anim. Ecol., 50:111–124.

TAPPER, S.   1979.   The effect of fluctuating vole numbers (*Microtus agrestis*) on a population of weasels (*Mustela nivalis*) on farmland.   J. Anim. Ecol., 48: 603–617.

TAYLOR, L. R., I. P. WOIWOOD, AND J. N. PERRY.   1978.   The density-dependence of spatial behaviour and the rarity of randomness. J. Anim. Ecol., 47: 383–406.

TERMAN, C. R. 1968. Population dynamics. Pp. 412–450, *in* Biology of *Peromyscus* (Rodentia) (J. A. King, ed.). Spec. Publ., Amer. Soc. Mamm., 2: 1–593.

———. 1973. Recovery of reproductive function by prairie deermice (*Peromyscus maniculatus bairdii*) from asymptotic populations. Anim. Behav., 21: 443–448.

———. 1980. Social factors influencing delayed reproductive maturation in prairie deermice (*Peromyscus maniculatus bairdii*) in laboratory populations. J. Mamm., 61:219–223.

TERTIL, R. 1972. The effect of behavioural thermoregulation on the daily metabolism of *Apodemus agrarius* (Pallas, 1771). Acta Theriol., 17:295–313.

THOMPSON, S. D. 1982. Structure and species composition of desert heteromyid rodent species assemblages: effects of a simple habitat manipulation. Ecology, 63:1313–1321.

VAN HORNE, B. 1981. Demography of *Peromyscus maniculatus* in seral stages of coastal coniferous forest in southeast Alaska. Canadian J. Zool., 59: 1045–1061.

———. 1982. Niches of adult and juvenile deer mice (*Peromyscus maniculatus*) in seral stages of coniferous forest. Ecology, 63:992-1003.

VAUGHAN, T. A. 1974. Resource allocation in some sympatric, subalpine rodents. J. Mamm., 55:764–795.

VENABLES, L. S. V., AND O. M. VENABLES. 1971. Mammal population changes in a young conifer plantation 1960–1970, Newborough Warren, Anglesey. Nature in Wales, 12:159–163.

VESTAL, B. M., AND J. J. HELLACK. 1978. Comparison of neighbor recognition in two species of deer mice (*Peromyscus*). J. Mamm., 59:339–346.

VISINESCU, N. 1967. The particularities of thermoregulation and their seasonal variations in *Clethrionomys glareolus* Schreb. (1870) and *Apodemus sylvaticus* L. (1758). Rev. Roum. Biol. Zoologie, 12:127–137.

———. 1970. Thermoregulation particularities in the transition period from hypothermia to normothermia. Rev. Roum. Biol. Zoologie, 15:347–353.

VLCEK, M. 1984. Spacial activity of small mammals (Rodentia) in areas of large-scale livestock production farms. Vest. Czech. Spolec. Zool., 48:69–80.

WALKER, E. P. 1975. Mammals of the World. Third ed. Johns Hopkins University Press, Baltimore, 2:647-1500.

WALTON, J. B., AND J. P. ANDREWS. 1981. Torpor induced by food deprivation in the wood mouse *Apodemus sylvaticus*. J. Zool., 194:260–263.

WATTS, C. H. S. 1968. The foods eaten by wood mice (*Apodemus sylvaticus*) and bank voles (*Clethrionomys glareolus*) in Wytham Woods, Berkshire. J. Anim. Ecol., 37:25–41.

———. 1969. The regulation of wood mouse (*Apodemus sylvaticus*) numbers in Wytham Woods, Berkshire. J. Anim. Ecol., 38:285–304.

———. 1970. Long distance movement of bank voles and wood mice. J. Zool., 161:217–256.

WECKER, S. C. 1963. The role of early experience in habitat selection by the prairie deer mouse, *Peromyscus maniculatus bairdii*. Ecol. Monogr., 33:307–325.

WENDLAND, V. 1975. Dreijahriger Rhythmus im Bestandwechsel der Gelbhalsmaus (*Apodemus flavicollis* Melchior). Oecologia, 20:301–310.

————. 1981. Cyclic population changes in three mouse species in the same woodland. Oecologia, 48:7–12.

WHITAKER, J. O., JR. 1966. Food of *Mus musculus*, *Peromyscus maniculatus bairdii* and *Peromyscus leucopus* in Vigo County, Indiana. J. Mamm., 47:473–486.

————. 1967. Habitat relationships of four species of mice in Vigo County, Indiana. Ecology, 48:867–872.

————. 1968. Parasites. Pp. 254–311, *in* Biology of *Peromyscus* (Rodentia) (J. A. King, ed.). Spec. Publ., Amer. Soc. Mamm., 2:1–593.

WHITAKER, R. H. 1975. Communities and ecosystems. Second ed. Macmillan Publishing Co., Inc., New York, 385 pp.

WHITAKER, W. L. 1940. Some effects of artificial illumination on reproduction in the white-footed mouse, *Peromyscus leucopus noveboracensis*. J. Exp. Zool., 83:33–60.

WHITFORD, W. G. 1976. Temporal fluctuation in density and diversity of desert rodent populations. J. Mamm., 57:351–369.

WHITSETT, J. M., L. E. GRAY, JR., AND G. M. BEDIZ. 1979. Gonadal hormones and aggression toward juvenile conspecifics in prairie deer mice. Behav. Ecol. Sociobiol., 6:165–168.

WIRTZ, W. O., II, AND P. G. PEARSON. 1960. A preliminary analysis of habitat orientation in *Microtus* and *Peromyscus*. Amer. Midland Nat., 63:131–142.

WOLFF, J. O. 1985. Maternal aggression as a deterrent to infanticide in *Peromyscus leucopus* and *P. maniculatus*. Anim. Behav., 33:117–123.

WOLFF, J. O., M. H. FREEBERG, AND R. D. DUESER. 1983. Interspecific territoriality in two sympatric species of *Peromyscus* (Rodentia: Cricetidae). Behav. Ecol. Sociobiol., 12:237–242.

WOLFF, J. O., AND B. HURLBUTT. 1982. Day refuges of *Peromyscus leucopus* and *Peromyscus maniculatus*. J. Mamm., 63:666–668.

WOLTON, R. J. 1983. The activity of free-ranging wood mice. J. Anim. Ecol., 52:781–794.

————. 1985. The ranging and nesting behaviour of wood mice, *Apodemus sylvaticus* (Rodentia:Muridae), as revealed by radio-tracking. J. Zool., (A) 206:203–224.

WOLTON, R. J., AND J. R. FLOWERDEW. 1985. Spatial distribution and movements of wood mice, yellow-necked mice and bank voles. Symp. Zool. Soc. London, 55:249–275.

YAHNER, R. H. 1983. Population dynamics of small mammals in farmstead shelterbelts. J. Mamm., 64:380–386.

YALDEN, D. W. 1984. The yellow-necked mouse, *Apodemus flavicollis*, in Roman Manchester. J. Zool., 203:285–288.

YALDEN, D. W., P. A. MORRIS, AND J. HARPER. 1973. Studies on the comparative ecology of some French small mammals. Mammalia, 37:257–276.

YOSHIDA, H. 1970. Small mammals of Mt. Kiyomizu, Fukuoka Pref. I. Ecological distribution of the small mammals. J. Mamm. Soc. Japan, 5:8–14.

————. 1971. Small mammals of Mt. Kiyomizu, Fukuoka Pref. J. Mamm. Soc. Japan, 5:123–129.

ZATSEPINA, R. A. 1960. Data on the ecological morphology of mice of the genus *Apodemus*. Zhur. Biol., No. 100254.

ZEGERS, D. A., AND J. C. HA. 1981. Niche separation of *Peromyscus leucopus* and *Blarina brevicauda*. J. Mamm., 62:199–201.

ZEJDA, J., AND V. HOLISOVA. 1970. On the prebaiting of small mammals in the estimation of their abundance. Zool. Listy, 19:103–118.

ZEMANEK, M. 1972. Food and feeding habits of rodents in a deciduous forest. Acta Theriol., 15:315–325.

ZIMMERMAN, K. 1962. Die Untergattungen der Gattung *Apodemus*. Bonn. Zool. Beitr., 13:198–208.

CONTRIBUTORS

MICHAEL D. CARLETON
Division of Mammals, National Museum of Natural History, Smithsonian Institution, Washington, DC 20560, USA

THEODORE GARLAND, JR.
Department of Ecology and Evolutionary Biology, University of California—Irvine, Irvine, California 92717, USA; Present address *Department of Zoology, University of Wisconsin, Madison, Wisconsin 53706, USA*

DONALD W. KAUFMAN
Division of Biology, Kansas State University, Manhattan, Kansas 66506, USA

GLENNIS A. KAUFMAN
Division of Biology, Kansas State University, Manhattan, Kansas 66506, USA

GORDON L. KIRKLAND, JR.
Vertebrate Museum, Shippensburg University, Shippensburg, Pennsylvania 17257, USA

JAMES N. LAYNE
Archbold Biological Station, Lake Placid, Florida 33852, USA

RICHARD E. MACMILLEN
Department of Ecology and Evolutionary Biology, University of California—Irvine, Irvine, California 92717, USA

JOHN S. MILLAR
Department of Zoology, The University of Western Ontario, London N6A 5B7, Ontario, Canada

W. I. MONTGOMERY
Department of Zoology, The Queen's University of Belfast, Belfast BT7 1NN, Northern Ireland, U.K.

JERRY O. WOLFF
Department of Biology, Villanova University, Villanova, Pennsylvania 19085, USA